Brazilian Marine Biodiversity

Series Editor
Alexander Turra, Unesco Chair for Ocean Sustainability,
Oceanographic Institute and Advanced Studies Institute
University of São Paulo
São Paulo, Brazil

The book series Brazilian Marine Biodiversity was designed to communicate to abroad and international readership the diversified marine and coastal habitats along the large Brazilian coast.

***

The diversity of marine habitats found in Brazil is astonishing and includes estuaries, coral reefs, rocky shores, sandy beaches, rhodolith beds, mangroves, salt marshes, deep-sea habitats, vegetated bottoms, and continental shelf. These habitats are addressed from an ecosystem perspective across the series, and characterized in terms of distribution and peculiarities along the Brazilian coast, records of relevant species, and information on the prevailing structuring ecological and oceanographic processes governing biodiversity.

The series also presents an analysis of the role of biodiversity and the importance of ecosystem services, and discusses the threats to each habitat, such as pollution, habitat loss, invasive species, overfishing, and global environmental changes. Conservation efforts are also considered as well as gaps in scientific knowledge and science-policy interface.

This series is an initiative of the UNESCO Chair for Ocean Sustainability, hosted by the Oceanographic Institute and the Advanced Studies Institute of the University of São Paulo, within the Brazilian Network for Monitoring Coastal Benthic Habitats (ReBentos; rebentos.org), which is supported by the Brazilian National Council for Scientific and Technological Development (CNPq), the Research Program on Biodiversity Characterization, Conservation, Restoration and Sustainable Use of the São Paulo Research Foundation (BIOTA-FAPESP), the Coordination for the Improvement of Higher Education Personnel (CAPES) and the Brazilian Innovation Agency (FINEP). ReBentos is part of the Brazilian Network on Global Climate Change Research (Rede Clima) and the Science and Technology National Institute on Climate Changes (INCT Mudanças Climáticas) at the Ministry of Science, Technology, and Innovation (MCTI).

Paulo Antunes Horta • Marina Nasri Sissini
Editors

Brazilian Rhodolith Beds

Oasis of Marine Biodiversity to be protected

Springer

Editors
Paulo Antunes Horta
Laboratório de Ficologia
Departamento de Botânica
Universidade Federal de Santa Catarina
Florianópolis, SC, Brasil

Marina Nasri Sissini
Programa de Pós-Graduação em Ecologia
Universidade Federal de Santa Catarina
Florianópolis, SC, Brazil

ISSN 2520-1077 ISSN 2520-1085 (electronic)
Brazilian Marine Biodiversity
ISBN 978-3-031-61448-4 ISBN 978-3-031-61449-1 (eBook)
https://doi.org/10.1007/978-3-031-61449-1

This Springer imprint is published by the registered company Springer Nature Switzerland AG
The registered company address is: Gewerbestrasse 11, 6330 Cham, Switzerland

Foreword

The world and humanity are facing the most delicate moment in recent history, with the acceleration of climate change and the imminent collapse of numerous points of support for the balance of the biosphere, society and our economies. It is increasingly clear and evident that environments dominated by algae, especially calcareous algae, are fundamental elements for sustaining these balances. In our book, we present a little of the history of studies on the group in Brazil and important elements for understanding the functioning of species and ecosystems structured by them.

Rhodolith beds, oases of biodiversity on the Brazilian continental shelf, can be found from north to south of our coast, providing contributions to our society that encompass environmental, sociocultural and economic aspects. Despite their importance and strong presence on our coast, calcareous algae and associated biodiversity are threatened by direct exploitation, pollution, ocean warming and acidification.

In our book, part of Brazilian Network for Monitoring Coastal Benthic Habitats (REBENTOS) contributions, the reader will have access to the diagnosis of the current state and to solutions that point to alternatives for managing coastal marine environments. May diving into the pink powerful world of these submerged oases, a natural heritage of humanity, awaken new passions for these unique representatives of our sociobiodiversity. University students and researchers interested in different fields of marine sciences will find, besides additional information, new arguments to foster conservation and regenerative development in our blue planet.

Contents

1 **Coralline Red Algae and Rhodolith Beds: A World Natural Heritage in the Southwestern Atlantic** 1
Marina Nasri Sissini and Paulo Antunes Horta

2 **A Historical Perspective of the Brazilian Rhodolith Beds: Discovering a Hidden Ecosystem** 15
Flávio Berchez, Gilberto Dias, Marina N. Sissini, Paulo A. Horta, and Cláudio Gonçalves Tiago

3 **Rhodolith Beds: Occurrence, Diversity and Evolution** 29
Maria Beatriz B. Barros-Barreto, Luana M. Coutinho, Fernanda P. Gomes, Iara O. Costa, Maria Carolina Henriques, Talita Vieira-Pinto, Marina N. Sissini, Mariana C. Oliveira, and Paulo A. Horta

4 **Biodiversity Associated with Rhodolith Beds in Brazil** 67
Frederico T. S. Tâmega, Guido Grimaldi, André Macedo, Sulamita O. Barbosa, Antônio B. Anderson, Eduardo Bastos, Vanessa Freire, Anderson Moreira, Celso Fernandes, Marcia A. O. Figueiredo, Pedro Bastos Carneiro, Lidiane P. Gouvêa, and Paulo A. Horta

5 **Mollusks Associated with Rhodolith Beds from the Southwestern Atlantic** 105
Aramys Rocchetti de Mello Cesar, Izabela Felice Paes, Ana Clara Becker, Flávio Dias Passos, and Guilherme Henrique Pereira-Filho

6 **Fossil and Recent Coralline Algal Bioconstructions and Rhodoliths in the Brazilian Shelf** 127
Frederico T. S. Tâmega, Alex C. Bastos, Davide Bassi, Julio Aguirre, and Juan C. Braga

7 Rhodolith Beds Mapping Along Brazilian Continental Shelf 145
Vanessa Freire de Carvalho, Lidiane P. Gouvêa, Eduardo Bastos, Marina Nasri Sissini, and Paulo A. Horta

8 Major Threats to Rhodolith Beds: Ocean Acidification, Global Warming, and Local Stressors 163
Gabrielle Koerich, Marina Nasri Sissini, and Paulo Antunes Horta

9 Contributions of Rhodolith Beds to People 185
Tainá Luchese Gaspar, Marina Nasri Sissini, Jorge Luiz Rodrigues Filho, and Paulo A. Horta

10 The Legacy of Gilberto M. Amado-Filho to South Atlantic Rhodolith Beds Knowledge: Memories of His Students 203
Guilherme H. Pereira-Filho, Ricardo da Gama Bahia, Fernando C. Moraes, and Leonardo T. Salgado

Index 211

About the Editors

Paulo Antunes Horta graduated in Biological Sciences in the University of São Paulo (1996), Ph.D. in Biological Sciences in the University of São Paulo (2000), and PostDoc. in Marine Ecology in Plymouth University, UK (2013). He was a professor at the Federal University of Paraíba (2004) and a visiting professor at the University of Algarve (2018) and at the University of Salento (2021). Since 2006 he has been professor at the Federal University of Santa Catarina. He is interested in marine ecology, studying the impacts of climate change and local stressors on marine biodiversity.

Marina Nasri Sissini graduated as biologist at Federal University of Santa Catarina (2011) and Ph.D. in Ecology at the same university. She is particularly interested in the Biodiversity, Evolution, and Conservation of rhodolith beds in the South Atlantic.

Contributors

Julio Aguirre Departamento de Estratigrafía y Paleontología, Universidad de Granada, Granada, Spain

Antônio B. Anderson Laboratório de Ichthyologia, Departamento de Oceanografia, Universidade Federal do Espírito Santo, Vitória, Brazil

Sulamita O. Barbosa Programa de Pós-Graduação em Botânica, Museu Nacional – UFRJ, Rio de Janeiro, RJ, Brazil

Maria Beatriz B. Barros-Barreto Laboratório de Ficologia, Universidade Federal do Rio de Janeiro, Rio de Janeiro, RJ, Brazil

Davide Bassi Dipartimento di Fisica e Scienze della Terra, Università degli Studi di Ferrara, Ferrara, Italy

Alex C. Bastos Departamento de Oceanografia, Universidade Federal do Espírito Santo, Espírito Santo, Vitória, ES, Brazil

Eduardo Bastos Departamento de Botânica, Universidade Federal de Santa Catarina, Florianópolis, Santa Catarina, Brazil

Ana Clara Becker Instituto do Mar, Laboratório de Ecologia e Conservação Marinha, Universidade Federal de São Paulo, São Paulo, Brazil

Flávio Berchez Departamento de Botânica, Universidade de São Paulo, São Paulo, SP, Brazil

Cape Horn International Center (CHIC), Puerto Williams, Chile

Juan C. Braga Departamento de Estratigrafía y Paleontología, Universidad de Granada, Granada, Spain

Pedro Bastos Carneiro Universidade Federal do Delta do Parnaíba (UFDPar), Parnaíba, Brazil

Iara O. Costa Programa de Pos-Graduacao em Botanica, Universidade Estadual de Feira de Santana, Feira de Santana, BA, Brazil

Departamento de Ciencias Humanas e Tecnologias (DCHT) do Campus XXIV, Universidade do Estado da Bahia (UNEB), Xique-Xique, Brazil

Luana M. Coutinho Laboratório de Ficologia, Universidade Federal do Rio de Janeiro, Rio de Janeiro, Brazil

Ricardo da Gama Bahia Instituto de Pesquisas Jardim Botânico do Rio de Janeiro, Rio de Janeiro, Brazil

Vanessa Freire de Carvalho Bióloga, Departamento de Pesquisa e Conservação de Fauna (MAPCF), Prefeitura Municipal de Curitiba, Florianópolis, SC, Brazil

Programa de Pós-Graduação em Ecologia, Universidade Federal de Santa Catarina, Florianópolis, SC, Brazil

Aramys Rocchetti de Mello Cesar Departamento de Biologia Animal, Instituto de Biologia, Universidade Estadual de Campinas (UNICAMP), São Paulo, Brazil

Gilberto Dias Lagemar, Departamento de Geologia, Universidade Federal Fluminense, Niterói, RJ, Brazil

Celso Fernandes Laboratório de Meios Porosos e Propriedades Termofísicas—EMC, CTC, Universidade Federal de Santa Catarina, Florianópolis, SC, Brazil

Marcia A. O. Figueiredo Instituto de Pesquisas Jardim Botânico do Rio de Janeiro, Rio de Janeiro, RJ, Brazil

Jorge Luiz Rodrigues Filho Laboratório de Ecologia Aplicada e Conservação, Universidade do Estado de Santa Catarina, Laguna, SC, Brazil

Vanessa Freire Departamento de Botânica, Centro de Ciências Biológicas, Universidade Federal de Santa Catarina, Florianópolis, SC, Brazil

Tainá Luchese Gaspar Programa de Pós-Graduação em Ecologia, Universidade Federal de Santa Catarina, Florianópolis, SC, Brazil

Laboratório de Ecologia de Ambientes Recifais, Departamento de Ecologia e Zoologia, Universidade Federal de Santa Catarina, Florianópolis, SC, Brazil

Fernanda P. Gomes Laboratório de Ficologia, Universidade Federal do Rio de Janeiro, Rio de Janeiro, RJ, Brazil

Lidiane P. Gouvêa Centre of Marine Sciences, University of Algarve, Faro, Portugal

Guido Grimaldi Departamento de Botânica, Centro de Ciências Biológicas, Universidade Federal de Santa Catarina, Florianópolis, SC, Brazil

Maria Carolina Henriques Laboratório de Ficologia, Universidade Federal do Rio de Janeiro, Rio de Janeiro, RJ, Brazil

Paulo Antunes Horta Laboratório de Ficologia, Departamento de Botânica, Universidade Federal de Santa Catarina, Florianópolis, SC, Brazil

Gabrielle Koerich Phycology Laboratory, Botanical Department, Federal University of Santa Catarina, Florianópolis, Brazil

Federal University of Santa Catarina, Florianópolis, Brazil

André Macedo Departamento de Botânica, Centro de Ciências Biológicas, Universidade Federal de Santa Catarina, Florianópolis, SC, Brazil

Fernando C. Moraes Instituto de Pesquisas Jardim Botânico do Rio de Janeiro, Rio de Janeiro, Brazil

Anderson Moreira Laboratório de Meios Porosos e Propriedades Termofísicas—EMC, CTC, Universidade Federal de Santa Catarina, Florianópolis, SC, Brazil

Mariana C. Oliveira Departamento de Botânica, Instituto de Biociências, Universidade de São Paulo, São Paulo, SP, Brazil

Izabela Felice Paes Departamento de Biologia Animal, Instituto de Biologia, Universidade Estadual de Campinas (UNICAMP), São Paulo, Brazil

Flávio Dias Passos Departamento de Biologia Animal, Instituto de Biologia, Universidade Estadual de Campinas (UNICAMP), São Paulo, Brazil

Guilherme Henrique Pereira-Filho Instituto do Mar, Laboratório de Ecologia e Conservação Marinha, Universidade Federal de São Paulo, São Paulo, Brazil

Leonardo T. Salgado Instituto de Pesquisas Jardim Botânico do Rio de Janeiro, Rio de Janeiro, Brazil

Marina Nasri Sissini Programa de Pós-Graduação em Ecologia, Universidade Federal de Santa Catarina, Florianópolis, SC, Brazil

Frederico T. S. Tâmega Departamento de Biotecnologia Marinha, Instituto de Estudos do Mar Almirante Paulo Moreira, Arraial do Cabo, RJ, Brazil

Cláudio Gonçalves Tiago Centro de Biologia Marinha, Universidade de São Paulo, São Sebastião, SP, Brazil

Talita Vieira-Pinto Departamento de Botânica, Instituto de Biociências, Universidade de São Paulo, São Paulo, SP, Brazil

Chapter 1
Coralline Red Algae and Rhodolith Beds: A World Natural Heritage in the Southwestern Atlantic

Marina Nasri Sissini and Paulo Antunes Horta

Abstract In this book, we will discuss different aspects of the coralline red algae (CRA) and their rhodolith beds. Besides being the main builders of these biogenic reefs, they are also an important constituent of coral reefs, with important interconnections and interdependencies among these systems. Recent advancements in the characterization of the community structure of biogenic reefs in the South Atlantic revealed unique reef formations as the Rocas Atoll and the Abrolhos "chapeirões," where CRA are majorly responsible for the construction and maintenance of these formations. In order to contextualize the long evolutionary process of calcareous algae, as well as the origin of the coralline marine flora in the Southwestern Atlantic, we briefly present the main geological events and the role of ecophysiological barriers and filters to understand the distribution pattern of the current flora. We discussed some of the main threats to this group of organisms and highlighted the unquestionable ecological importance of CRA and their role in the reefs of the Southwestern Atlantic. Finally, we reinforce their contribution to people and our needs to foster their conservation in the Blue Amazon as a world natural heritage.

Keywords Reef builders · Rhodolith · Reef evolution · Biogenic reefs

---

M. N. Sissini
Programa de Pós-Graduação em Ecologia, Universidade Federal de Santa Catarina, Florianópolis, SC, Brazil

P. A. Horta (✉)
Laboratório de Ficologia, Departamento de Botânica, Universidade Federal de Santa Catarina, Florianópolis, SC, Brasil

P. A. Horta, M. N. Sissini (eds.), *Brazilian Rhodolith Beds*, Brazilian Marine Biodiversity, https://doi.org/10.1007/978-3-031-61449-1_1

1.1 Introduction

Biogenic reefs[1] are among the most diverse and threatened biological formations on the planet, yet the high-value ecosystem services and functions they provide are still underestimated for coastal regions (Worm et al., 2006; Paoli et al., 2016; Amado-Filho et al., 2017; Hall-Spencer & Harvey, 2019). Considering coral bleaching and changes in the composition of reef environments, diversity losses have been identified, quantified, and monitored across the globe (Bellwood et al., 2004; Mcmanus & Polsenberg, 2004; Maliao et al., 2008), while biogenic algal reefs have received little attention.

With the help of recent advancements in the characterization of the community structure of these formations, especially in the south Atlantic (Moura et al., 2016; Mazzei et al., 2017; Aued et al., 2018; Bastos et al., 2018), some particularities have been revealed, such as unique reef formations (mushroom-shaped coral pinnacles called "chapeirões," Leão et al., 2003) and a significant contribution of coralline red algae[2] (CRA) to the process of construction and maintenance of these reefs (review in Horta et al., 2016; Mazzei et al., 2017; Aued et al., 2018; Bastos et al., 2018). These primary producers often play the largest role in the formation of present-day reefs, as is the case for the Rocas Atoll (Gherardi & Bosence, 2001), whereas in

[1] To conceptualize the word "reef" is quite an audacious task that must be done from the perspective of the different organisms that structure them and contribute to their biogeochemical roles. The term reef encompasses a set of marine biosedimentary structures resulting from evolutionary events at different spatial and temporal scales. Many authors define reefs as "confined structures built by the growth or metabolic activity of sessile benthic aquatic organisms" (Kiessling, 2003). However, contemporary coral banks in deep, cold waters can also be described as reefs (Roberts et al., 2006; Kiessling, 2009), even though they do not fit under the generic definition. Riding (2002) highlighted the flaws and consequences of defining reefs based only on superficial quantitative aspects instead of essential functional characteristics. A comparison based solely on biological aspects misses valuable information about the characteristics and great diversity of different types of reefs. For this reason, we believe that the structural (physical) classification of biogenic reefs proposed by Riding (2002) is the most comprehensive way to classify biogenic reefs, including rhodolith beds, in the Southwestern Atlantic.

[2] In the present book, the term "coralline red algae," or CRA, is used for the group of red algae that deposit calcium carbonate in their cell walls. The deposition of calcium carbonate in these algae can occur: (1) throughout the whole thallus in non-articulated, or non-geniculate, CRA, and (2) in an intercalated fashion, where non-calcified regions occur between calcified portions of the thallus in articulated, or geniculate, CRA. Non-articulated CRA are classified by their growth-form as either (1) growing attached to the substrate, such as rocks, shells, or corals, which are called crustose coralline algae (or encrusting calcareous algae), or (2) growing freely on unconsolidated substrate as nodules, which are commonly known as rhodoliths. In the past two decades, the systematics of this group has been changing, especially due to the use of molecular tools. Currently, CRA are in the subclass Corallinophycidae, which is morphologically characterized by pit connections with dome-shaped pit plugs and a mineralized thallus via the deposition of calcite (Le Gall & Saunders, 2007). There are five recognized orders within the subclass: Rhodogorgonales, Sporolithales (Le Gall et al., 2010), Corallinales, Hapalidiales (Nelson et al., 2015), and Corallinapetrales (Jeong et al., 2021), with only the later four representing CRA.

other reef environments, they act as cementers, providing rigidity and adding volume to the reef structure (Bosence, 1983; Riding, 2002).

The main biogeochemical and ecological roles of CRA are related somehow to growth and calcification patterns. These processes are directly or indirectly related to photosynthesis/calcification they can provide. At the geological scale, these processes resulted in abundant sinks for CO_2. However, in the short term, they can represent eventual sources, depending on the balance of calcification/photosynthesis, region, and community structure (Carvalho et al., 2020b). Calcification typically demands two bicarbonate ions, incorporating one carbonate into the $CaCO_3$ mineral structure and releasing the other into the environment as CO_2. CRA photosynthesis reduces the availability of CO_2, increases pH, and can lead to carbonate oversaturation. Therefore, calcification of CRA, calcifier photoautotrophs, poses a "chicken and egg" dilemma, as highlighted by McConnaughey and Whelan (1997). Photosynthesis elevates carbonate saturation and fosters calcification, or does calcification provide CO_2 and stimulate photosynthesis? CRA have calcification occurring on the same cell or thallus region where photosynthetic carbon uptake is occurring. Photosynthetic rates, aquatic chemistry, and diffusion within the boundary layer surrounding the organism determine carbonate saturation levels and potential calcification rates. Sometimes, calcification also occurs away from the site of carbon uptake, due to ion transport through the organism, pumps determine ion concentrations, $CaCO_3$ saturation levels, and calcification rates. Controlling the ion pumps allows the organism to control calcification (McConnaughey & Whelan, 1997). Together, these two processes will result in complex and diverse morphologies, growth patterns, and eco-evolutionary roles of organisms conspicuous in many marine ecosystems around the world.

In the Southwestern Atlantic, CRA are present in virtually all marine environments (Fig. 1.1), from shallow to mesophotic reefs, columnar reefs ("chapeirões"), rocky reefs (or rocky shores), and even on unconsolidated substrates building free nodules, known as rhodoliths (Gherardi & Bosence, 2001; Amado-Filho et al., 2012; Pereira-Filho et al., 2012; Meirelles et al., 2015; Aued et al., 2018; Bastos et al., 2018). Based on the current knowledge of their distribution, rhodoliths are estimated to occur over an area from 167,000 to 230,000 km^2 along the Brazilian coast (Carvalho et al., 2020a; Santos et al., 2023), which includes registries of the most extensive rhodolith beds in the world (Amado-Filho et al., 2012). The ecological importance of these organisms and biogenic formations is strategic to the Atlantic. Rhodolith beds are structurally complex habitats that house assemblages of diverse groups of marine animals (Simon et al., 2016; Carvalho et al., 2020b; Veras et al., 2020; Moura et al., 2021) and plants (Amado-Filho et al., 2010, Brasileiro et al., 2016; Carvalho et al., 2020a; Fig. 1.2.) influencing ecological and evolutionary processes, as well as the biogeographical patterns (Anderson et al., 2023).

Nevertheless, the functional role and biodiversity of non-articulated CRA on reefs are influenced by physical (Adey & Steneck, 2001; Wilson et al., 2004), biochemical (Bjork et al., 1995; Delgado & Lapointe, 1994; Wilson et al., 2004), and biological factors (Steneck, 1983) that have varied dramatically since the Cambrian

Fig. 1.1 Examples of coastal and oceanic reefs along the Brazilian coast. (**a**) Pedra do Silva, in Rio Grande do Norte; (**b**) Rocas Atoll; (**c**) Abrolhos, Bahia; (**d**) Guarapari, Espíritu Santo; (**e**) Trindade Island; (**f**) Arvoredo Island, Santa Catarina. (Credits: (**a**, **b**, **e**) M.N. Sissini; (**c**, **d**) SISBIOTA Mar; (**f**) E.O. Bastos)

Period, when the first CRA crusts are registered to have emerged (*c.* 540 Million years ago, Ma, Johnson & Hoeg, 1961; Steneck, 1983). CRA have evolved throughout geologic time and became abundant during the Pennsylvanian Era (*c.* 323 Ma; Steneck, 1983), but it was only at the beginning of the Jurassic Period (*c.* 201 Ma) that this group became more abundant and diversified on reefs. Adaptive radiation took place at the end of the Cretaceous Period and continued during the Cenozoic Era, which resulted in the geologic period with the greatest diversity of CRA in all evolutionary history. Regarding the role of CRA in the structuring of reefs, from the end of the Carboniferous Period to the beginning of the Permian Period (*c.* 298 Ma), they proved to be important in the construction of these environments, not for forming a rigid structure, but for trapping sediment between branches (Wray, 1977;

Fig. 1.2 The complex habitat built by the rhodoliths, forming the rhodolith beds, is home to a huge diversity of marine fauna and flora. It functions as a refuge for macroinvertebrates and also a feeding area for larger species, such as: (**a**) groupers, (**b**) turtles, and (**c**) rays. Also contain a rich marine fauna (**d**), in different stages of the life cycle. (Credits: (**a**, **c** and **d**) Daniel Venturini; (**b**) Fábio Borges)

Kiessling, 2009). More recently, these organisms were consolidated as cementing bioengineers of key formations that promote and maintain marine biodiversity in every ocean on the planet.

Reefs containing non-articulated CRA and corals, as we observe them today, emerged during the Upper Cretaceous (*c.* 80 Ma). CRA diversified at the beginning of the Cenozoic Era after recovering from the mass extinction at the end of the Cretaceous (*c.* 65 Ma, Kiessling, 2008). However, they began to decline at the end of the Paleocene (*c.* 56 Ma; Kiessling et al., 2002) and continued declining throughout the Eocene (*c.* 56 Ma; Kiessling, 2008). Nevertheless, the appearance of ancestral CRA is associated with the emergence and diversification of herbivores throughout the Phanerozoic (*c.* 542 Ma), due to the ability of herbivores to remove competitors for light and nutrients (e.g., epiphytes), excavate the limestone substrate, and predate on and disperse reef-builders (Steneck, 1983; Bellwood, 2003; Fig. 1.3). However, various morpho-anatomical characteristics and high regeneration capacity, such as depositing calcite (a crystalline form of carbonate) in the cell wall and sloughing of epithelial and peri-epithelial cells, allowed CRA to be less vulnerable to predation and epibiont cover. Their anatomical structure, which is

Fig. 1.3 Herbivores such as parrotfish (pictured, a *Sparisoma amplum*) played a fundamental role in the evolution and diversification of calcareous algae (Steneck, 1983; Bellwood, 2003). Herbivores remove competitors for light and nutrients (e.g., epiphytes), excavate the limestone substrate (**a**), and predate on and disperse reef-builders (**b**). (Credits: Daniel Venturini)

characterized by cell connections and fusions, permits the efficient translocation of resources (nutrients and gases) throughout the hypothallus, even when important photosynthetic regions are damaged (Steneck, 1983; Pitlik & Paul, 1997).

South Atlantic reefs have been developing since the Cretaceous (*c.* 145 Ma), when the continents were still joined as a supercontinent, Gondwana (Pittman et al., 1993). With the initial separation of the supercontinent, the South and Central Atlantic oceans began to form, and the American and African plates separated around 100 Ma ago (Müller et al., 2016), which resulted in the formation of new biogeographic connections and barriers. The new conformation of the tectonic plates allowed for the colonization of species not only from the Indo-Pacific and Central Atlantic but also those from the East Pacific through the Panama Channel, which opened approximately 3.1–3.5 Ma ago (Coates & Obando, 1996; Floeter et al., 2008). The closure of the Tethys Sea about 12–19 Ma ago was also an important event that imposed a new physical barrier to the connectivity between the Atlantic and Indo-Pacific oceans (Steininger & Rögl, 1984).

Aside from the new landmasses that have blocked passages, another important aspect to consider in the process of reef formation and the origin of reef organisms is the "soft" barriers, which are known as ecophysiological filters. Ecophysiological filters are oceanographic processes, such as currents or long distances of deep open ocean, which, even though transposable, impose obstacles to the dispersion of organisms (Floeter et al., 2008; Luiz et al., 2012; Cowman & Bellwood, 2013). Some examples are the formation of the deep sea in the Atlantic (*c.* 60 Ma, Pittman et al., 1993), the large contribution of freshwater from the Amazon River (Amazonian barrier, *c.* 6–10 Ma, Hoorn et al., 1995), and the formation of the Benguela current (*c.* 2 Ma, Shannon, 1985; Marlow et al., 2000). Considering these scenarios, many hypotheses have emerged about the origin of Brazilian marine flora (Oliveira, 1977; Horta et al., 2001) based on aspects of diversity and geologic and/or oceanographic processes (Taylor, 1955; Joly, 1964, 1965; Oliveira, 1977; Horta et al., 2001).

Taylor (1955) considered the Caribbean to be a huge center of dispersion in the tropical Atlantic due to its species richness and endemism. On the other hand, based on current oceanographic aspects, Joly (1964, 1965) proposed that Brazilian marine flora were the center of dispersion for the Caribbean via the Guiana Current. However, Oliveira (1977) considered geologically older and more species-rich locations to be the centers of dispersion, yet that the Caribbean would be a source of species for the South Atlantic. Based on an expanded database and multivariate analyses, Horta et al. (2001) incorporated historic oceanographic processes when considering the theory of continental drift and formation of the South Atlantic Ocean. These authors suggested that the Brazilian flora originated via different routes and at different moments. The southern portion of the Atlantic, south of the last connection between Africa and South America, would have had its original colonization influenced by the Indo-Pacific while the northern portion represented by the present-day states east of Rio Grande do Norte, by the Caribbean. These authors theorized that regardless of the complete separation of America and Africa, the oceanic currents along the Brazilian coast ~100 Ma ago did not come from the North Atlantic but moved from the southern to the northern hemisphere. These currents also had influences from the circulation of the Indian Ocean, which could potentially bring propagules and spores that could colonize the South African coast and the American coast in the South Atlantic (Berggren, 1980; Horta et al., 2001).

Coralline red algae, as well as many other marine organisms, are affected by climate change and ocean acidification (Martin & Gattuso, 2009; Martin & Hall-Spencer, 2017), in addition to regional and local threats, such as the introduction of non-treated sewage into the marine environment (Schubert et al., 2019), deforestation, runoff of fertilizers and pesticides from agro-industrial production, marine mining, oil spills (Sissini et al., 2020), direct exploitation (Paiva et al., 2023), and predatory fishing (Steller et al., 2003) with special attention to the bottom trawling (Fragkopoulou et al., 2021). These activities put the health and ecosystem services provided by these organisms at risk, which, although fundamental, still need to be better described and publicized, so that their conservation status or importance can be recognized (Moura et al., 2021) and determined (Halpern et al., 2019; Magris et al., 2020). The growing number of new, cryptic and endemic taxa being discovered in rhodolith beds suggests that their associated biodiversity is still unknown, as is their distribution, extent, and health status (Tuya et al., 2023).

This book brings these organisms and environments present all over the Brazilian coast (Fig. 1.4), within the focus of a broad and diverse group of marine scientists who are providing important international contributions. It is organized into nine additional chapters that describe different aspects of rhodolith beds, from a historical perspective, passing through their biodiversity, functioning, contributions to people, paleontological perspective, and main threats in the Brazilian Southwestern Atlantic. After this introduction, Berchez and collaborators bring back a bit of the history of surveys in these environments on our coast encompassing the pioneer

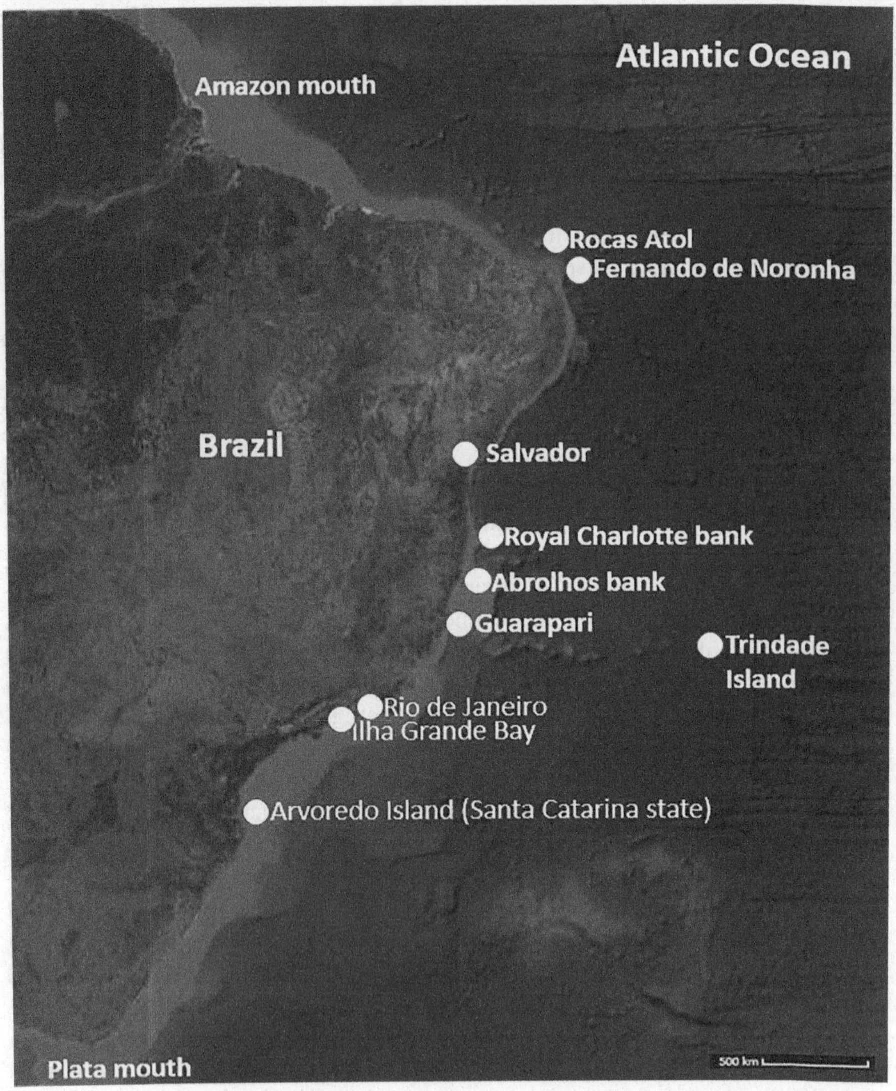

Fig. 1.4 Brazilian southwestern Atlantic highlighting main toponymies utilized in this book

Foslie identification up to the last molecular and physiological evaluations of coralline red algae and the rhodolith beds built by their species. In Chap. 3, Barros-Barreto and colleagues offer the first synthesis of the taxonomy and evolution of CRA considering the molecular information available up to now. In recent decades, there have been substantial efforts toward describing the biodiversity of CRA along the Brazilian coast (Tâmega & Figueiredo, 2005, 2007; Villas-Bôas et al., 2009, 2014; Bahia, 2014a, b, c, 2015; Sissini et al., 2014; Costa et al., 2014, 2019; Tâmega et al., 2014, 2015; Henriques et al., 2014a,b; Vieira-Pinto et al., 2014; Torrano-Silva et al., 2014, 2018; Borges et al., 2014; Villas-Boas et al., 2015; Jesionek et al., 2016, 2020). Barros-Barreto et al. put all this together and provide a very comprehensive scenario of diversity and future needs.

However, the associated biodiversity (Chaps. 4 and 5), paleontological perspectives (Chap. 6), mapping and modeling (Chap. 7), and functioning and main threats (Chap. 8) and their contributions to the people and to the planet (Chap. 9) in the Southwestern Atlantic presented in the last years important advances. In the last chapter is presented a tribute to our friend and missed colleague Dr. Gilberto Amado Filho, a Brazilian phycologist that dedicated his life to the Rhodolith bed studies, producing pioneer surveys and understanding. Gilberto left us in March 2019, leaving his immeasurable legacy to knowledge development, formation of young scientists, and science diffusion about marine science.

This compilation reinforces the idea that these organisms and environments play a key biogeochemical role to the homeostasis of our ocean, reinforcing the argument that they should be considered a world natural heritage. In addition, this book is dedicated to all those researchers that dedicate their efforts and lives, to bring rhodolith beds to the light of science, with a special tribute to Gilberto Amado in its last segment.

Acknowledgments This book is dedicated to the oceanographer Thiony Emanuel Simon as well as to the Professor Gilberto Amado Filho and students, researchers, and technicians dedicated, in the last decades, to the understanding of coralline red algae and rhodolith bed biology, ecology, evolution, and conservation. We thank all authors who contributed to this book, as well as the Brazilian research agencies that provided financial support to the production of all results and discussions summarized here, highlighting the federal (CNPq, Capes), state (FAPs), and private organizations. Finally, we thank Prof. Alexander Turra for leading Rebentos initiative, a product of the Brazilian Network for the Monitoring of Coastal Benthic Biodiversity (SISBIOTA-CNPq/FAPESP).

References

Adey, W. H., & Steneck, R. S. (2001). Thermogeography over time creates biogeographic regions: A temperature/space/time-integrated model and an abundance-weighted test for benthic marine algae. *Journal of Phycology, 37*, 677–698.

Amado-Filho, G. M., Maneveldt, G. W., Pereira-Filho, G. H., et al. (2010). Seaweed diversity associated with a Brazilian tropical rhodolith bed | Diversidad de macroalgas asociada con un manto de rodolitos tropical de Brasil. *Ciencias Marinas, 36*, 371–391.

Amado-Filho, G. M., Moura, R. L., Bastos, A. C., et al. (2012). Rhodolith beds are major $CaCO_3$ BIO-factories in the tropical south West Atlantic. *PLoS One, 7*, e35171.

Amado-Filho, G. M., Bahia, R. G., Pereira-Filho, G. H., & Longo, L. L. (2017). South Atlantic rhodolith beds: Latitudinal distribution, species composition, structure and ecosystem functions, threats and conservation status. In *Rhodolith/Maërl beds: A global perspective*. Springer.

Anderson, A. B., Pinheiro, H. T., Batista, M. B., et al. (2023). Biogeographic patterns of marine fishes associated with rhodolith beds in the southwestern Atlantic reveal an ecotone of biodiversity. *Biodiversity and Conservation, 32*(2), 821–837.

Aued, A. W., Smith, F., Quimbayo, J. P., et al. (2018). Large-scale patterns of benthic marine communities in the Brazilian province. *PLoS One, 13*, e0198452.

Bahia, R. G., Amado-Filho, G. M., & Maneveldt, G. W. (2014a). *Sporolithon molle* (Heydrich) Heydrich (Sporolithales, Corallinophycidae, Rhodophyta): An addition to the Atlantic flora found on a remote oceanic Island. *Cryptogamie Algologie, 35*, 7–14.

Bahia, R. G., Amado-Filho, G. M., Maneveldt, G. W., et al. (2014b). Sporolithon tenue sp. nov. (Sporolithales, Corallinophycidae, Rhodophyta): A new rhodolith-forming species from the tropical southwestern Atlantic. *Phycological Research, 62*, 44–54.

Bahia, R. G., Amado-Filho, G. M., Azevedo, J., et al. (2014c). *Porolithon improcerum* (Porolithoideae, Corallinaceae) and *Mesophyllum macroblastum* (Melobesioideae, Hapalidiaceae): New records of crustose coralline red algae for the Southwest Atlantic Ocean. *Phytotaxa, 190*, 38–44.

Bahia, R. G., Amado-Filho, G. M., Maneveldt, G. W., et al. (2015). *Sporolithon yoneshigueae* sp. *nov*. (Sporolithales, Corallinophycidae, Rhodophyta), a new rhodolith-forming coralline alga from the Southwest Atlantic. *Phytotaxa, 224*, 140–158.

Bastos, A. C., Moura, R. L., Moraes, F. C., et al. (2018). Bryozoans are major modern builders of South Atlantic oddly shaped reefs. *Scientific Reports, 8*, 9638.

Bellwood, D. R. (2003). Origins and escalation of herbivory in fishes: A functional perspective. *Paleobiology, 29*(1), 71.

Bellwood, D. R., Hughes, T. P., Folke, C., et al. (2004). Confronting the coral reef crisis. *Nature, 429*, 827–833.

Berggren, W. A. (1980). Les courants de 1'Atlantique. *La Recherche, 113*, 786–795.

Bjork, M., Mohammed, S. M., Bjorklund, M., et al. (1995). Coralline algae, important coral-reef builders threatened by pollution. *Ambio, 24*, 502–505.

Borges, V. P., Bastos, E., Batista, M. B., et al. (2014). The genus *Melobesia* (Corallinales, Rhodophyta) from the subtropical South Atlantic, with the addition of *M. rosanoffii* (Foslie) Lemoine. *Phytotaxa, 190*, 268–276.

Bosence, D. W. J. (1983). Coralline algal reef frameworks. *Journal of the Geological Society of London, 140*, 365–376.

Brasileiro, P. S., Pereira-Filho, G. H., Bahia, R. G., et al. (2016). Macroalgal composition and community structure of the largest rhodolith beds in the world. *Marine Biodiversity, 46*, 407–420.

Carvalho, V. F., Assis, J., Serrão, E. A., et al. (2020a). Environmental drivers of rhodolith beds and epiphytes community along the South Western Atlantic coast. *Marine Environmental Research, 154*, 104827.

Carvalho, V. F., Silva, J., Kerr, R., et al. (2020b). When descriptive ecology meets physiology: A study in a South Atlantic rhodolith bed. *Journal of the Marine Biological Association of the United Kingdom, 100*, 347–360.

Coates, A. G., & Obando, J. A. (1996). The geologic evolution of the Central American isthmus. In *Evolution and environments in tropical America*. University of Chicago Press.

Costa, I. A., Horta, P. A., Bergstrom, E. R., et al. (2014). Taxonomic study of crustose coralline algae off the northeastern Brazilian coast. *Phytotaxa, 190*, 130–161.

Costa, I. A., Jesus, P. B., Jesus, T. S., et al. (2019). Reef-building coralline algae from the Southwest Atlantic: Filling gaps with the recognition of *Harveylithon* (Corallinaceae, Rhodophyta) on the Brazilian coast. *Journal of Phycology, 55*, 1370–1385.

Cowman, P. F., & Bellwood, D. R. (2013). Vicariance across major marine biogeographic barriers: Temporal concordance and the relative intensity of hard versus soft barriers. *Proceedings of the Royal Society of London, 280*, 1–8.

Delgado, O., & Lapointe, B. E. (1994). Nutrient-limited productivity of calcareous versus fleshy macroalgae in a eutrophic, carbonate-rich tropical marine environment. *Coral Reefs, 13*, 151–159.

Floeter, S. R., Rocha, L. A., Robertson, D. R., et al. (2008). Atlantic reef fish biogeography and evolution. *Journal of Biogeography, 35*, 22–47.

Fragkopoulou, E., Serrão, E. A., & Horta, P. A. (2021). Bottom trawling threatens future climate refugia of rhodoliths globally. *Frontiers in Marine Science, 7*, 1246.

Gherardi, D. F. M., & Bosence, D. W. J. (2001). Composition and community structure of the coralline algal reefs from Atol das Rocas, South Atlantic, Brazil. *Coral Reefs, 19*, 205–219.

Hall-Spencer, J. M., & Harvey, B. P. (2019). Ocean acidification impacts on coastal ecosystem services due to habitat degradation. *Emerging Topics in Life Sciences, 3*, 197–206.

Halpern, B. S., Cottrell, R. S., Blanchard, J. L., et al. (2019). Putting all foods on the same table: Achieving sustainable food systems requires full accounting. *PNAS, 116*, 18152–18156.

Henriques, M. C., Coutinho, L. M., Riosmena-Rodríguez, R., et al. (2014a). Three deep water species of *Sporolithon* (Sporolithales, Rhodophyta) from the Brazilian continental shelf, with the description of *Sporolithon elevatum* sp. *nov*. *Phytotaxa, 190*, 320–330.

Henriques, M. C., Riosmena-Rodríguez, R., Coutinho, L. M., et al. (2014b). Lithophylloideae and Mastophoroideae (Corallinales, Rhodophyta) from the Brazilian continental shelf. *Phytotaxa, 190*, 112–129.

Hoorn, C., Guerrero, J., Sarmiento, G. A., et al. (1995). Andean tectonics as a cause for changing drainage patterns in Miocene northern South America. *Geology, 23*, 237–240.

Horta, P. A., Amancio, E., Coimbra, C. S., et al. (2001). Considerações sobre a distribuição e origem da flora de macroalgas marinhas brasileiras. *Hoehnea, 28*, 243–265.

Horta, P. A., Riul, P., Amado-Filho, G. M., et al. (2016). Rhodoliths in Brazil: Current knowledge and potential impacts of climate change. *Brazilian Journal of Oceanography, 64*, 117–136.

Jeong, S. Y., Nelson, W. A., Sutherland, J. E., Peña, V., Le Gall, L., Diaz-Pulido, G., ... & Cho, T. O. (2021). Corallinapetrales and Corallinapetraceae: A new order and family of coralline red algae including Corallinapetra gabrielii comb. nov. *Journal of Phycology, 57*(3), 849–862.

Jesionek, M. B., Bahia, R. G., Hernández-Kantún, J., et al. (2016). A taxonomic account of non-geniculate coralline algae (Corallinophycidae, Rhodophyta) from shallow reefs of the Abrolhos Bank, Brazil. *Algae, 31*, 317–340.

Jesionek, M. B., Bahia, R. G., Lyra, M. B., et al. (2020). Newly discovered coralline algae in Southeast Brazil: *Tectolithon fluminense gen. et* sp. *nov.* and *Crustaphytum atlanticum* sp. *nov.* (Hapalidiales, Rhodophyta). *Phycologia, 59*, 1–15.

Johnson, J. H., & Hoeg, A. O. (1961). Studies of Ordovician algae. *Colorado School of Mines, 56*, 1–120.

Joly, A. B. (1964). Extensão da Fiora Marinha Tropical no Sul do Brasi1. *Boletim do Instituto de Biologia marinha de Mar Del Plata, 7*, 1–15.

Joly, A. B. (1965). Marine flora of the tropical and subtropical Western South Atlantic. *Anais da Academia Brasileira de Ciências, 37*, 279–282.

Kiessling, W., Flügel, E., & Golonka, J. (2002). From patterns to processes: the future of reef research. *SSEPM Special Publications 72*, 735–743. https://doi.org/10.2110/pec.02.72.0735

Kiessling, W. R. (2003). *Encyclopedia of sediments and sedimentary rocks*. Kluwer Academic.

Kiessling, W. (2008). Sampling-standardized expansion and collapse of reef building in the Phanerozoic. *Foss Rec, 11*, 7–18.

Kiessling, W. (2009). Geologic and biologic controls on the evolution of reefs. *Annual Review of Ecology, Evolution, and Systematics, 40*, 173–192.

Leão, Z. M. A. N., Kikuchi, R. K. P., & Testa, V. (2003). Corals and coral reefs of Brazil. In *Latin American coral reefs*. Elsevier Science.

Le Gall, L., & Saunders, G. W. (2007). A nuclear phylogeny of the Florideophyceae (Rhodophyta) inferred from combined EF2, small subunit and large subunit ribosomal DNA: establishing the new red algal subclass Corallinophycidae. *Molecular Phylogenetics and Evolution, 43*(3), 1118–1130.

Le Gall, L., Payri, C. E., Bittner, L., & Saunders, G. W. (2010). Multigene phylogenetic analyses support recognition of the Sporolithales ord. nov. *Molecular Phylogenetics and Evolution, 54*(1), 302–305.

Luiz, O. J., Madin, J. S., Robertson, D. R., et al. (2012). Ecological traits influencing range expansion across large oceanic dispersal barriers: Insights from tropical Atlantic reef fishes. *Proceedings of the Royal Society of London, 279*, 1033–1040.

Magris, R. A., Costa, M. D. P., Ferreira, C. E. L., et al. (2020). A blueprint for securing Brazil's marine biodiversity and supporting the achievement of global conservation goals. *Diversity and Distributions, 24*, 1–18.

Maliao, R. J., Turingan, R. G., & Lin, J. (2008). Phase-shift in coral reef communities in the Florida Keys National Marine Sanctuary (FKNMS), USA. *Journal of Marine Biology, 154*, 841–853.

Marlow, J. R., Lange, C. B., Wefer, G., et al. (2000). Upwelling intensification as part of the Pliocene-Pleistocene climate transition. *Science, 290*, 2288–2291.

Martin, S., & Gattuso, J. P. (2009). Response of Mediterranean coralline algae to ocean acidification and elevated temperature. *Global Change Biology, 15*, 2089–2100.
Martin, S., & Hall-Spencer, J. M. (2017). Effects of ocean warming and acidification on rhodolith/maërl beds. In *Rhodolith/Maërl beds: A global perspective*. Springer.
Mazzei, E. F., Bertoncini, A. A., Pinheiro, H. T., et al. (2017). Newly discovered reefs in the southern Abrolhos Bank, Brazil: Anthropogenic impacts and urgent conservation needs. *Marine Pollution Bulletin, 114*, 123–133.
McConnaughey, T. A., & Whelan, J. F. (1997). Calcification generates protons for nutrient and bicarbonate uptake. *Earth-Science Reviews, 42*(1–2), 95–117.
Mcmanus, J. W., & Polsenberg, J. F. (2004). Coral–algal phase shifts on coral reefs: Ecological and environmental aspects. *Progress in Oceanography, 60*, 263–279.
Meirelles, P. M., Amado-Filho, G. M., Pereira-Filho, G. H., et al. (2015). Baseline assessment of mesophotic reefs of the Vitória-Trindade Seamount Chain based on water quality, microbial diversity, benthic cover and fish biomass data. *PLoS One, 10*, e0130084.
Moura, R. L., Amado-Filho, G. M., Moraes, F. C., et al. (2016). An extensive reef system at the Amazon River mouth. *Science Advances, 2*, 1–12.
Moura, R. L., Abieri, M. L., Castro, G. M., et al. (2021). Tropical rhodolith beds are a major and belittled reef fish habitat. *Scientific Reports, 11*, 794.
Müller, R. D., Seton, M., Zahirovic, S., et al. (2016). Ocean basin evolution and global-scale plate reorganization events since Pangea breakup. *Annual Review of Earth and Planetary Sciences, 44*, 107–138.
Nelson, W. A., Sutherland, J. E., Farr, T. J., Hart, D. R., Neill, K. F., Kim, H. J., & Yoon, H. S. (2015). Multi-gene phylogenetic analyses of New Zealand coralline algae: Corallinapetra Novaezelandiae gen. et sp. nov. and recognition of the Hapalidiales ord. nov. *Journal of Phycology, 51*(3), 454–468.
Oliveira, E. C. (1977). *Algas marinhas bentônicas do Brasil*. Tese de Livre-Docência, Universidade de São Paulo, São Paulo, 406 f.
Paiva, S. V., Carneiro, P. B. M., Garcia, T. M., et al. (2023). Marine carbonate mining in the southwestern Atlantic: Current status, potential impacts, and conservation actions. *Marine Policy, 148*, 105435.
Paoli, C., Morten, A., Biachi, C. N., et al. (2016). Capturing ecological complexity: OCI, a novel combination of ecological indices as applied to benthic marine habitats. *Ecological Indicators, 66*, 86–102.
Pereira-Filho, G. H., Amado-Filho, G. M., Moura, R. L., et al. (2012). Extensive rhodolith beds cover the summits of southwestern Atlantic Ocean seamounts. *Journal of Coastal Research, 28*, 261–269.
Pitlik, T. J., & Paul, V. J. (1997). Effects of toughness, calcite level, and chemistry of crustose coralline algae (Rhodophyta, Corallinales) on grazing by the parrotfish *Chorurus sordidus*. In *Proceeding of the 8th international coral reefs symposium, Panama* (Vol. 1, pp. 701–706). Smithsonian Tropical Research Institute.
Pittman, W. C., Cande, S., LaBrecque, J., et al. (1993). Fragmentation of Gondwana: The separation of Africa from South America. In *Biological relationships between Africa and South America*. Yale University Press.
Riding, R. (2002). Structure and composition of organic reefs and carbonate mud mounds: Concepts and categories. *Earth-Science Reviews, 58*, 163–231.
Roberts, J. M., Wheeler, A. J., & Freiwald, A. (2006). Reefs of the deep: the biology and geology of cold-water coral ecosystems. *Science, 312*(5773), 543–547.
Santos, V. S., de Moura, R. L., Magdalena, U. R., et al. (2023). Spatial modeling reveals a growing threat to the world's largest rhodolith beds. *Ocean & Coastal Management, 232*, 106441.
Schubert, N., Salazar, V. W., Rich, W. A., et al. (2019). Rhodolith primary and carbonate production in a changing ocean: The interplay of warming and nutrients. *Science of the Total Environment, 676*, 455–468.
Shannon, L. V. (1985). The Benguela ecosystem. Part I. Evolution of the Benguela physical features and processes. *Annual Review of Marine Science, 23*, 105–182.
Simon, T. E., Pinheiro, H. T., Moura, R. L., et al. (2016). Mesophotic fishes of the Abrolhos shelf, the largest reef ecosystem in the South Atlantic. *Journal of Fish Biology, 89*, 990–1001.

Sissini, M. N., Oliveira, M. C., Gabrielson, P. W., et al. (2014). *Mesophyllum erubescens* (Corallinales, Rhodophyta)—So many species in one epithet. *Phytotaxa, 190*, 299–319.
Sissini, M. N., Berchez, F., Hall-Spencer, J., et al. (2020). Brazil oil spill response: Protect rhodolith beds. *Science, 367*, 156.
Steininger, F. F., & Rögl, F. (1984). Paleography and palinspastic reconstruction of the Neogene of the Mediterranean and Paratethys. In *The geological evolution of the eastern Mediterranean*. Blackwell Scientific Publications.
Steller, D. L., Riosmena-Rodríguez, R., Foster, M. S., et al. (2003). Rhodolith bed diversity in the Gulf of California: The importance of rhodolith structure and consequences of disturbance. *Aquatic Conservation, 13*, S5–S20.
Steneck, R. S. (1983). Escalating herbivory and resulting adaptive trends in calcareous algal crusts. *Paleobiology, 1*, 44–61.
Tâmega, F. T. S., & Figueiredo, M. A. O. (2005). Distribuição das algas calcárias incrustantes (Corallinales Rhodophyta) em diferentes habitats na Praia do Forno Armação dos Búzios Rio de Janeiro. *Rodriguésia, 56*, 123–132.
Tâmega, F. T. S., & Figueiredo, M. A. O. (2007). Distribution of crustose coralline algae (Corallinales Rhodophyta) in the Abrolhos reefs Bahia Brazil. *Rodriguésia, 58*, 941–947.
Tâmega, F. T. S., Riosmena-Rodríguez, R., Mariath, R., et al. (2014). Nongeniculate coralline red algae (Rhodophyta: Corallinales) in coral reefs from northeastern Brazil and a description of *Neogoniolithon atlanticum* sp. nov. *Phytotaxa, 190*, 277–298.
Tâmega, F. T. S., Riosmena-Rodríguez, R., Sportono-Oliveira, P., et al. (2015). Taxonomy and distribution of non-geniculate coralline red algae (Corallinales, Rhodophyta) on rocky reefs from Ilha Grande Bay, Brazil. *Phytotaxa, 192*, 267–278.
Taylor, W. R. (1955). Marine algal flora of the Caribbean and its extension into neighboring seas. In *Essays in the natural sciences in honour of Capitain Allan Hancock*. Creative Media Partners.
Torrano-Silva, B. N., Riosmena-Rodriguez, R., & Oliveira, M. C. (2014). Systematic position of *Paulsilvella* in Lithophylloideae (Corallinaceae, Rhodophyta) confirmed by molecular data. *Phytotaxa, 190*, 94–111.
Torrano-Silva, B. N., Vieira, B. R., Riosmena-Rodríguez, R., et al. (2018). Guidelines for DNA barcoding of coralline algae, focusing on Lithophylloideae (Corallinales) from Brazil. *Botanica Marina, 61*, 127–140.
Tuya, F., Schubert, N., Aguirre, J., et al. (2023). Levelling-up rhodolith-bed science to address global-scale conservation challenges. *Science of the Total Environment, 892*, 164818.
Veras, P. C., Pierozzi-Jr, I., Lino, J. B., et al. (2020). Drivers of biodiversity associated with rhodolith beds from euphotic and mesophotic zones: Insights for management and conservation. *Perspectives in Ecology and Conservation, 18*, 37–43.
Vieira-Pinto, T., Oliveira, M. C., Bouzon, J., et al. (2014). *Lithophyllum* species from Brazilian coast: Range extension of *Lithophyllum margaritae* and description of *Lithophyllum atlanticum sp. nov*. (Corallinales, Corallinophycidae, Rhodophyta). *Phytotaxa, 190*, 355–369.
Villas-Bôas, A. B., Riosmena-Rodriguez, R., Amado-Filho, G. M., et al. (2009). Rhodolith-forming species of *Lithophyllum* (Corallinales; Rhodophyta) from Espírito Santo State, Brazil, including the description of *L. depressum* sp. nov. *Phycologia, 48*, 237–248.
Villas-Bôas, A. B., Riosmena-Rodriguez, R., & Figueiredo, M. A. O. (2014). Community structure of rhodolith-forming beds on the central Brazilian continental shelf. *Helgoland Marine Research, 68*, 27–35.
Villas-Boas, A. B., Riosmena-Rodriguez, R., Tâmega, F. T. S., et al. (2015). Rhodolith-forming species of the subfamilies Neogoniolithoideae and Hydrolithoideae (Rhodophyta, Corallinales) from Espírito Santo State, Brazil. *Phytotaxa, 222*, 169–184.
Wilson, S., Blake, C., Berges, J. A., et al. (2004). Environmental tolerances of free-living coralline algae (maërl): Implications for European marine conservation. *Biological Conservation, 120*, 279–289.
Worm, B., Barbier, E. B., Beaumont, N., et al. (2006). Impacts of biodiversity loss on ocean ecosystem services. *Science, 314*, 787–790.
Wray, J. L. (1977). Calcareous algae and their environment. In *Calcareous algae*. Elsevier.

Chapter 2
A Historical Perspective of the Brazilian Rhodolith Beds: Discovering a Hidden Ecosystem

Flávio Berchez, Gilberto Dias, Marina N. Sissini, Paulo A. Horta, and Cláudio Gonçalves Tiago

Abstract This chapter discusses, from an historical perspective, the evolution of scientific knowledge since the description, by foreign naturalists, of the first types of rhodolith-forming species found in the Brazilian coast, in the late nineteenth century. During the second half of the twentieth century, geological and geophysical surveys demonstrated that calcareous algae beds along the Brazilian continental shelf possibly were the largest on the planet, which, consequently, attracted many commercial interests to the region. These commercial potential of rhodolith beds leveraged the first ecological studies on the Espirito Santo Continental Shelf, as a requirement for exploration activities. These studies revealed a remarkably diverse community, dominated by high brown algae biomass, attached to the relatively light rhodoliths, resulting in the constant displacement of the whole community, due to the wave action. A decrease of biomass was observed, probably related to the abrasion and burial during storms, resulting in a decrease from 2328 $g.m^{-2}$ f.w. in December 1986 to 65 $g.m^{-2}$ in May 1994. After the new millennium, new technical conditions allowed the increase of studies, but the knowledge is still incipient, with special concerns relying on their functional and economic roles. Besides, the

F. Berchez (✉)
Departamento de Botânica, Universidade de São Paulo, São Paulo, SP, Brazil

Cape Horn International Center (CHIC), Puerto Williams, Chile
e-mail: fberchez@ib.usp.br

G. Dias
Lagemar, Departamento de Geologia, Universidade Federal Fluminense, Niterói, RJ, Brazil
e-mail: gilbertotmd@id.uff.br

M. N. Sissini
Programa de Pós-Graduação em Ecologia, Universidade Federal de Santa Catarina, Florianópolis, SC, Brazil

P. A. Horta
Laboratório de Ficologia, Departamento de Botânica, Universidade Federal de Santa Catarina, Florianópolis, SC, Brazil

C. G. Tiago
Centro de Biologia Marinha, Universidade de São Paulo, São Sebastião, SP, Brazil
e-mail: clgtiago@usp.br

P. A. Horta, M. N. Sissini (eds.), *Brazilian Rhodolith Beds*, Brazilian Marine Biodiversity, https://doi.org/10.1007/978-3-031-61449-1_2

population and managers, in great part, do not even know the existence of such an ecosystem, precluding support to conservation.

Keywords Brazilian rhodolith beds · Calcareous reefs · Ocean literacy

2.1 The Beginning

Knowledge and perception of rhodolith beds is closely linked to prevailing historical concepts we have on the sea, to the paths followed by research on marine environments in Brazil, and existent seabed sampling and mapping technologies. The mid-twentieth century sense of marine and coastal environments was of a preserved nature with a great abundance of resources. Urban agglomeration were isolated, while large, practically unexplored, stretches of coastal sandy areas and rocky shore. Natural resources, such as fishing grounds, were extremely abundant, and urbanization for touristic purposes appeared as a promising economical alternative.

Within this context, the interest of Brazilian researchers on the marine environment began in a field that previously has been restricted to foreign naturalists (e.g., Harvey, 1947; Howe, 1928; Taylor, 1931, 1960). Regarding the rhodolith taxonomy, it was during one of these great scientific expeditions that were sampled and identified the first types of Brazilian rhodolith-forming species, as *Lithothamnion brasiliense* (Foslie, 1900) and *L. heteromorphum* (Foslie, 1907; species currently regarded as a synonym of *Lithothamnion crispatum*), both collected at the São Sebastião, São Paulo state, Brazil (Fig. 2.1).

The beginning of studies on taxonomy and systematics of marine algae began with Aylthon Brandão Joly (1957), who also contributed to the formation of many phycologists over the next decades (Oliveira, 1977) and the formation of an extremely productive working group, which was fundamental to the beginning of

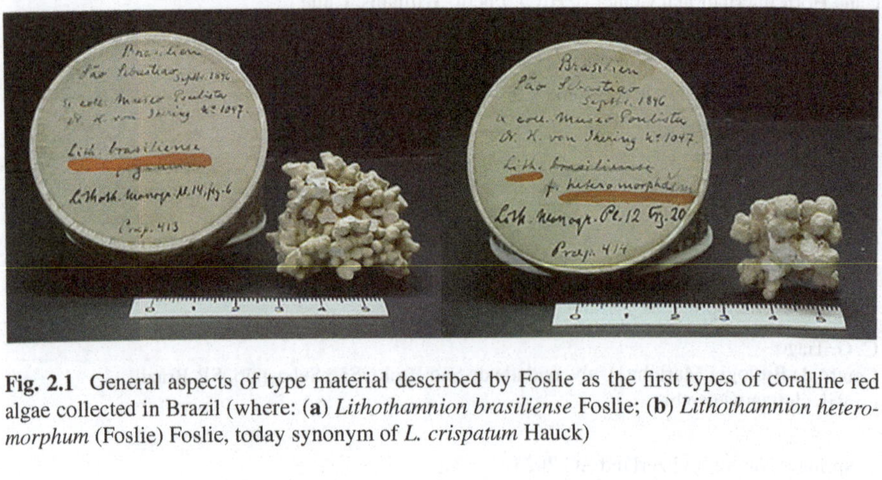

Fig. 2.1 General aspects of type material described by Foslie as the first types of coralline red algae collected in Brazil (where: (**a**) *Lithothamnion brasiliense* Foslie; (**b**) *Lithothamnion heteromorphum* (Foslie) Foslie, today synonym of *L. crispatum* Hauck)

knowledge about the structure and composition of the rhodolith beds. Although the vast majority of data related to this initial work refers mainly to the mid-littoral and shallow infralittoral region, occasional collections were also carried out on the Brazilian continental shelf by means of dredging. Studies based in these early collections, besides unveiling the taxonomic composition of these habitats, and their relationship with the midlittoral flora, also gave initial glimpses of their structure (Joly & Yoneshigue-Braga, 1966; Joly & Oliveira Filho, 1967).

The first ecological works that emerged from that time on (Nonato & Pérèz, 1961; Oliveira Filho & Mayal, 1976; Oliveira Filho & Paula, 1983) most focused on the intertidal zone, because of the limitations of Scuba Diving in Brazil. The first exploring the infralittoral, introducing the scientific SCUBA diving aiming to ecological/taxonomic purposes, are related to the works of Jacques Laborel and Françoise Laborel-Deguen, in the north east coast of Brazil (Laborel-Deguen et al., 2019). One decade later, the Oxford Diving Expedition to Cabo Frio (Maggs et al., 1979), followed by Eston et al. (1986), also studied the infralittoral.

Even with the local development of the first Scuba Diving techniques in the late 1970s, there still remained the challenge of diving at open sea, and in greater depths, this posing a delay in starting studies of these habitats.

In the 1970s, a research developed as part of the Remac Project (Vicalvi & Milliman, 1976; Milliman, 1977), and involving geological and geophysical surveys along the Brazilian continental shelf, revealed calcareous algae deposits occupying a large area, possibly the largest on the planet. Consolidated data also indicated a diverse structure and the occurrence at higher depths, when compared with the beds studied in other regions, showing a great economic potential, especially for marine limestone application in agriculture (Amaral, 1979).

The appearance, at this stage, of the first companies involved in commercial explorations resulted in the first ecological surveys, related to the Environmental Impact Studies, financed by Fermisa Mineração, who had received authorization to mine, experimentally, at the inside "Área Moleques" polygon, located in the State of Espírito Santo (ES) shelf, about 20 km from the coast, off the town of Marataízes (20°30′ S, 40°00′ W).

2.2 The Second Period

The first surveys carried out in the area were with geological bias, leading to the approval by the Departamento Nacional de Produção Mineral (DNPM), the Mineral Research Report, and the Mining Report and finally obtaining of the Mining Decree. A prior-license from Instituto Brasileiro do Meio Ambiente (IBAMA) was provided after the presentation of a preliminary biological diagnosis of the area, carried out in 1985 by a team from Universidade Federal do Rio de Janeiro (UFRJ), using qualitative samples, collected by dredging, in 16 stations. Initial studies relied on scarce bibliographic references to maërl-bottom surveys carried out in France. Attempts at comparison were difficult due to divergence in structural factors, such

Fig. 2.2 (**a**) View of the Fermisa's deck, with the clamshell dredge and a small amount of sediments already harvested. (**b**) Clam shell preserved sediment sample, showing a thin layer of live rhodoliths at the surface of the sea bottom. (**c**) Submarine image of live rhodoliths

as the depth of bioclastic sediments in ES. The dredging of bioclastic sediment, made on an experimental basis, was on a former ferryboat used in the Pacific during the second world war for disembarking British navy vehicles, the "FERMISA" (Fig. 2.2), using a clamshell equipment (100 × 200 cm). The Fermisa was able to purposefully stand on the beach and open the bow portal for entry of a small tractor that removed the dredged sediment. Curiously, on a small beach between the towns of Piúma and Anchieta, and bordering a small farm, on hearing the "Fermisa's" arrival, dairy cows would always go to the beach to munch very much appreciated rhodoliths.

At that time, environmental agencies at federal (IBAMA) and state (SEAMA/ES) levels required monitoring the area and an Environmental Impact Study (EIA/

RIMA). The team who conceived the research was linked to the Instituto de Biociências da Universidade de São Paulo (IB/USP), consisting of Verena Eston, Fábio Lang da Silveira, Cláudio Gonçalves Tiago and Flávio Berchez. Gilberto Dias (LAGEMAR-UFF) gave support to diving, as well as to geological observations and mapping. A Before-after-control-impact (BACI) design (Murray et al., 2006) was elected. A permanent buoy was anchored north of the mining area for location. Bearing in mind that the site was in the waters of the Brazilian Current, with an intensely predominating north-south flow, the control area was defined within a radius of 200 meters to the north of the buoy, with an opposing experimental area to the south, where the exploration would take place. The effect of the sediment plume resulting from the dredging was thus avoided on the control area.

This was one of the pioneering experiences of Brazilian scientific diving and particularly in the open sea, also incorporating unprecedented sampling techniques, related to the BACI Sampling Design and the use of random sampling elements (Fig. 2.3). In spite of sampling irregularity, up to now is the longest time series of a location related to rhodolith beds in the Southwestern Atlantic Ocean.

Observations were made over a decade, between 1986 and 1994. Part of the results were published (Berchez et al., 2009), but not in their entirety, due to limitations on the data domain by the contracting company. For the initial samplings (1986–1988), only data of the total biomass and the checklist of species were made available. The monitoring revealed that the initial biomass (in December 1986) was the highest with 2328 g.m^{-2} and dominated by *Sargassum filipendula C. Agardh* and *Sargassum hystrix J. Agardh* (Table 2.1). From this period on, a progressive

Fig. 2.3 Square 0.5 m^2 sampling element located at a place dominated by brown algae and sponges, during the 1986's summer campaign

Table 2.1 Pioneer monitoring program of rhodolith beds in Espírito Santo state according to before-after-control-impact (BACI) design (Murray et al., 2006) to evaluate the dredging impact

	Dec/1986	Feb/1987	Jul/1987	Sep/1988	Mar/1992	Oct/1992	Dec/1992	May/1994
Total biomass (g.m^{-2})	2328	1809	1094	703	1150	410	1585	65
Rhodolith (%)	–	–	–	–	54	17	24	56
Unconsolidated substratum (%)	–	–	–	–	16	69	38	33

Only total biomass (fresh weight), relative percent cover of rhodoliths, and percentage of unconsolidated substratum were evaluated from December 1986 to May 1984

reduction was observed, up to a minimum of 65 g.m^{-2} in the last sampling (in May 1994). Until March 1992, the Phaeophyta were dominant, and *Sargassum* spp. were the highest biomass (690 g.m^{-2}, Fig. 2.4), followed by *Dictyopteris* spp., *Zonaria tournefortii*, *Padina* spp., and *Stypopodium zonale*. The total biomass was 1150 g.m^{-2}, with a recuperation from the previous sampling (1988).

During the same year, a significant variation was observed in the total biomass, varying the dominant species. In October 1992, a reduction of total biomass was observed (410 g.m^{-2}), with those of *Sargassum* spp. still dominant, but lower (140 g.m^{-2}), closely followed by the red algae *Cryptonemia crenulata* (100 g.m^{-2}), a psamic species. On December 1992, although the total biomass (wet weight) was higher (1585 g.m^{-2}), the dominance was of the red algae *Halymenia* spp. and *Meristotheca gelidium* (699 g.m^{-2}), named in that moment as *Meristiella echinocarpum*, followed by *Sargassum* spp. (280 g.m^{-2}).

During the last sampling, in May 1994, the total biomass reduction was dramatic (65 g.m^{-2}), and the non-calcified were found in a scattered pattern. Under the gravel sediment rhodoliths with the remains of fleshy algae still attached to them could be found, in special *Sargassum* spp. holdfasts. *Cryptonemia crenulata*, *Stypopodium zonale,* and *Caulerpa sertularioides* were also found buried, but in a good condition, denoting a resistance to burial. This could explain the relative higher importance of the first during both periods of biomass reduction (October, 1992 and May, 1994).

Based on these observations, Berchez et al. (2009) concluded that one of the factors involved in the drop in both periods was the wave action, very often detected in the bottom by the divers, due to the low depth of the "Área Moleques'" polygon (15–20 m depth). Rhodolith pendular displacement was constant as result of the wave action on the attached epiphytic algae. This movement also reduced the precision of measurements based on a fixed sampling element. During periods of rough sea, much more intense and chaotic effects would be expected.

This dynamic also explains the origin of the great amount of rhodoliths found on adjacent quartzose sand beaches. Due to the epiphyte-reduced densities and thallus features such as *Sargassum* air vesicles, pendular movements combined with the strong drift SW current facilitate transport toward the beach. Over time, the leaf algae decompose on the emerged beach, thus leaving the clustered rhodolith intact.

Percent cover data allowed the evaluation of the relative proportion of rhodolith and unconsolidated sediment along the last four samplings (Table 2.1). The total percent cover of rhodolith plus unconsolidated substratum were high during the two periods of biomass reduction, October 1992 and May 1994, respectively, 88% and 89%. Lower values are related to the peak of red algae biomass (62%), which, unlike *Sargassum* spp., do not have their thalli erect and collapse over the rhodoliths and unconsolidated substratum. The percent cover of rhodolith alone varied from 54% to 17%, without a clear correlation with other factors. Throughout sampling from March 1992 to May 1994, species richness decreased from 20.6 to 14.7, Shannon Diversity from 2.9 to 1.6, and Simpson Dominance increased from 0.2 to 0.4.

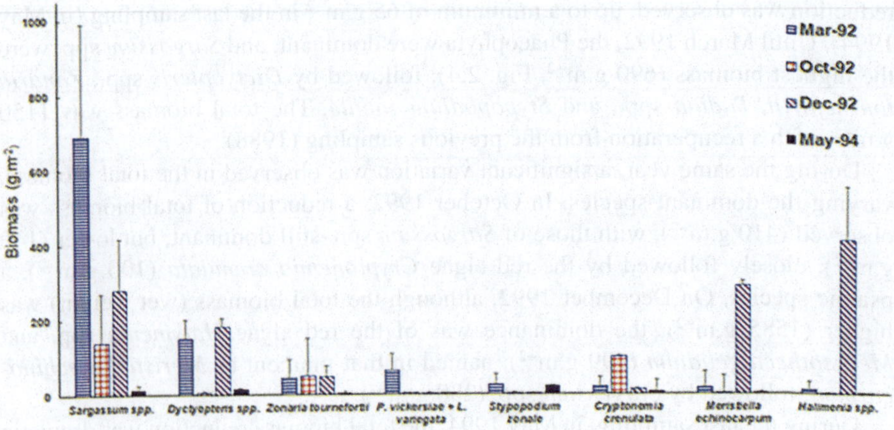

Fig. 2.4 Biomass (fresh weight) of the dominant species from March 1992 to May 1994. $N = 10$ square 0.5 m² sampling elements. Vertical lines: confidence intervals ($p = 0.05$)

In compliance with IBAMA's requirements, activities continued between March 2002 and October 2003 but under a new team led by Gilberto Dias and Roberto Villaça (Dep. of Marine Biology at UFF). The results (Dias & Villaça, 2012) placed in evidence the geological pattern of the bed, evidencing a layer of calcareous sediments over 10 m thick, formed mainly by the fragmentation of coralline algae. In terms of the species distribution and abundance, although only presenting observational results, without quantitative data, the authors indicate a pattern similar to the previous studies, with the establishment of foliose macroalgae on the rhodoliths during the summer. Later, during the winter, these became senescent and disturbed by the storms, which altered the sea-bottom, simultaneously burying the rhodoliths and removing epiphyte algae. Around 10–20% of rhodoliths became permanently buried and are incorporated into the sediment mass. Additionally, the authors found a recurrent bloom of an opportunistic Bivalvia (*Pinctata pectinata*) in the springs.

Fermisa was sold and became ALGAREA. The Public Hearing for the presentation of the Algarea Project took place in June 2005. The IBAMA Operating License (LO), valid for 4 years, and referring to an area of 600 m² within the "Moleques" polygon, was issued in 2007. In 2010, the license was renewed for another 8 years. However, the company was unable to extract a significant amount of material. During the eight trips made from 2007 to 2010, only approximately 6000 m³ were extracted. By comparison, 600,000 m³ alone were extracted every year in France during the 1970s.

2.3 Modern Surveys

With the arrival of the new millennium, the number of researches related to rhodolith beds had increased. Access to appropriate technologies has opened up new opportunities for studying benthic communities. Underwater digital photography facilitates obtaining a large number of high-quality images, which could be achieved and checked from the surface, by using underwater cables or remote operated vehicles (ROVs). Side-scan sonar has facilitated mapping extensive areas, then showing the distribution of habitats along the Brazilian continental shelf with great accuracy, and revealing the largest rhodolith bed in the world, at the Abrolhos shelf (Amado-Filho et al., 2012). There has also been an increase in marine ecologists trained for scientific diving, which, together with novel ecological approaches and new forms of analysis, has greatly facilitated obtaining answers to questions about these habitats.

Despite the significant advances made since the first ecological articles, which date back to a decade and a half ago, the knowledge about the rhodolith beds is still incipient with special concerns relying on their functional and economic roles. Besides the improvement of our knowledge about the biological diversity related to these key environments (Horta et al., 2016), much more attention is necessary to describe the microbiomes associated with these builders (Calegario et al., 2021), as well as their responses to complex scenarios related with the interaction pollution, latitudinal gradient, deep, ocean warming, and acidification (Schubert et al., 2019). Despite advancing our understanding of their taxonomy, distribution, and ecological drivers (Sissini et al., 2020), we still have these organisms and environments threatened, with most of their biological and ecological particularities among priorities areas to discuss and implement marine protected areas (MPA, Magris et al., 2021). Recent pieces of evidence point to the threats related to predatory fishery impacting trophic chains (Moura et al., 2021) and the niche suitability when these stressors, as trawling, interact with ocean warming and acidification (Fragkopoulou et al., 2021). A recent episode of oil spill, which constitutes the biggest ecological accident on the Brazilian continental shelf, evidenced, in principle, inaction, lack of government coordination, and the absence of contingency plans for various possible risk situations, not only for the rhodolith beds (Sissini et al., 2020) but also for several marine ecosystems (Brum et al., 2020; Soares et al., 2020). More recently, the governmental discussions promoted to discuss a widespread and rapid exploitation of rhodolith resources highlighted a similar problem (Berchez et al., 2022). It became evident the lack of knowledge about rhodoliths beds, since possible impacts were not even cited, at any time, by different actors, such as the press, local agencies, or government managers, and very wrong concepts were disseminated along the discussions. This observation coincides with several empirical ones and preliminary data, which shows that even the existence of the Rhodolith Beds is unknown by the vast majority of the Brazilian population.

Concern is a consequence of knowledge. Invisible ecosystems, such as the rhodolith beds, need stronger efforts to be recognized and thus gain support for conservation as other coastal ecosystems, as tropical reefs, kelp forest, mangroves, and seagrasses (Tuya et al., 2023). The lack of ecological knowledge (Soares et al., 2020) prevents the assessment of impacts on the ecosystem and makes it difficult to take possible actions for mitigation and for recovery assessment. Furthermore, social surveys, in order to evaluate perceptions and knowledge of this habitat by various categories of public, are essential to direct policymakers actions toward conservation.

Within the proclamation of the 2021–2030, an international Decade of Ocean Science for Sustainable Development by IOC/UNESCO is a clear understanding that, to allow nations, stakeholders, and citizens to build conceptions and substantiate decisions in scientific knowledge, it is necessary a "transparent and accessible" ocean, where data obtained from research projects are available to all actors, using current information technologies (Ryabinin et al., 2019). The main topics of concern in Ocean Literacy are a better understanding of the importance of the oceans, the human-ocean interactions, and how to act sustainably and thus reduce human impacts (Santoro et al., 2017; Borja et al., 2020).

Within this scenario, the first priority to Brazilian policymakers is to improve understanding on rhodolith beds, including their structural diversity, based on hypotheses related to geographic and depth gradients, their functional roles, and consequent ecological and economic importance. Clearly, the establishment of long-term monitoring stations at several and more sensitive points along the coast and the Brazilian oceanic islands is primordial. Simultaneously, bold projects to spread knowledge about their importance and sensibilities to all audiences, improving communication based on social survey's data (such as the documentary produced by the Darwin 200 and PELD ILOC Projects, https://youtu.be/1-XLuEvOMlw?si=-bCx6s3sGk5WYlFQ), are essential to induce conservation concern and support. Regarding the diverse and persistent threats related with anthropogenic factors (Costa et al., 2023), policies related with mitigation of local stressors are urgent to avoid cascading effects when these interact with ocean warming and acidification (Venegas et al., 2023).

Brazil has many marine protected areas (MPA, Rylands & Brandon, 2005). In the last few decades, many marine and coastal MPA have been created, part of them including rhodolith beds. The Abrolhos National Marine Park is a prominent example. However, there are no MPAs directly associated with the large rhodolith beds on the continental shelf of northeastern or northern Brazil.

2.4 Final Remarks

Due to implicit biodiversity, unique characteristics worldwide, and sensitivity to phenomena linked to local stressors, as trawling impacts, and to global climate change, the Brazilian rhodolith beds merit MPAs of various categories and sizes. Together with other forthcoming benefits would be the provision of wildlife refuges, stimulation of studies and long-term monitoring, coordination of the use of natural resources, installation of warning systems against natural and man-made impacts, and incitement to environmental education actions. However, on considering the only slight MPA efficacy for reducing impacts linked to major global threats (Hilborn 2015), of which rhodolith beds are particularly sensitive to global warming and acidification consequences, other strategies for their protection should not be neglected.

References

Amaral, C. A. B. (1979). *Recursos Minerais da Margem Continental Brasileira e das Áreas Oceânicas Adjacentes*. Projeto REMAC. Vol. 10. Petrobras. Rio de Janeiro, 112p.

Amado-Filho, G. M., Moura, R.L., Bastos, A.C., Salgado, L.T., Sumida, P. Y., Guth, A., Francini-Filho, R.B., Pereira-Filho, G.H., Abrantes, D., Brasileiro, P., Bahia, R., Leal, R.N., Kaufman, L., Kleypas, J. Farina, M. & Tompson, F. (2012). Rhodolith beds are major $CaCO_3$ bio-factories in the tropical South West Atlantic. *PloS One, 7(4), e35171.*

Berchez, F., Tiago, C. G., Rosso, S., Dias, G., & Oliveira, E. C. (2009). Structure of a coralline algal bed on southeastern Brazil. *Brazilian Journal of Ecology, 13*, 49–57.

Berchez, F., Horta, P. A., Sissini, M. N., Tiago, C. G., Turra, A., Schubert, N., Santos, I., & Cabral, A. (2022). Brazil: Heed price of marine mining for an alternative fertilizer. *Nature, 607*(7918), 239.

Borja, A., Santoro, F., Scowcroft, G., Fletcher, D., & Strosser, P. (2020). Connecting people to their oceans: Issues and options for effective ocean literacy (Editorial). *Frontier in Marine Science, 6*, 4–5.

Brum, H. D., Campos-Silva, J. V., & Oliveira, E. G. (2020). Brazil oil spill response: Government inaction. *Science, 367*(6474), 155–156.

Calegario, G., Freitas, L., Appolinario, L. R., Venas, T., Arruda, T., Otsuki, K., et al. (2021). Conserved rhodolith microbiomes across environmental gradients of the great Amazon reef. *Science of the Total Environment, 760*, 143411.

Costa, D. A., Dolbeth, M., Christoffersen, M.L., Zúñiga-Upegui, P.T., Venâncio, M., Lucena, R. F. P. (2023) An Overview of Rhodoliths: Ecological Importance and Conservation Emergency. *Life, 13, 1556.*

Dias, G., & Villaça, R. (2012). Coralline algae depositional environments on the Brazilian central–south-eastern shelf. *Journal of Coastal Research., 28*, 270–279.

Eston, V. R., Migotto, A. E., Oliveira Filho, E. C., Rodrigues, S. A., & Freitas, J. C. (1986). Vertical distribution of benthic marine organisms on rocky coasts of the Fernando de Noronha Archipelago (Brazil). *Boletim do Instituto Oceanográfico, 34*, 37–54.

Foslie, M. (1900). New or critical calcareous algae. *Det Kongelige Norske Videnskabers Selskabs Skrifter, 1899*(5), 1–34.

Foslie, M. (1907). Algologiske notiser IV. *Det Kongelige Norske Videnskabers Selskabs Skrifter, 1907*(6), 1–30.

Fragkopoulou, E., Serrão, E. A., Horta, P. A., Koerich, G., & Assis, J. (2021). Bottom trawling threatens future climate refugia of rhodoliths globally. *Frontiers in Marine Science, 7*, 1246.

Harvey, H. W. (1947). Manganese and the growth of phytoplankton. *Journal of the Marine Biological Association of the United Kingdom, 26(4), 562–579.*

Hilborn, R. (2015). Marine protected areas miss the boat. *Science, 350*, 1326.

Horta, P. A., Riul, P., Amado Filho, G. M., Gurgel, C. F. D., Berchez, F., Nunes, J. M. C., Scherner, F., Pereira, S., Lotufo, T., Peres, L., Sissini, M., Bastos, E. O., Rosa, J., Munoz, P., Martins, C., Gouvêa, L., Carvalho, V., Bergstrom, E., Schubert, N., Bahia, R. G., Rodrigues, A. C., Rörig, L., Barufi, J. B., & Figueiredo, M. (2016). Rhodoliths in Brazil: Current knowledge and potential impacts of climate change. *Brazilian Journal of Oceanography (Online), 64*, 117–136.

Howe, M. A. (1928). Notes on some marine algae from Brazil and Barbados. *Journal of the Washington Academy of Sciences, 18(7), 186–194.*

Joly, A. B. (1957). *Contribuição ao conhecimento da flora ficológica marinha da Baia de Santos e arredores*. Bolm Fac. Filos. Cienc. e Letras da Univ. S. Paulo 217 (Botânica 14), 198pp.

Joly, A. B., & Oliveira Filho, E. C. (1967). Notes on Brazilian algae. I – New findings confirming uncertain records. *Boletim da Faculdade de Filosofia, Ciências e Letras da USP, 305*, 313–320.

Joly, A. B., & Yoneshigue-Braga, Y. (1966). *Primeira nota sobre algas coletadas durante as viagens do navio oceanográfico "Almirante Saldanha"* (Vol. 34, pp. 1–12). Publicação do Instituto de Pesquisas Marinhas do Rio de Janeiro.

Laborel-Deguen, F., Nunes, F., Castro, C. & Pires, D. (2019). Recifes brasileiros: o legado de Laborel. Série Livros do Museu Nacional, Rio de Janeiro, 64. ISBN 978-85-7427-075-3., 376p

Maggs, C. A., Milner, A., Watts, W., & Whittle, M. R. (1979). The Oxford diving expedition to Cabo Frio, Brazil. *Bulletin of the Oxford University Exploration Club, New Series, 4*, 13–40.

Magris, R. A., Costa, M. D., Ferreira, C. E., Vilar, C. C., Joyeux, J. C., Creed, J. C., et al. (2021). A blueprint for securing Brazil's marine biodiversity and supporting the achievement of global conservation goals. *Diversity and Distributions, 27*(2), 198–215.

Marina Nasri, Sissini Gabrielle, Koerich Maria Beatriz, de Barros-Barreto Luana M., Coutinho Fernanda P., Gomes William, Oliveira Iara O., Costa José Marcos, de Castro Nunes Maria Carolina, Henriques Talita, Vieira-Pinto Beatriz N., Torrano-Silva Mariana Cabral, Oliveira Line, Le Gall Paulo Antunes, Horta (2022). Diversity distribution and environmental drivers of coralline red algae: the major reef builders in the Southwestern Atlantic *Coral Reefs, 41(3) 711–725.* https://doi.org/10.1007/s00338-021-02171-1

Milliman, J. D. (1977). Role of calcareous algae in Atlantic continental margin sedimentation. In E. Flügel (Ed.), *Fossil algae*. Springer. https://doi.org/10.1007/978-3-642-66516-5_26

Moura, R. L., Abieri, M. L., Castro, G. M., Carlos-Júnior, L. A., Chiroque-Solano, P. M., Fernandes, N. C., et al. (2021). Tropical rhodolith beds are a major and belittled reef fish habitat. *Scientific Reports, 11*, 794.

Murray, S. N., Ambrose, R. A., & Dethier, M. N. (2006). *Monitoring rocky shores*. University of California Press.

Nonato, E. F., & Pérèz, J. (1961). Observations sur quelques peuplements intertidaux de substrat dur dans la région d'Ubatuba (État de Sao Paulo). *Cahiers de Biologie Marine, 2*, 263–270.

Oliveira Filho, E. C. (1977). Algas marinhas bentônicas do Brasil. São Paulo, 407p.

Oliveira Filho, E. C., & Mayal, E. M. (1976). Seasonal distribution of intertidal organismos at Ubatuba, São Paulo, Brazil. *Revista Brasileira de Biologia, 36*, 3050–3316.

Oliveira Filho, E. C., & Paula, E. J. (1983). *Aspectos da distribuição vertical e variação sazonal de comunidades da zona de marés em costões rochosos do litoral norte do Estado de São Paulo. I Encontro de Macrófitas Marinhas* (p. 147). Boletim do Instituto de Pesquisas da Marinha.

Ryabinin, V., Barbière, J., Haugan, P., Kullenberg, G., Smith, N., McLean, C., Troisi, A., Fischer, A., Aricò, S., Aarup, T., Pissierssens, P., Visbeck, M., Enevoldsen, H., & Rigaud, J. (2019). The UN decade of ocean science for sustainable development. *Frontiers in Marine Science, 6*, 470.

Rylands, A. B., & Brandon, K. (2005). Brazilian protected areas. *Conservation Biology, 19*, 612–618.

Santoro, F., Santin, S., Scowcroft, G., Fauville, G., & Tuddenham, P. (2017). *Ocean literacy for all – A toolkit*. IOC/UNESCO & UNESCO Venice office Paris. IOC Manuals and Guides, 80 revised in 2018, 136p.

Schubert, N., Salazar, V. W., Rich, W. A., Bercovich, M. V., Saá, A. A., Fadigas, S. D., et al. (2019). Rhodolith primary and carbonate production in a changing ocean: The interplay of warming and nutrients. *Science of the Total Environment, 676*, 455–468.

Sissini, M. N., Berchez F., Hall-Spencer, J., Ghilardi-Lopes, N., Carvalho, V.F., Schubert, N., Koerich, G., Diaz-Pulido, G., Silva, J., Serrão, E., Assis J., Santos, R., Floeter, S. R., Rorig, L., Barufi, J. B., Bernardino, A.F., Francini-Filho, R., Turra, A., Hofmann, L. C., Aguirre, J., le Gall, L., Peña, V., Nash, M.C., Rossi, S., Soares, M., Pereira-Filho, G., Tâmega, F., Horta, P. A. (2020). Brazil oil spill response: protect rhodolith beds. *Science, 367:156.*

Soares, M. O., Teixeira, C. E. P., Bezerra, L. E. A., Rossi, S., Tavares, T., & Cavalcante, R. M. (2020). Brazil oil spill response: Time for coordination. *Science, 367*, 155.

Taylor, W. R. (1931). A synopsis of the marine algae of Brazil. Laboratoire de Cryptogamie du Museum National d'Histoire Naturelle. *Brittonia, 1(1), 7–33.*

Taylor, W. R. (1960). Marine algae of the eastern tropical and subtropical coasts of the Americas. University of Michigan. Ann Arbor, USA.

Tuya, F., Schubert, N., Aguirre, J., et al. (2023). Levelling-up rhodolith-bed science to address global-scale conservation challenges. *Science of the Total Environment, 892*, 164818.

Venegas, R. M., Acevedo, J., & Treml, E. A. (2023). Three decades of ocean warming impacts on marine ecosystems: A review and perspective. *Deep Sea Research Part II: Topical Studies in Oceanography, 212*, 105318.

Vicalvi, M. A., & Milliman, J. D. (1976). *Carbonate sedimentation on the continental shelf off southern Brazil, with special reference to the benthonic foraminifera*. American Association of Petroleum Geologist, Special Paper. The University of Texas, Austin.

Chapter 3
Rhodolith Beds: Occurrence, Diversity and Evolution

Maria Beatriz B. Barros-Barreto, Luana M. Coutinho, Fernanda P. Gomes, Iara O. Costa, Maria Carolina Henriques, Talita Vieira-Pinto, Marina N. Sissini, Mariana C. Oliveira, and Paulo A. Horta

Abstract Within Rhodophyta, with nearly 7100 known living species, the Corallinophycidae deserve main attention. Besides the crustose form, the coralline red algae are responsible for the rhodolith bed formation. With five orders, these algae present a complex taxonomy and a sub-estimated richness. This chapter describes their anatomical and reproductive characters, the diversity and distribution along the Brazilian Blue Amazon.

Keywords Corallinophycidae · Corallinales · Hapalidiales · Sporolithales · Systematic

---

M. B. B. Barros-Barreto (✉) · L. M. Coutinho · F. P. Gomes · M. C. Henriques
Laboratório de Ficologia, Universidade Federal do Rio de Janeiro, Rio de Janeiro, RJ, Brazil
e-mail: beatrizdebb@biologia.ufrj.br

I. O. Costa
Programa de Pós-Graduação em Botânica, Universidade Estadual de Feira de Santana, Feira de Santana, BA, Brazil

Departamento de Ciências Humanas e Tecnologias (DCHT) do Campus XXIV, Universidade do Estado da Bahia (UNEB), Xique-Xique, Brazil

T. Vieira-Pinto · M. C. Oliveira
Departamento de Botânica, Instituto de Biociências, Universidade de São Paulo, São Paulo, SP, Brazil
e-mail: mcdolive@ib.usp.br

M. N. Sissini
Programa de Pós-Graduação em Ecologia, Universidade Federal de Santa Catarina, Florianópolis, SC, Brazil

P. A. Horta
Laboratório de Ficologia, Departamento de Botânica, Universidade Federal de Santa Catarina, Florianópolis, SC, Brasil

P. A. Horta, M. N. Sissini (eds.), *Brazilian Rhodolith Beds*, Brazilian Marine Biodiversity, https://doi.org/10.1007/978-3-031-61449-1_3

3.1 Introduction

Within the red algae phylum (Rhodophyta), with nearly 7100 known living species, among the calcareous groups, the Corallinophycidae, characterised by calcification in their cellular walls, deserve special attention (Cabioch & Giraud, 1986). This class is composed of five orders: Corallinales, Hapalidiales, Sporolithales and more recently Corallinapetrales (Jeong et al., 2021) and Rhodogorgonales (Fredericq & Norris, 1995; Peña et al., 2020). The Corallinales can be geniculated with uncalcified segments and calcified segments, or non-geniculated with entirely calcified thalli. Hapalidiales and Sporolithales are non-geniculate, growing as crustose coralline algae (CCA) or free-living forms. Non-geniculate coralline algae are present in practically all marine environments, where they can grow firmly attached to a consolidated substrate, such as rocky shores and biological reefs or unattached to the substrate, by forming calcareous structures of free life, known as rhodoliths (Foster, 2001).

Structurally, a rhodolith can be exclusively structured of one or more CCA species or contain other organisms, such as bryozoans, foraminifers and gastropods (Steneck, 1986; Foster, 2001; Holz et al., 2020). Rhodoliths can develop around a nucleus (e.g. shell or rock fragments), or through spore settlements on hard substrates (Freiwald & Henrich, 1994; Foster, 2001) or reef fragmentation (Halfar & Riegl, 2013). When grouped, they form rhodolith beds. The composition of these beds has many variations in the oceans and can be monospecific (Riosmena-Rodriguez et al., 1999) or multispecific (Birkett et al., 1998). Most of the rhodolith beds in the world are comprised of two or three species of CCA (Konar et al., 2006). Even though at Ilha do Arvoredo, Santa Catarina and Brazil, Pascelli et al. (2013) found six different species.

The rhodoliths, as well as the CCA, are broadly distributed, both in latitude and longitude, and are present from the poles to the tropics (Steneck, 1986; Foster, 2001), as well as from the intertidal zones to a depth of 268 m (Littler et al., 1986). Until 2016, the calcareous algae were thought to be exclusively marine; however, Žuljević et al. (2016) registered *Pneophyllum cetinense* in freshwater environments, in a river in Croatia, and this species is genetically close to marine *Pneophyllum* (91–95%).

The Brazilian continental platform possesses the largest area of rhodoliths in the world, ranging from Amazon Reefs, Amapá state (Bahia, 2014; Moura et al., 2016; Richards et al., 2019) to the state of Santa Catarina, at Arvoredo and Deserta Islands (Sissini et al., 2014). The beds can also be found on oceanic islands, such as Rocas Atoll, Fernando de Noronha Archipelago, along the Vitoria-Trindade Chain, and São Pedro e São Paulo Archipelago (Villas-Bôas et al., 2005; Figueiredo et al., 2007; Amado-Filho et al., 2012b; Bahia, 2014; Crespo et al., 2014).

3.2 Vegetative and Reproductive Characters

The morphology of the CCA is quite varied. To identify them, we must observe the external and internal characteristics, with some diagnostic characteristics to Brazilian taxa discussed by Horta (2002).

3.2.1 *Vegetative Characters*

3.2.1.1 External Characters

Regarding their external aspect, we can analyse four main characteristics: (1) colouring, (2) growth forms, (3) types of reproductive structures and (4) position of the reproductive structures on the thallus surface and form a disposition of epithelial cells. Only characteristics 1 and 2 can be observed with the naked eye and on the field; for all others, it is required the use of a stereo or scanning electron microscopy.

1. *Colouring:* Varies from rose and goes to other shades of red to purple, blue, greenish grey or brown (Littler & Littler, 2013). Colouring, however, is not considered an aspect for taxonomic determination (Fig. 3.1).

Fig. 3.1 Colouring. (**a**) Pale rose. (**b**) Brownish rose. (**c**) Grey to rose. (**d**) Rose. (**e**). Greenish grey. Scale bar = 1 cm

Table 3.1 Types of growth according to Woelkerling et al. (1993)

Growth forms	Description	Image
Encrusting	Plants crustose without protuberances and branches	
Warty	Plants with warty (verrucose) protuberances that are usually <3 mm long and unbranched	
Lumpy	Plants with irregular lumpy, usually nodular and contiguous, rarely may be branched	
Fruticose	Plants with cylindrical to compressed protuberances that are mostly >3 mm long	

2. *Growth forms:* The way they growth (i.e. external aspect) has an important role in the taxonomic history of the CCA because this character was used to distinguish genera and species in the first taxonomic works to be published (e.g. Linnaeus, 1767; Lamouroux, 1816; Philippi, 1837; Areschoug, 1852). Currently, this character is no longer diagnostic for genera or species because of the high morphological plasticity (Table 3.1).

   Due to the absence of a pattern referring to the different types of growth, in 1993, Woelkerling et al. reviewed the names of the types used since 1895 and grouped them into 10 different types: (1) not consolidated, (2) encrusting, (3) warty, (4) lumpy, (5) fruticose, (6) discoid, (7) in layers, (8) foliose, (9) ribbon-like and (10) arborescent. These terms are still used in modern taxonomy to describe the types of growth of the CCA. The types of growth found in this work are presented in Table 3.1.
3. *Reproductive structures:* CAs present their reproductive cells developed into calcified chambers isolated or fused. When fused, these chambers are called conceptacles, which can be opened by one or multiple pores (Fig. 3.2).
4. *Position of the reproductive structures on the thallus surface:* This characteristic is used mainly to improve the description of the types of reproductive structures. We can position the reproductive structure as flush, in the same plane or raised above the surrounding thallus surface.

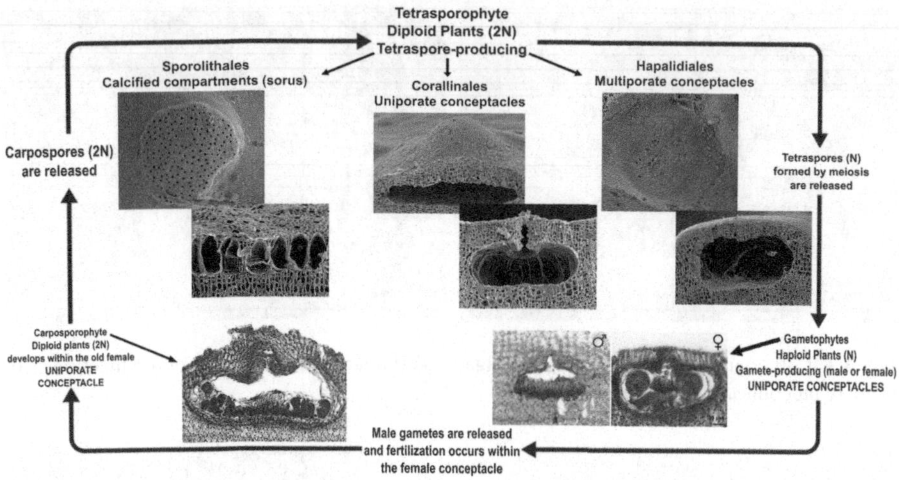

Fig. 3.2 Corallinophycidae life cycle. Tetrasporophyte (diploid phase-2n), meiosis occurs to form tetraspores (n) that can be developed into calcified chambers in Sporolithales, or fused forming conceptacles, that can be opened by one pore in Corallinales, or multiple pores in Hapalidiales. Tetraspores are released and will grow into Gametophytes (haploid phase-n), and gametes (male or female, haploid) are produced in uniporate conceptacles. Male gametes are released, and fertilisation occurs within the female conceptacle. After gametes fusion, a third phase, called Carposporophyte (diploid phase-2n), develops into the female, uniporate conceptacle, which will produce and release carpospores (2n). Modified from Harvey (2005)

3.2.1.2 Internal Characters

As to their internal characteristics, we can analyse three characteristics: (1) cell connections, (2) epithelial and subepithelial cells form and (3) thallus organisation. For all the above characteristics, the use of optical microscopy and/or scanning electron microscopy (SEM) is required

1. *Cell connections: As in all other rhodophytes*, primary pit connections are present between cells in the same filament. However, between cells of adjacent filaments, we will observe secondary pit connection (Fig. 3.3a) or cell fusions (Fig. 3.3b). In some groups, secondary pit and cell fusion can be present at the same specimen.
2. *Thallus construction:* The thallus of the coralline algae is made up of filaments that can be organised in two patterns: dimerous or monomerous (coaxial and non-coaxial).

   - *Dimerous:* Two groups of filaments oriented perpendicularly, one being parallel to thallus surface (Fig. 3.4a).
   - *Monomerous:* A single system of filaments that contribute to a ventral or central core and a peripheral region that runs outwards towards the thallus surface.

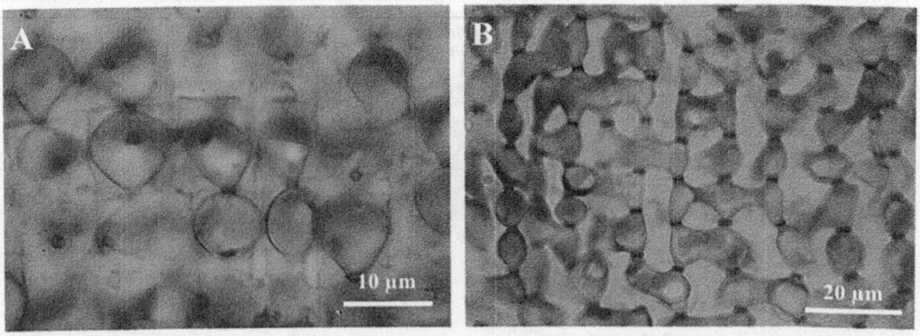

Fig. 3.3 Cell connections in coralline red algae. (**a**) Primary and secondary pit connections. (**b**) Primary pit connections and cell fusions

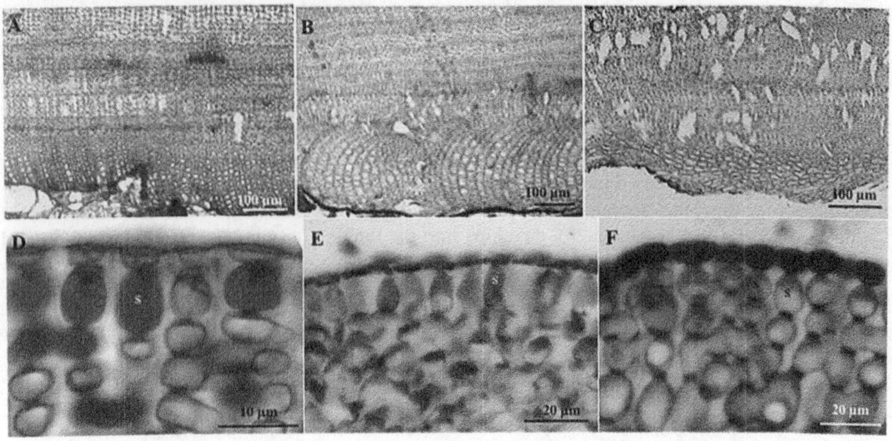

Fig. 3.4 Thallus construction. (**a**). Dimerous. (**b**) Monomeorus coaxial. (**c**) Monomerous non coaxial. (**d**) Flared epithelial cell. Size of the subepithelial cell (s) initials longer than the subtending ones. (**e**) Flattened epithelial cell. Size of the subepithelial cell (s) initials longer than the subtending ones. (**f**) Rounded epithelial cell. Size of the subepithelial cell (s) initials equal to the subtending ones

- The internal organisation of these CAs can also be organised in a perithallial region, an internal cortex, and hypothallus. This more internal region can present their filaments with different organisations. In the **coaxial** structure, core filaments are aligned in arching tiers (Fig. 3.3b), while in the **non-coaxial** (plumose), the core filaments are more or less parallel (Fig. 3.4C).

3. *Epithelial and subepithelial cells form:* Epithelial cells can be flared (Fig. 3.4d), flattened (Fig. 3.4e) or elliptical to rounded (Fig. 3.4f). The subepithelial cells can be as long as, longer than or shorter than the subtending cells (Fig. 3.4d–f).

3.2.2 *Reproductive Characters*

A. *Types of reproductive structures:* Reproductive structures hold most of the diagnostic features utilised in the taxonomy of CCA. In each life stage of its triphasic cycle, different types of reproductive structures can be found: Tetrasporophyte, diploid plants; Gametophytes, haploid plants; and Carposporophyte, diploid spores which develop within the old female conceptacle (Fig. 3.2).

Gametophyte (n): Present uniporate conceptacles, where sexual cells, or female and male gametes develop. Female conceptacles, with carpogonia and trycogine, are called carpogonial conceptacles, while the male ones are spermatangial conceptacles. The **carposporophyte (2n)**, resulted from the carpogonia (female gamete) fertilisation by the spermatio (male gamete), is observed in the uniporate female conceptacle that after fertilisation can be called carposporangial conceptacle.

In the **Sporophyte (2n)**, the Sporolithales present their tetraspores in calcified chambers aggregated in sorus. These chambers can be fused forming a conceptacle that can be uniporate in Corallinales or multiporate in Hapalidiales (Fig. 3.2).

3.3 The Molecular Revolution in Systematics Rhodolith Forming Corallinophycidae Species

The taxonomic study of coralline algae is considered complex and problematic (Horta, 2000, 2002). The carbonatic structure that imposes further difficulties in the search of anatomical diagnostic features, plus the morphological plasticity, where traits can change with changes of environmental conditions, can be raised as arguments to explain why the taxonomy of this important group have been neglected (Irvine & Chamberlain, 1994). Due to morpho-anatomical limitation to clarify the adaptive strategies utilised by this group to occupy so diverse niches in the global scale, molecular analyses have been allied to classical taxonomy to clarify doubts related to the correct position of taxa (Huisman et al., 2004).

Molecular studies have brought a considerable turnaround in Corallinophycideae. The leading molecular markers currently used in this group are SSU, *psb*A and COI. Bailey and Chapman (1998) conducted the first phylogenetic study in Corallinales based on the small subunit ribosomal RNA gene (SSU rDNA). After this work, the SSU rDNA marker is being used in Corallinophycideae studies (Vidal et al., 2003; Bittner et al., 2011; Vieira-Pinto et al., 2014; Hernández-Kantún et al., 2015; Nelson et al., 2015). The gene that transcribes the D1 protein of photosystem II (*psb*A), which is more variable than SSU rDNA, is an important phylogenetic marker in this group to rebuild relationships between families, genera and even species (Broom et al., 2008; Bittner et al., 2011; Sissini et al., 2014; Carro et al., 2014; Peña et al., 2015; Nelson et al., 2015; Hernández-Kantún et al., 2015). Other markers known as DNA barcodes are being used. The 5′ to 3′ region of the mitochondrial cox1, which codifies subunit 1 of the cytochrome oxidase enzyme (COI), is one of the main markers used as DNA barcode (Hebert et al., 2003). Saunders (2005) has used it in Rhodophyta species and concluded that it was a good marker for species identification.

In 2010, Le Gall et al. proposed the Sporolithales order based on molecular markers and diagnostic characters: calcified tetrasporangial compartments and cruciate tetraspores. At that time, Corallinales was composed of two families Corallinaceae and Hapalidiaceae with zonate tetraspores (Harvey & Woelkerling, 2007).

The family Corallinaceae had, then, four subfamilies: Corallinoideae, Lithophylloideae, Masthophoroideae and Metagoniolithoideae (Harvey et al., 2003). Bailey et al. (2004) and Broom et al. (2008), based on SSU rDNA and *psb*A, showed that three of these subfamilies were monophyletic and Mastophoroideae was polyphyletic. Based on molecular phylogeny, Kato et al. (2011) proposed the segregation of *Neogoniolithon* to the new subfamily Neogoniolithoideae, and Rösler et al. (2016) included *Spongites* into this subfamily.

In 2015, Nelson et al. elevated Hapalidiaceae to Hapalidiales based on concatenated molecular markers and very distinct morphological characteristics such as having multiporate tetrasporangial conceptacles, mainly. In this order, only Melobesioideae is represented on the Brazilian coast. This subfamily is characterised by adjacent cells connected by cell fusion and apical plugs blocking the porous channel before the liberation of spores (Woelkerling, 1996; Harvey et al., 2003).

3.4 Species Records in Brazil

Currently, 76 living and one fossil CCA taxa are recognized in Brazil, building rhodoliths (Appendix 1); among them, 31 Corallinales taxa, nine genera are recognized in this order, the most speciose one being *Lithophyllum* (9 spp.); 17 Sporolithales taxa, being 16 *Sporolithon* species and *Heydrichia* (?) *poignantii*; and 29 taxa within Hapalidiales, encompassing 8 genera, *Lithothamnion* was the most speciose one (9 spp.), now, sharing its diversity with a recently proposed new genus, *Roseolithon* (9 spp.).

Villas-Boas et al. (2009) described *Lithophyllum depressum* Villas-Boas, Figueiredo and Riosmena-Rodriguez. Farias et al. (2010) re-evaluated *L. brasiliense* and *L. heteromorphum* in Brazil, proposed *L. heteromorphum* as a heterotypic synonym of *L. superpositum*, and extended its distribution to the Western Atlantic. Mariath et al. (2012) described *Lithothamnion steneckii* Mariath and Figueiredo and expanded the occurrence of *Pneophyllum conicum* to the Atlantic Ocean. Comparative taxonomic analyses revealed the conspecificity of *Sporolithon dimotum* to *S. ptychoides*, reported for the first time to the Atlantic Ocean (Bahia et al., 2011).

Riosmena-Rodriguez et al. (2014) have published a special issue on Phytotaxa about systematics and biogeography of the Corallinophycideae of the Atlantic

Ocean, where several articles contemplating some non-geniculate coralline algae of Brazil were published. Bahia et al. (2014b) described *Porolithon improcerum* and *Mesophyllum macroblastum* as new occurrences for the South Atlantic Ocean; Costa et al. (2014) studied the taxonomy of CCA of the northeastern littoral of Brazil; Henriques et al. (2014a) studied the Lithophylloideae and Mastophoroideae sub-families of the Brazilian continental platform; Henriques et al. (2014b) described three species of *Sporolithon* of the Brazilian continental platform, including the description of a new species *Sporolithon elevatum*; Tâmega et al. (2014) studied non-geniculate corallines from the coral reefs of Brazil's northeast and described *Neogoniolithon atlanticum,* also a new species; and Crespo et al., 2014 studied the floristic composition of the CCA of the São Pedro and São Paulo archipelago. Vieira-Pinto et al. (2014) described the Lithophylloideae species from the South of Brazil, proposing the new species *L. atlanticum.* Borges et al. (2014) described the genus *Melobesia* to the subtropical southwestern Atlantic coast, with the addition of *M. rosanoffii.* Sissini et al. (2014) re-evaluated *Mesophyllum erubescens* (Corallinales, Rhodophyta) based on molecular information.

Woelkerling et al. (2014) described for the first time the *Heydrichia* (?) *poignantii*, a 100-year-old fossil species, from the northeastern region of Brazil. *Sporolithon molle* was added to the Atlantic flora (Bahia et al., 2014a).

3.4.1 Occurrence and Distribution

The world's broadest extensive area of rhodolith beds is located in the Brazilian coast, distributed in a discontinued way, from the Amazon Reefs of Amapá to Santa Catarina (Fig. 3.4) covering the entire extension of the Brazilian continental platform (Richards et al., 2019; Moura et al., 2016; Sissini et al., 2014; Fig. 3.5), as well as oceanic islands (Rocas Atoll, Fernando de Noronha Archipelago, Vitoria-Trindade Chain, Trindade Island and São Pedro e São Paulo; Villas-Boas et al., 2005; Amado-Filho et al., 2012b; Crespo et al., 2014. Bahia et al., 2014b). There is a relevant concentration in the region where the continental platform widens up—from the south of Bahia to the north of Espírito Santo (Amado-Filho & Pereira-Filho, 2012; Horta et al., 2016). The world's largest rhodolith bed was recently discovered in the region of the Abrolhos continental platform (Amado-Filho et al., 2012b). This bed occupies an area of more than 20,000 km and has global importance in the production of calcium carbonate, comparable to the coralline reefs of the Caribbean and the Great Reef Barrier of Australia (Amado-Filho et al., 2012b; Moura et al., 2013). Another recent discovery was the carbonate reef of approximately 10,000 km, in the region between the French Guiana and Brazil, encompassing the Amazon River Delta region; this is the first reef found under the plume of a large river (Moura et al., 2016). The most recent discover described the occurrence of rhodoliths in the São Paulo coast, filling an important gap of the rhodolith bed distribution in southwestern Atlantic coast (Pereira Filho et al., 2019).

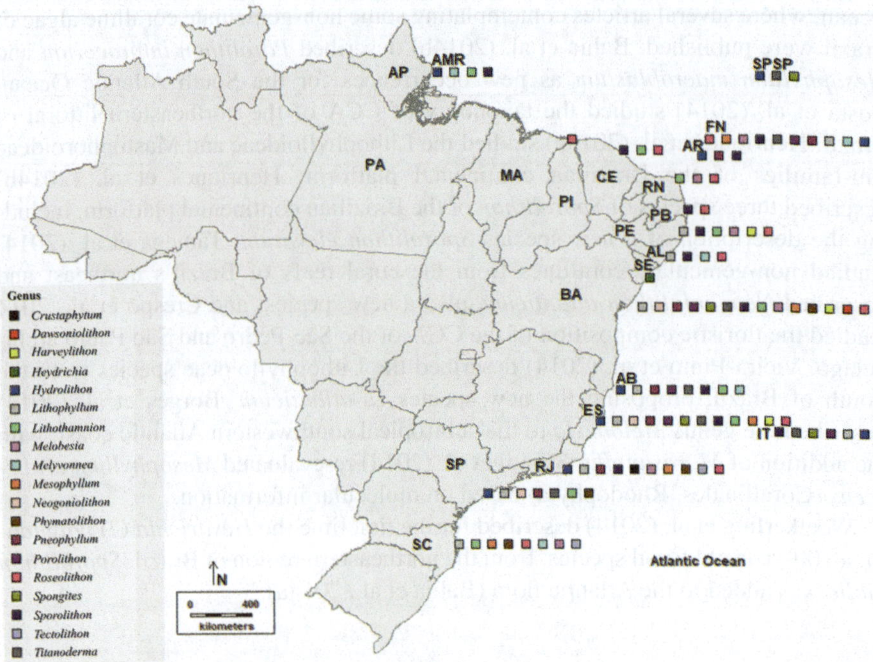

Fig. 3.5 Distribution of non-geniculate calcareous algae on the Brazilian coast. Brazilian states: *AP* Amapá, *PA* Pará, *MA* Maranhão, *CE* Ceará, *RN* Rio Grande do Norte, *PB* Paraíba, *PE* Pernambuco, *AL* Alagoas, *SE* Sergipe, *BA* Bahia, *ES* Espírito Santo, *RJ* Rio de Janeiro, *SP* São Paulo, *SC* Santa Catarina, *RS* Rio Grande do Sul. Oceanic Islands Reefs and Archipelago: *AMR* Amazon Reef, *SPSP* São Pedro and São Paulo Archipelago, *FN* Fernando de Noronha, *AR* Rocas Atoll, *AB* Abrolhos Archipelago, *IT* Trindade Island

3.5 Brazilian Diversity Based on Integrative Taxonomy

Since the pioneer works of Foslie (1900, 1905), when were described, respectively, *Lithothamnion brasiliense* and *L. heteromorphum* (species currently regarded as a synonym of *Lithothamnion crispatum*), to São Sebastião, São Paulo coast, the taxonomy of coralline red algae was based on morpho-anatomical evaluations (Horta, 2002). Since the last decade, the number of studies using molecular markers has been increasing significantly for the Brazilian coast (e.g. Vieira-Pinto, 2011; Bahia et al., 2013; Sissini et al., 2014; Henriques, 2016; Jesionek et al., 2016; Vieira-Pinto, 2016). So far, 11 studies have been published, including monographies and theses. Bahia et al. (2013, 2015) described two species: *Sporolithon tenue* and *S. yoneshigueae*, with morpho-anatomic and molecular approaches, using *psb*A as a molecular marker. Recently, comparative analyses of DNA sequences from Brazilian *S. ptychoides* were circumscribed to a new species *S. amadoi* (Richards et al., 2019). Vieira-Pinto et al. (2014) described a new species for Brazil: *Lithophyllum atlanticum* using the SSU rDNA, CO1 and UPA molecular markers.

In a Master degree study, Coutinho (2014) identified seven taxa, of the Corallinales and Hapalidiales orders, in one single island of Armação dos Búzios, Rio de Janeiro state, with molecular (CO1 and *psb*A) and morphological data. Jesionek et al. (2016) molecularly confirmed the presence of six known taxa on the Brazilian coast and two unknown taxa—*Lithophyllum* sp. and a Hapalidiales species (Fig. 3.6).

At the moment, the Melobesioideae subfamily possesses 18 genera, seven of them distributed in Brazil (*Crustaphytum, Lithothamnion, Melobesia, Mesophyllum, Melyvonnea, Phymatolithon, Roseolithon* and *Tectolithon*) (Guiry & Guiry, 2023; Flora do Brasil, 2020). Sissini et al. (2014) sequenced the type species, *Mesophyllum erubescens*, collected at Fernando de Noronha in 1887, with the *rbc*L-3P molecular marker, indicating the existence of several cryptic species with this same specific epithet. The *M. erubescens* species, however, is restricted to warm tropical and

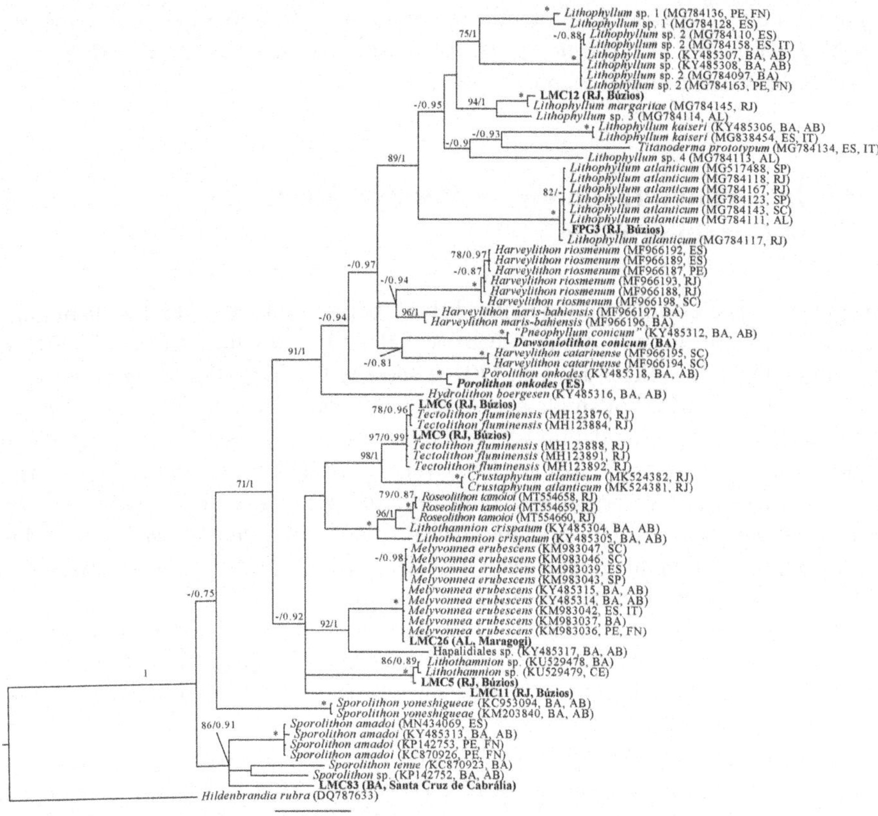

Fig. 3.6 Bayesian Inference, tree inferred with *psb*A marker for Brazilian samples. Bootstrap support over 70% is shown on branches for ML tree (GTR + I + G, −ln=, processed in PhyML-Guindon et al., 2010) with 1000 replicates. *Posterior probability* over 0.75 is shown on branches for BI tree (Cipres server, mixed models, rates = invgamma, with one million generations and 0.25 burnin). *Total support for branches. Newly generated sequences are shown in bold

temperate regions of the Western Atlantic Ocean, including the Caribbean Sea and the South of the Gulf of Mexico. Although Athanasiadis and Ballantine (2014) limited *Mesophyllum* to the North Hemisphere and proposed *Melyvonnea* to the South, DNA analyses integrated with morpho-anatomical data are needed to clarify the circumscription of this genus (Gabrielson et al., 2018). Recently, Coutinho (2019) sequenced samples from Brazil, with different molecular markers; one SSU sequence from this clade has 0% divergence in comparison to U61257 shown in Athanasiadis and Ballantine (2014) cladogram. Therefore, after confirmation with molecular data, we will consider the name *Melyvonnea erubescens* for the Brazilian samples. Until now, *Melobesia* and *Phymatolithon* have no molecular data.

Tectolithon fluminensis Jesionek was described, and *Crustaphytum* L.-C. Liu and S.-M. Lin is now extended to the Atlantic Ocean, as the new species C. *atlanticum* Jesionek (Jesionek et al., 2020; Fig. 3.6). Coutinho et al. (2021) proposed *Roseolithon* to circumscribe samples characterized by flared epithelial cells and pitted aspect in multiporate conceptacles given by the depressed rosette cells surrounding the pore canal; *Lithothamnion crispatum* from Abrolhos, BA (Jesionek et al., 2016), were included in Roseolithon (Fig. 3.6).

3.6 Integrative Taxonomy to Resolve Species in Coralline Algae

At present, species-delimitation methods have been an important tool to distinguish species, mainly in red algae (Jesus et al. 2016; Gomes et al. 2020). In Corallinophycidae, Richards et al. (2016) showed that *Lithothamnion* is polyphyletic based on analyses of single (*psb*A and COI) and concatenated markers (*psb*A, COI and UPA). Species delimitation analyses performed on *psb*A and COI with Automatic Barcode Gap Species Discovery (ABGD) and General Mixed Yule Coalescence (GMYC) segregated nine *Lithothamnion* spp. This study included three samples from Brazil, and these were placed in *Lithothamnion* sp. I. The recently proposed genus *Roseolithon* (Coutinho et al., 2021, Fig. 3.6) solved, partially, the polyphyly of *Lithothamnion*.

Torrano-Silva et al. (2018) have shown the usefulness of multiple-marker (COI-5P, UPA, *psb*A and *rbc*L-3P) and two algorithmic approaches [automatic barcode gap discovery (ABGD), and Poisson tree process (PTP)] for the species delimitation of Lithophylloideae in Brazil.

Based on three markers (*psb*A, *rbc*L and COI), morpho-anatomical characters and species delimitation methods (mBGD, ABGD, SPN and PTP), Costa et al. (2019) described three new species as a result of this approach. The well-known *Hydrolithon samoense* turned out to be three new species (Fig. 3.6).

Several Brazilian taxa are still not defined. This can be seen in the Bayesian inferred (BI) tree with *psb*A (Fig. 3.6). Not published sequences are shown, and illustrations are given (Figs. 3.6, 3.7, 3.8 and 3.9).

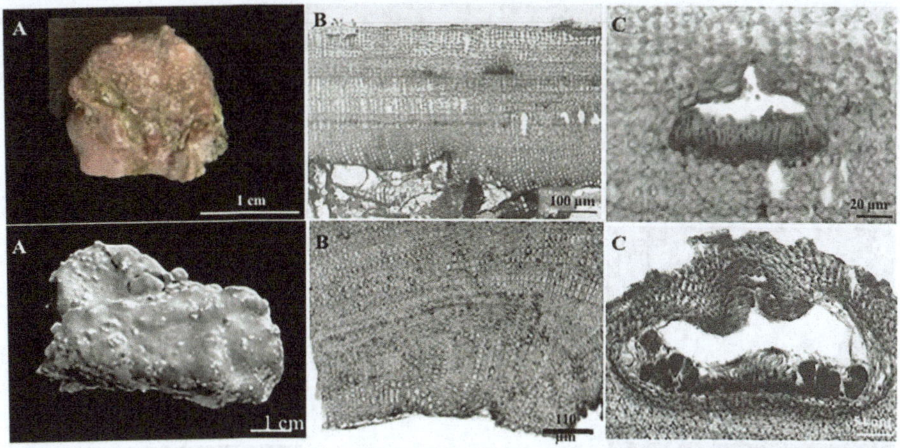

Fig. 3.7 *Lithophyllum margaritae* (LMC12). (**a**) Encrusting growth-form. (**b**) Vertical section through the vegetative thallus showing the dimerous thallus construction. (**c**) Vertical section showing buried spermatangial conceptacle. Spermatangial systems confined to the conceptacle floor. *Lithophyllum atlanticum* (FPG3). (**d**) Encrusting growth-form. (**e**) Vertical section through the vegetative thallus showing the dimerous thallus construction. (**f**) Vertical section through a uniporate carpogonial conceptacle showing carpogonial branches at the centre of the conceptacle floor

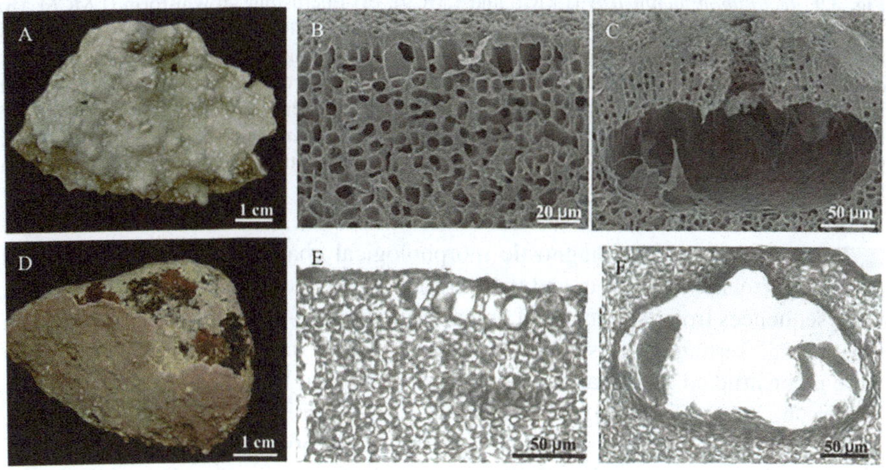

Fig. 3.8 *Dawsoniolithon conicum* (ALCB125758). (**a**) Encrusting growth-form. (**b**) Longitudinal fracture, SEM, showing trichocytes occurring in horizontal fields, and cells of adjacent filaments linked by cell fusions. (**c**) Longitudinal fracture, SEM, showing uniporate tetrasporangial conceptacle. *Porolithon onkodes* (ALCB125642). (**d**) Encrusting growth-form. (**e**) Longitudinal section through the thallus showing trichocytes occurring in horizontal fields, and cells of adjacent filaments linked by cell fusions. (**f**) Longitudinal section through uniporate tetrasporangial conceptacle with sporangium zonately divided

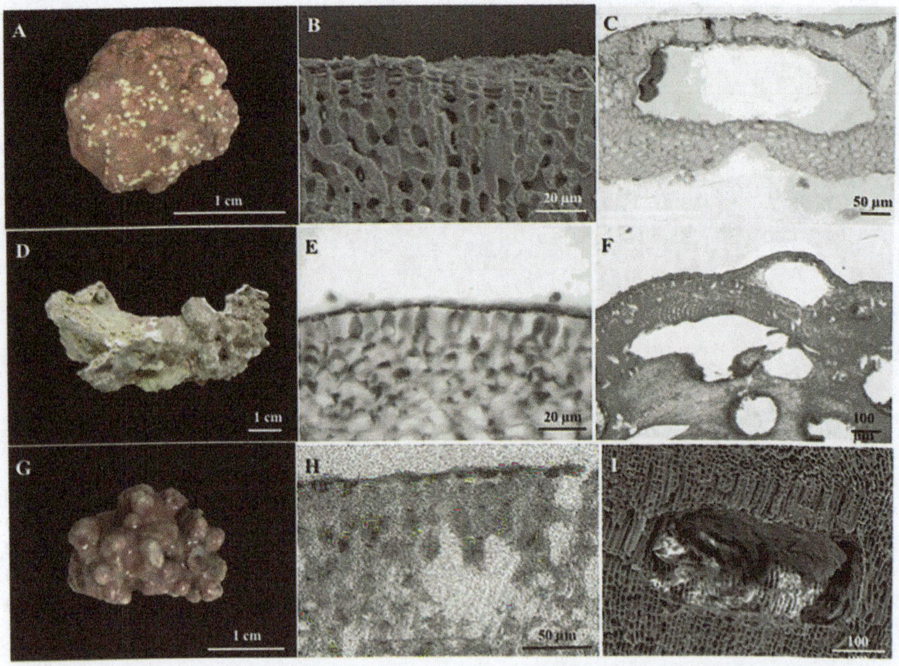

Fig. 3.9 *Tectolithon fluminensis* (LMC6 and LMC9). (**a**) Encrusting growth-form (LMC6). (**b**) Vertical section showing one to four layers of flared epithelial cells. (LMC9). (**c**) Section through a multiporate conceptacle showing pore plugs and tetrasporangium (LMC6). *Melyvonnea erubescens* (LMC26). (**d**) Warty growth-form. (**e**) Section of epithelial cells flattened to rounded, but not flared, and subepithelial cells more elongated than the subsequent ones. (**f**) Section through the thallus showing the coaxial with a multiporate conceptacle. *Lithotahmnion* sp. (LMC5 e IBC1907). (**g**) Lumpy growth-form. (LMC5). (**h**) Section showing flared epithelial cells. (LMC5). (**i**) Section showing a putative buried multiporate conceptacle (IBC1907)

To analyse the CCA, diagnostic morphological characters were reconstructed over the *psb*A Bayesian tree, and the most informative ones are shown (Table 3.3). Only sequences from type material or taxa with morpho-anatomical characters from Operational Terminal Units (OTU) were included in the analyses. The characters were reconstructed with maximum parsimony (MP) in Mesquite 3.6 (Maddison & Madisson, 2019) (Fig. 3.10).

Tetrasporangial compartments clearly corroborate the definition of Corallinophycidae orders (Fig. 3.10a). Apical plugs are present in Hapalidiales and Sporolithales (Fig. 3.10b). Thallus construction can be present within different specimens of the same species (Fig. 3.10c), as shown. This character is useful only for descriptions and cannot be used to define taxa. Cell fusions are observed in all taxa, except *Lithophyllum* and *Titanoderma*, in which cells of adjacent filaments are joined by a secondary pit connection and *Sporolithon*, which is characterized by both features—secondary pit connections and cell fusions (Fig. 3.10d). Trichocytes are commonly used to differentiate species but are not informative, since all

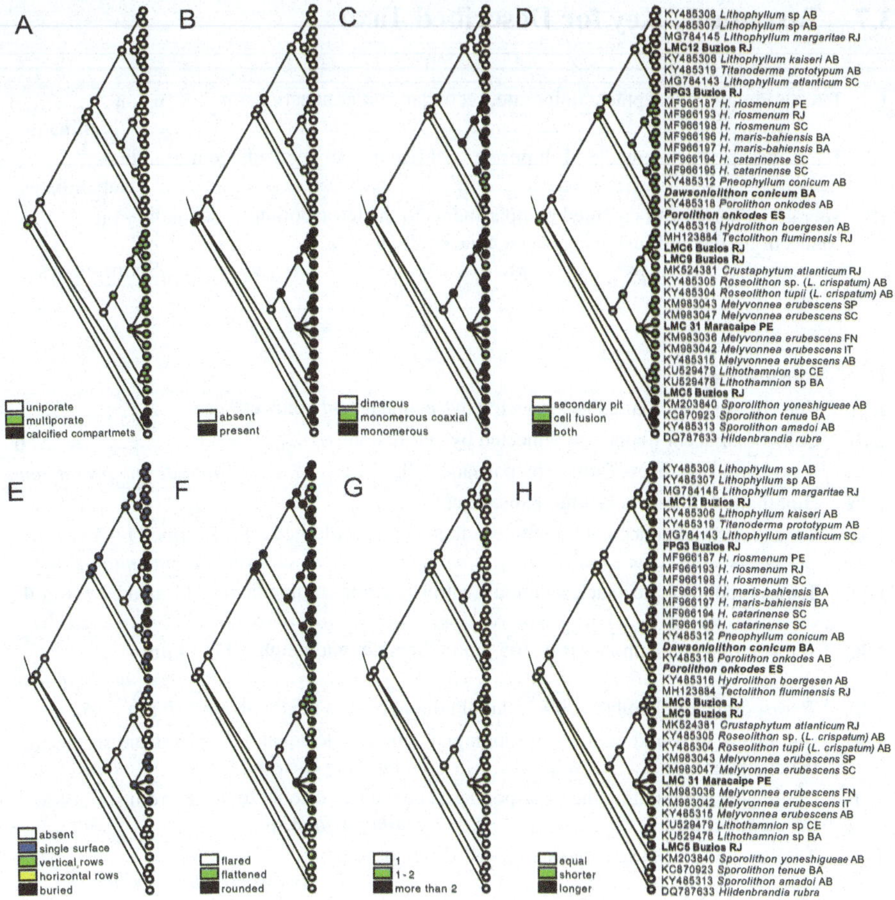

Fig. 3.10 Ancestral character reconstruction under MP over the Bayesian tree (*psb*A) Areas in the pie diagrams represent the relative support for each character reconstructed. (**a**) Tetrasporangial conceptacle type. (**b**) Apical plugs. (**c**) Thallus construction. (**d**) Cells connections between adjacent filaments. (**e**) Trichocytes. (**f**) Epithelial cells form. (**g**) Epithelial cells layers. (**h**) Subepithelial initials size

different states are present in *Harveylithon riosmenun* (Fig. 3.10e). As it is already known, one diagnostic character to distinguish *Lithothamnion* and *Mesophyllum* is the epithelial cell form that solely is not informative in the actual circumscription of these genera showing the necessity to circumscribe the Hapalidiales genera (Fig. 3.10f). Epithelial cell layers (two or more) can be a diagnostic character for *Tectolithon* and *Crustaphytum* (Fig. 3.8 and 3.10g). The size of the subepithelial cells and the subtending ones are not an informative character (Fig. 3.10h). Although these characters revealed not being informative, the combination of some of them can contribute to the taxa morpho-anatomical circumscription.

3.7 Taxonomic Key for Described Taxa

I.	Tetrasporangial conceptacle uniporate, secondary pit connections or cell fusion .... Corallinales
II.	Tetrasporangial conceptacle Multiporate, cell fusions between adjacent filaments .... Hapalidiales
III.	Sporangial sorus with calcified compartments to the tetrasporangia, secondary pit connections and cell fusions between adjacent filaments ....Sporolithales

I.	**Corallinales**
I.1a.	Cells of adjacent filaments connected by secondary pit connections 2
I.1b.	Cells of adjacent filaments connected by cell fusion 6
I.2a.	Hipothallial filaments formed by palisade cells .... *Titanoderma prototypum*
I.2b.	Hipothallial filaments lacking palisade cells 3
I.3a.	Tetrasporangial conceptacles with calcified central columella, roof formed by 3–4 cells .... *Lithophyllum margaritae*
I.3b.	Tetrasporangial conceptacles without central calcified columella, roof formed by 4–6 cells 4
I.4a.	Tetrasporangial chambers 196–365 µm in diameter and height 61–114 µm .... *Lithophyllum atlanticum*
I.4b.	Tetrasporangial chambers 200-350um in diameter and height above 115 µm … 5
I.5a.	Cell layers surrounding the tetrasporangial conceptacle equal to 12–16 (from surface to the chamber base) .... *Lithophyllum kaiseri*
I.5b.	Cell layers surrounding the tetrasporangial conceptacle equal to 9–12 (from surface to the chamber base) .... *Lithophyllum* sp.
I.6a.	Thallus dimerous, trichocytes arranged singly or in pairs .... *Hydrolithon boergesen*
I. 6b.	Thallus monomerous 7
I.7a.	Tetrasporangial pore canals lined by cells oriented more-or-less parallel or at a sharp angle to the conceptacle roof surface .... *Dawsoniolithon conicum*
I.7b.	Tetrasporangial pore canals lined by a ring of conspicuous and enlarged cells that are orientated more or less perpendicularly (vertically orientated) to the roof surface 8
I.8a.	Trichocytes arranged in horizontal fields becoming buried in the thallus .... *Porolithon onkodes*
I.8b.	Trichocytes arranged singly on surface, buried within the thallus and/or in short vertical rows.. 9
I.9a.	Roof of tetrasporangial conceptacle with 4–8 cell layers .... *Harveylithon catarinense*
I.9b.	Roof of tetrasporangial conceptacle with 2–4 cell layers 10
I.10a.	Intercalary meristematic cells of the roof of the tetrasporangial conceptacle generally columnar, >10 µm in height .... *Harveylithon riosmenum*
I.10b.	Intercalary meristematic cells of the roof of the tetrasporangial conceptacle generally ovoid, <10 µm in height .... *Harveylithon maris-bahiensis*

(continued)

II.	**Hapalidiales**
II.1.a.	Rounded or flattened epithelial cells 2
II.1.b.	Flared epithelial cells 4
II.2.a.	Non coaxial monomerous hypothallus .... *Tectolithon fluminensis*
II.2.b.	Coaxial monomerous hypothallus 3
II.3.a.	Filaments lining the pore canals are composed of 6–9 cells (the subepithelial initial is shorter than those comprising other roof filaments) .... *Crustaphytum atlanticum*
II.3.b.	Filaments lining the pore canals are composed of 3–5 cells (the basal cells being elongate) .... *Melyvonnea erubescens*
II.4.a.	Subepithelial initials either larger than, approximately equal in size to, or smaller than their immediate inward derivatives. Rosette cells surrounding the multiporate pore canal in depressions giving the surface a pitted appearance... 5
II.4.b.	Subepithelial initials either larger than, approximately equal in size to, or smaller than their immediate inward derivatives........... *Lithothamnion* sp.
II.5.a.	Subepithelial initials usually as long as or longer than immediate inward derivatives .... *Lithothamnion crispatum* (*Roseolithon* OTU3)
II.5.b.	Subepithelial initials shorter than or approximately equal to those subtending them. Multiporate conceptacle bordered by 6–7 rosette cells with raised rims and slightly sunken pore plates .... *Roseolithon tamoioi*

III.	**Sporolithon**
III.1a.	Thallus composed by less than 20 cell layers, senescent tetrasporangial chambers shedding without elongated cells at the base ......*Sporolithon tenue*
III.1b.	Thallus composed by more than 20 cell layers, senescent tetrasporangial chambers buried on the thallus, with elongated cells at the base 2
III.2a.	Tetrasporangial compartments pore with 35–43 μm in diameter surrounded by 19–24 rosette cells .... *Sporolithon yoneshigueae*
III.2b.	Tetrasporangial compartments pore with 8.5–16.5 μm in diameter surrounded by 9–13 rosette cells .... *Sporolithon amadoi*

3.8 Final Remarks

Based on presented data (Appendix 2), it is possible to conclude that DNA analyses are essential in the identification of calcareous algae species and current knowledge of CCA species based on morpho-anatomical characteristics (Appendix 3) is not robust and does not reflect the true diversity of species and, therefore, should be updated by adding molecular data information.

Cryptic diversity of Brazilian flora is being revealed with the use of integrative taxonomy. This approach shows that molecular markers have been fundamental to help grouping closely related organisms and frequently pointed out hidden morphological characters not seen before. Therefore, more efforts regarding rhodolith bed composition will clarify if all CCA can form rhodoliths and if all rhodolith species present representatives under the crustose form.

The resolution of Corallinophycidae taxa is still confusing, not only at the species level but also at genus level. To clarify the relationship within CCA, an integrative taxonomy, combining multiple markers or phylogenomics with species delimitation methods and morpho-anatomical analyses, is needed. It is also relevant to explore poorly studied localities as the northern part of the Brazilian coast and deeper areas, including the southern region. It is important to invest more efforts in the recognition of dispersal mechanisms, connectivity and intra- and inter-specific genetic diversity of rhodolith populations. This information will improve our knowledge about the representativity of these free living CCAs to foster further discussions about conservation of this group in tropical and warm temperate southwestern Atlantic coast.

Appendices

Appendix 1: Non-geniculate Calcareous Algae on the Brazilian Coast

Taxa	Distribution	Literature	Markers and Genbank accessed codes
Corallinales P.C. Silva & H.W. Johansen			
Dawsoniolithon conicum	BA	Costa et al. (2019)[a]	*psb*A, SSU
Harveylithon catarinense I.O. Costa, P.A. Horta & J.M.C. Nunes	SC	Costa et al. (2019)[a]	*psb*A (MF966195, MF966194), *rbc*L, SSU
Harveylithon maris-bahiensis I.O. Costa, P.A. Horta & J.M.C. Nunes	BA	Costa et al. (2019)[a]	*psb*A (MF966197 MF966196), *rbc*L, SSU
Harveylithon riosmenum I.O. Costa, P.A. Horta & J.M.C. Nunes	PE, ES, RJ, SC	Costa et al. (2019)[a]	*psb*A (MF966188, MF966192, MF966189, MF966187, MF966193, MF966198), *rbc*L, SSU
Hydrolithon boergesenii (Foslie) Foslie	BA	Figueiredo and Steneck (2002) and Jesionek et al. (2016)[a]	*psb*A (KY485316)

(continued)

Taxa	Distribution	Literature	Markers and Genbank accessed codes
Hydrolithon breviclavium (Foslie) Foslie	IT	Henriques et al. (2014b)	–
Hydrolithon farinosum (J.V. Lamouroux) D. Penrose & Y.M. Chamberlain	RJ, SP	Taylor (1960), Oliveira Filho (1977), and Horta (2000)	–
Hydrolithon rupestre (Foslie) Penrose	AR, BA, ES, FN, IT, SPSP	Villas-Bôas (2008), Amado-Filho et al. (2012a, b, 2016), Pereira-Filho et al. (2012), Crespo et al. (2014), Bahia (2014), Villas-Bôas et al. (2015), and Moura et al. (2016)	–
Hydrolithon samoënse (Foslie) Keats & Y.M. Chamberlain	RJ	Tâmega and Figueiredo (2005)	–
Lithophyllum atlanticum Vieira-Pinto, Oliveira & Horta	AL, RJ, RS, SC, SP	Vieira-Pinto et al. (2014)[a], Torrano-Silva (2015)[a], Torrano-Silva et al. (2018)[a], and Henriques (2016)[a]	COI, *psb*A (**FPG3**, MG517488, MG784111, MG784117, MG784118, MG784123, MG784143, MG784167), *rbc*L, SSU, UPA
Lithophyllum congestum (Foslie) Foslie	BA	Figueiredo and Steneck (2002)	–
Lithophyllum corallinae (P.L. Crouan & H.M. Crouan) Heydrich	AL, BA, ES, FN, RJ	Villas-Bôas (2008), Villas-Bôas et al. (2009), Amado-Filho et al. (2012b), Bahia (2014), Henriques et al. (2014b), and Torrano-Silva (2015)[a]	–
Lithophyllum depressum Villas-Bôas, Figueiredo & Riosmena-Rodriguez	ES	Villas-Bôas et al. (2009)	–
			–
Lithophyllum cf. *depressum* Villas-Bôas, Figueiredo & Riosmena-Rodriguez	ES, BA, AB, IT	Torrano-Silva (2015)[a]	COI, *psb*A, rbcL-3P, UPA
Lithophyllum johansenii Woelkerling & Campbell	ES	Villas-Bôas et al. (2009)	–
Lithophyllum kaiseri (Heydrich) Heydrich	BA	Jesionek et al. (2016)[a] and Torrano-Silva (2015)[a], Torrano-Silva et al. (2018)[a]	COI, *psb*A (KY485306, MG784114), *rbc*L SSU, UPA

(continued)

Taxa	Distribution	Literature	Markers and Genbank accessed codes
Lithophyllum margaritae (Hariot) Heydrich	AL, ES, RJ, SC, SP	Horta (2000), Pascelli et al. (2013), Vieira-Pinto et al. (2014)[a], and Torrano-Silva (2015)[a], Torrano-Silva et al. (2018)[a], Henriques (2016)[a]	COI[b] (KP192397, KP192398), *psb*A (**LMC12**, MG784145), *rbc*L SSU, UPA
Lithophyllum rugosum (Foslie) Me. Lemoine	SC	Pascelli et al. (2013)	
Lithophyllum stictiforme (Areschoug) Hauck	BA, SC, RJ, ES, PE	Nunes et al. (2008), Villas-Boas et al. (2009), Burgos (2011), Amado-Filho et al. (2012b), Pascelli et al. (2013), Bahia (2014), Costa et al. (2014), Henriques et al. (2014b), and Figueiredo and Steneck (2002)	–
Lithophyllum spp.	FN, BA, ES	Torrano-Silva et al. (2018)[a]	*psb*A (MG784136, MG784128, MG784110, MG784158, KY485307, KY485308, MG784097, MG784163, MG784115, MG784113)
Neogoniolithon accretum (Foslie & Howe) Setch & Mason	BA	Figueiredo and Steneck (2002)	–
Neogoniolithon atlanticum Tâmega, Riosmena-Rodriguez, Mariath & Figueiredo	BA	Tâmega et al. (2014)	–
Neogoniolithon brassica-florida (Harvey) Setchell & Mason	BA, ES	Villas-Bôas (2008), Amado-Filho et al. (2012b), and Villas-Bôas et al. (2015)	–
Neogoniolithon fosliei (Heydrich) Setchell & Mason	BA, SP	Taylor (1960), Oliveira Filho (1977), and Bahia (2014)	–
Pneophyllum conicum (Dawson) Keats, Chamberlain & Baba	BA	Mariath et al. (2012) and Jesionek et al. (2016)[a]	*psb*A (KY485312)
Pneophyllum fragile KŸtzing	BA, AR, ES, FN, RJ, SP	Taylor (1960), Oliveira Filho (1977), and Bahia (2014)	–
Porolithon improcerum (Foslie & Howe) M. Howe	IT	Bahia et al. (2014b)	–

(continued)

Taxa	Distribution	Literature	Markers and Genbank accessed codes
Porolithon onkodes (Heydrich) Foslie	BA, FN, ES, IT	Taylor (1960), Oliveira Filho (1977), Henriques et al. (2014b), Jesionek et al. (2016)[a], and Costa et al., (2019)[a]	*psb*A (KY485316), SSU
Spongites fruticulosa Kutzing	ES, IT, SPSP	Crespo et al. (2014), Bahia (2014), and Henriques et al. (2014b)	–
Spongites yendoi (Foslie) Chamberlain	BA, ES	Henriques et al. (2012) and Costa et al. (2014)	–
Titanoderma prototypum (Foslie) Woelkerling, Y.M. Chamberlain & P.C. Silva	BA, IT	Pereira-Filho et al. (2012), Bahia (2014),Torrano-Silva (2015)[a], Torrano-Silva et al. (2018)[a], and Jesionek et al. (2016)[a]	COI, *psb*A (MG784134), *rbc*L, UPA
Titanoderma pustulatum (J.V. Lamouroux) Nägeli	BA, ES, FN, SP, SPSP	Taylor (1960), Oliveira Filho (1977), Horta (2000), Crespo et al. (2014), Bahia (2014), Henriques et al. (2014b), and Torrano-Silva (2015)[a], Torrano-Silva et al. (2018)[a]	*rbc*L, UPA
Sporolithales Le Gall & G.W. Saunders			
Heydrichia (?) *poignantii* Woelkerling, Granier & Dias-Brito	SE	Woelkerling et al. (2014)	–
Sporolithon africanum (Foslie) Afonso-Carillo	AL	Tomita (1976) and Oliveira Filho (1977)	–
Sporolithon australasicum (Foslie) N. Yamaguishi-Tomita ex M.J. Wynne	ES	Tomita (1976) and Oliveira Filho (1977)	–
Sporolithon durum (Foslie) R.A. Townsend & Woelkerling	ES e BA	Henriques et al. (2009)	–
Sporolithon elevatum M.C. Henriques & R. Riosmena-Rodriguez	RJ	Henriques et al. (2014b)	–
Sporolithon episoredion (Adey, Townsend & Boykins) Verheij	ES, IT	Henriques et al. (2014a, b)	–
Sporolithon episporum (M. Howe) E.Y. Dawson	BA, CE, PB e FN	Tomita (1976), Nunes et al. (2008), Bahia et al. (2010, 2015), Burgos (2011), Amado-Filho et al. (2012b), Costa (2013), and Costa et al. (2014)	–

(continued)

Taxa	Distribution	Literature	Markers and Genbank accessed codes
Sporolithon erythraeum (Rothpletz) Kylin	PE, BA, ES	Tomita (1976) and Oliveira Filho (1977)	–
Sporolithon howei (Me. Lemoine) N. Yamaguishi-Tomita ex M.J. Wynne	ES	Tomita (1976), Oliveira Filho (1977)	–
Sporolithon molle (Heydrich) Heydrich	ES	Bahia et al. (2014b)	–
Sporolithon pacificum E.Y. Dawson	ES, PE	Tomita (1976) and Oliveira Filho (1977)	–
Sporolithon ptychoides Heydrich	BA, ES, FN, IT, PA, AR	Tomita (1976) (como *S. mediterraneum*), Bahia et al. (2011), Pereira-Filho et al. (2012), and Amado-Filho et al. (2012b)	–
Sporolithon amadoi J.L. Richards & R.G. Bahia	ES, FN, AP	Bahia et al. (2013[a], 2015)[a], Adey et al. (2015), Jesionek et al. (2016)[a], and Richards et al., 2019[a]	*psb*A (KC870926, KY485313, KP142753, KC870926)
Sporolithon franciscanum L.A.S. Leão & R.G. Bahia	SE, BA	Leão et al. (2020)	
Sporolithon tenue R.G. Bahia, G.M.Amado-Filho, Maneveldt et W.H. Adey	BA	Bahia et al. 2013[a], 2014a, 2015)[a]	*psb*A (KC870923), SSU
Sporolithon yoneshigueae Bahia, Amado-Filho, Maneveldt & W.H. Adey	BA, ES, IT	Bahia et al. (2015)[a]	*psb*A (KC953094, KM203840), SSU
Sporolithon sp.	BA	Jesionek et al. (2016)[a]	*psb*A (**LMC83**, KP142752)
Hapalidiales W.A. Nelson, J.E. Sutherland, T.J. Farr & H.S. Yoon			
Crustaphytum atlanticum	RJ	Jesionek et al. (2020)[a]	*psb*A (MK524382, MK524381)
Lithothamnion brasiliense Foslie	SP, BA	Oliveira Filho (1977), Horta (2000), Bahia et al. (2010), and Costa et al. (2014)	–
Lithothamnion crispatum Hauck	ES, RN, BA, SC, SP	Farias et al. (2010), Bahia et al. (2010), Amado-Filho et al. (2012b), Andrade (2017), Bahia (2014), and Jesionek et al. (2016)[a]	*psb*A (KY485304, KY485305)

(continued)

Taxa	Distribution	Literature	Markers and Genbank accessed codes
Lithothamnion glaciale Kjellman	ES	Henriques et al. (2012)	–
Lithothamnion muelleri Lenormand ex Rosanoff	BA, ES	Amado-Filho et al. (2010), Henriques et al. (2012), and Bahia (2014)	–
Lithothamnion occidentale (Foslie) Foslie	PE, RN, BA	Oliveira Filho (1977) and Figueiredo and Steneck (2000)	–
Lithothamnion sejunctum Foslie	BA	Figueiredo and Steneck (2000)	–
Lithothamnion steneckii Mariath & Figueiredo	BA	Mariath et al. (2012)	–
Lithothamnion superpositum Foslie	BA, RN, SC, SP	Farias et al. (2010)	–
Lithothamnion sp.	BA, CE	Richards et al. (2016)	*psb*A (**LMC5**, KU529478, KU529479)
Melobesia membranacea (Esper) J.V. Lamouroux	PE, BA, FN, RJ	Taylor (1960), Oliveira Filho (1977), and Amado-Filho (1991)	–
Melobesia rosanoffi (Foslie) Lemoine	RN, PE, BA, SC	Borges et al. (2014)	–
Melyvonnea erubescens (Foslie) Me. Lemoine	PE, FN, AL, BA, AB, ES, IT, SP, SC	Taylor (1960), Figueiredo and Steneck (2002), Rocha et al. (2006), Nunes et al. (2008), Bahia et al. (2010), Burgos (2011), Henriques et al. (2012), Horta et al. (2011), Sissini et al. (2014), Jesionek et al. (2016)[a], and Sissini et al. (2022)[a]	*psb*A (**LMC26**, KM983036, KM983037, KM983039, KM983042, KM983043, KM983046, KM983047, KY485314, KY485315), *rbc*L
Mesophyllum engelhartii (Foslie) W.H. Adey	BA, ES, FN, SC	Farias (2009), Marins (2009), Amado-Filho et al. (2010, 2012b), Tâmega et al. (2013), and Bahia (2014)	–
Mesophyllum macroblastum (Foslie) W.H. Adey	RJ	Bahia et al. (2014b)	–
Phymatolithon masonianum Wilks & Woelkerling	SC, BA	Horta (2000) and Costa et al. (2014)	–
Phymatolithon calcareum (Pallas) W.H. Adey & D.L. McKibbin ex Woelkering & L.M. Irvine	SC, RJ e SP	Taylor (1960), Oliveira (1977), Horta (2000), and Costa et al. (2014)	–

(continued)

Taxa	Distribution	Literature	Markers and Genbank accessed codes
Roseolithon goytacae L.M. Coutinho, F.P. Gomes & Barros-Barreto	RJ	Coutinho et al. (2021)[a]	*psb*A (MT554688) SSU 18S (MT512022)
Roseolithon karaiborum L.M. Coutinho & Barros-Barreto	PE	Coutinho et al. (2021)[a]	*psb*A (MT554677), COI-5P (MT512002)
Roseolithon occidentaleatlanticum L.M. Coutinho & Barros-Barreto	RJ, ES	Richards et al. (2022)[a] and Coutinho et al. (2021)[a]	*psb*A (MT554683, MT554684, MT554685), COI-5P (MT512003)
Roseolithon potiguarae L.M. Coutinho & Barros-Barreto	PI, CE, RN, PE	Coutinho et al. (2021)[a]	*psb*A (MT512014-19), COI-5P (MT511999-MT512001), SSU 18S (MT512017-19), *rbc*L (MW316627)
Roseolithon purii L.M. Coutinho & Barros-Barreto	PE, ES	Coutinho et al. (2021)[a]	*psb*A (MT554686, MT554687, MT554685), COI-5P (MT512004), *rbc*L (MW316628)
Roseolithon tamoioi L.M. Coutinho & Barros-Barreto	RJ	Coutinho et al. (2021)[a]	*psb*A (MT554658, MT554660) COI-5P (MT554659), SSU 18S (MT512011, MT5120121)
Roseolithon tremembei L.M. Coutinho & Barros-Barreto	PI	Coutinho et al. (2021)[a]	*psb*A (MT554663), SSU 18S (MT512013)
Roseolithon tupii L.M. Coutinho & Barros-Barreto	PI, CE, RN, PB, PE, AL, BA, ES, RJ	Coutinho et al. (2021)[a]	*psb*A (MT554642-57), COI-5P (MT511993-6), SSU 18S (MT512005-10), rbcL (MW316624-6)
Roseolithon sp.	BA, SC	Coutinho et al. (2021)[a]	*psb*A (MT554678, MT554679, MT554680, MT554681, MT554682), SSU 18S (MT512020, MT512021)
Tectolithon fluminensis Bahia, Jesionek and Amado-Filho	RJ, SC	Jesionek et al. (2020)[a]	*psb*A (**LMC6**, **LMC9**, MH123876, MH123884, MH123888, MH123891, MH123892)
HAPALIDIALES	BA, RJ	Jesionek et al. (2016)[a]	*psb*A (**LMC11**, KY485317)

New sequences are in bold

[a]References with molecular data

[b]Samples without *psb*A sequences, in this case we used similar samples having COI-5P sequences searched with Basic Local Alignment Tool BLAST

Appendix 2: New Sequences Analysed in this Study

Taxa	Sample	Locality	Voucher	Genbank accessed codes
Dawsoniolithon conicum (E.Y. Dawson) Caragnano, Foetisch, Maneveldt & Payri	-	Praia de Itapuã, Salvador, Bahia Brazil. 29.09.2011	ALCB125758	PP820529
Lithophyllum atlanticum Vieira-Pinto, Oliveira & Horta	FPG3	Ilha Feia, Armação dos Búzios, RJ, Brazil. 30.04.2013	R242044	MW672044
Lithophyllum margaritae (Hariot) Heydrich	LMC12	Ilha Feia, Armação dos Búzios, RJ, Brazil. 30.04.2013	RB609177	MW672043
Porolithon onkodes (Heydrich) Foslie	–	Praia de Parati, Anchieta, ES, Brazil. 21.08.2013	ALCB125642	PP820528
Sporolithon sp.	LMC83	Santa Cruz de Cabrália, Praia de Coroa Vermelha, BA, Brazil. 03.01.2018	RFA47286	PP820530
Lithothamnion sp.	LMC5	Ilha Feia, Armação dos Búzios, RJ, Brazil. 30.04.2013	RB609182	PP820524
Melyvonnea erubescens (Foslie) Me.Lemoine	LMC26	Alagoas, Maragogi, Aquário de Maragogi, Maragogi, AL, Brazil. 26.05.2016	R238879	MW916197
M. erubescens	LMC31	Pontal de Maracaípe, Ipojuca, PE, Brazil. 27.05.2016	R238881	MW916199
Tectolithon fluminense Bahia, Jesionek & Amado-Filho	LMC6	Ilha Feia, Armação dos Búzios, RJ, Brazil. 30.04.2013	RB609178	PP820526
T. fluminense	LMC9	Ilha Feia, Armação dos Búzios, RJ, Brazil. 30.04.2013	R223094	PP820525
HAPALIDIALES	LMC11	Ilha Feia, Armação dos Búzios, RJ, Brazil. 30.04.2013	RB609180	PP820527

Appendix 3: Morphological Characters and States

	Eco-regions	Tetrasporangial	Tetraspores arrangement	thallus construction	Cells connections	Epithelial cells form	Epithelial cells layers	Apical plugs	Trichocytes	Subepithelial initials	Columella
LMC5 Buzios RJ	1	1	0	2	1	0	0	1	0	0 & 2	?
LMC9 Buzios RJ	1	1	0	2	1	1	2	1	0	0	?
LMC12 Buzios RJ	1	0	0	0	0	1 & 2	1	0	1	?	?
KY485307 Lithophyllum sp AB	1	0	0	0	0	2	?	0	1	0 & 2	?
KY485304 L. crispatum	0	1	0	2	1	0 &1	?	1	0	0 & 2	?
KY485305 L. crispatum	0	1	0	2	1	0 & 1	?	1	0	0 & 2	?
KU529478 Lithothamnion sp BA	0	1	0	2	1	0	0	1	0	0	?
KU529479 Lithothamnion sp CE	0	1	0	2	1	0	?	1	0	0	?
KM983047 M. erubescens	1	1	0	2	1	2	0	1	1	?	?
KY485315 M. erubescens	0	1	0	1	1	1 & 2	?	1	0	0 & 2	?

KM983042 M. erubescens	0	1	0	2	1	2	0	1	1	?	0
KM983036 M. erubescens	0	1	0	2	1	2	0	1	1	?	0
KY485316 H. boergesen	0	0	0	0	1	1 & 2	?	0	1	0	?
KY485318 P. onkodes AB	0	0	0	2	1	2	?	0	1	0 & 2	?
KY485312 P. conicum AB	0	0	0	1	1	2	?	0	1	0 & 2	?
KY485306 L. kaiseri	0	0	0	?	0	2	?	0	1	0	?
KY485308 Lithophyllum sp AB	0	0	0	0	0	2	?	0	1	0 & 2	?
MG784143 L. atlanticum	1	0	0	2	0	1	1	0	0	?	?
KY485313 S. ptychoides	0	2	1	2	2	0	?	1	0	0	?
KM203840 S. yoneshigueae	0	2	1	2	2	0 & 1	?	1	?	0	?
KM983043 M. erubescens	1	1	0	2	1	2	0	1	1	?	0
LMC6 Buzios RJ	1	1	0	2	0	1	2	1	0	0	?
LMC 31 Maracaipe PE	0	1	0	1	1	1 & 2	1	1	1	2	?
FPG3 Buzios RJ	1	0	0	0	0	2	1	0	0	0	?
MF966193 H riosmenum RJ	1	0	0	2	1	2	0	0	4	0 & 2	0

(continued)

	Eco-regions	Tetrasporangial	Tetraspores arrangement	thallus construction	Cells connections	Epithelial cells form	Epithelial cells layers	Apical plugs	Trichocytes	Subepithelial initials	Columella
MF966187 H riosmenum PE	0	0	0	2	1	2	0	0	4	0 & 2	0
MF966197 H marisbahiensis BA	0	0	0	2	1	2	0	0	1	0 & 2	0
MF966196 H marisbahiensis BA	0	0	0	2	1	2	0	0	1	0 & 2	0
MF966195 H catarinense SC	1	0	0	2	1	2	0	0	4	0 & 2	0
MF966194 H catarinense SC	1	0	0	2	1	2	0	0	4	0 & 2	0
MF966198 H riosmenum SC	1	0	0	2	1	2	0	0	1, 2 & 4	0 & 2	0
Porolithon onkodes ES	0	0	0	0	1	2	2	0	3 & 4	0	0
Dawsoniolithon conicum BA	0	0	0	1	1	2	2	0	3 & 4	2	1
DQ787633 H. rubra	?	?	?	?	?	?	?	?	?	?	?
MK524381 C. atlanticum	1	1	0	?	?	?	0	1	0	0 & 2	?
MH123884 T. fluminensis RJ	1	1	0	?	?	?	?	?	1	?	?
KC870923 S. tenue BA	0	2	1	2	2	0 & 1	?	1	0	?	?
MG784145 L. margaritae	1	0	0	0 & 2	0	1	0	0	0	?	1

? = Unknown Data

Morphological characters and states.

Characters/states		State = 0	State = 1	State = 2	State = 3	State = 4
1	Ecoregions	TSA	WTSA			
2	Tetrasporangial	Uniporate	Multiporate	Calcified compartments		
3	Tetraspores arrangement	Zonate	Cruciate			
4	Thallus construction	Dimerous	Monomerous coaxial	Monomerous		
5	Cells connections	Secondary pit	Vell fusion	Both		
6	Epithelial cells	Flared	Flattened	Rounded		
7	Epithelial cell layers	1	1–2	More than 2		
8	Apical plugs	Absent	Present			
9	Trichocytes	Absent	Single surface	Vertical rows	Horizontal rows	Buried
10	Subepithelial initials	Equal	Shorter	Longer		
11	Columella	Absent	Central			

References

Adey, W. H., Hernandez-Kantun, J. J., Johnson, G. & Gabrielson, P. W. (2015). DNA sequencing, anatomy, and calcification patterns support a monophyletic, subarctic, carbonate reef-forming *Clathromorphum* (Hapalidiaceae, Corallinales, Rhodophyta). *Journal of Phycology, 51*, 189–203.

Amado-Filho, G. M. (1991). Algas Marinhas Bentônicas do litoral de Saquarema a Itacoatiara (RJ). Dissertação de Mestrado, Universidade Federal do Rio de Janeiro, Rio de Janeiro, 323p.

Amado-Filho, G., Maneveldt ,G., Pereira-Filho, G., Manso, R., Bahia, R., Barros-Barreto, M. & Guimarães, S. (2010). Seaweed diversity associated with a Brazilian tropical rhodolith bed. *Ciencias Marinas, 36, 371–391.*

Amado-Filho, G. M., & Pereira-Filho, G. H. (2012). Rhodolith beds in Brazil: A new potential habitat for marine bioprospection. *Revista Brasileira de Farmacognosia, 22*, 782–788.

Amado-Filho G., Moura, R. L., Bastos, A. C., Salgado, L. T., Sumida P. Y. G., Guth, A. Z., Francini-Filho, R. B., Pereira-Filho, G. H., Abrantes, D. P., Brasileiro, P. S., Bahia, R. G., Leal, R. N., Kaufman, L., Kleypas, J., Farina, M. & Thompson, F. L. (2012a). Rhodolith beds are major CaCO3 bio-factories in the Tropical South West Atlantic. *Plos One, 7(4), e35171.*

Amado-Filho G., Pereira-Filho, G. H., Bahia, R. G., Abrantes, D. P. Veras, P. C. & Matheus, Z. (2012b). Occurrence and distribution of rhodolith beds on the Fernando de Noronha Archipelago of Brazil. *Aquatic Botany, 101, 41–45.*

Amado-Filho, G. M., Moura, R. L., Bastos, A. C., Frnacini-Filho, R. B., Pereira-Filho, G. H., Bahia, R. G., Moraes, F. C., & Motta, F. S. (2016). Mesophotic cosystems of the unique South Atlantic atoll are composed by rhodolith beds and scattered consolidated reefs. *Mar Biodivers* 46:407–420

Andrade, V. M. S. (2017). Algas Calcárias da região entremarés da costa brasileira: uma descricmão qualitativa e quantitativa dos carbonatos. Dissertacmão (Mestrado) - Universidade Federal de Santa Catarina. 102p.

Areschoug, J. E. (1852). Ordo XII. Corallineae. In J. G. Agardh (Ed.), *Sepcies genera et ordines algarum. Volumen secundum: Algas florideas complectens* (pp. 506–576). CWK Gleerum.

Athanasiadis, A., & Ballantine, D. L. (2014). The genera *Melyvonnea* gen. nov. and *Mesophyllum* ss (Melobesioideae, Corallinales, Rhodophyta) particularly from the Central Atlantic Ocean. *Nordic Journal of Botany, 32*(4), 385–436.

Bahia, R. G. (2014). *Algas coralináceas formadoras de rodolitos da plataforma continental tropical e ilhas oceânicas do Brasil: levantamento florístico e taxonomia.* Tese de doutorado. Instituto de Pesquisas Jardim Botânico do Rio de Janeiro/Escola Nacional de Botânica Tropical, 221p.

Bahia, R. G., Abrantes, D. P., Brasileiro, P. S., Pereira-Filho, G. H. & Amado-Filho, G. M. (2010). Rhodolith bed structure along a depth gradient on the northern coast of Bahia State, Brazil. *Brazilian Journal of Oceanography, 58, 323–337.*

Bahia, R. G., Riosmena-Rodriguez, R., Maneveldt, G. W., & Amado Filho, G. (2011). First report of *Sporolithon ptychoides* (Sporolithales, Corallinophycidae, Rhodophyta) for the Atlantic Ocean. *Phycological Research, 59*, 64–69.

Bahia, R. G., Amado-Filho, G. M., Maneveldt, G. W., Adey, W. H., Johnson, G., Marins, B. V., & Longo, L. L. (2013). *Sporolithon tenue* sp. nov. (Sporolithales, Corallinophycidae, Rhodophyta): A new rhodolith-forming species from the tropical southwestern Atlantic. *Phycological Research, 62*(1), 44–54.

Bahia, R. G., Filho, G. M. A., & Maneveldt, G. W. (2014a). *Sporolithon molle* (Heydrich) Heydrich (Sporolithales, Corallinophycidae, Rhodophyta): An addition to the Atlantic flora found on a remote oceanic Island. *Cryptogamie, Algologie, 35*(1), 7–14.

Bahia, R. G., Amado Filho, G. M., Azevedo, J., & Maneveldt, G. W. (2014b). *Porolithon improcerum* (Porolithoideae, Corallinaceae) and *Mesophyllum macroblastum* (Melobesioideae, Hapalidiaceae): New records of crustose coralline red algae for the Southwest Atlantic Ocean. *Phytotaxa, 190*(1), 38–44.

Bahia, R. G., Amado-Filho, G. M., Maneveldt, G. W., Adey, W. H., Johnson, G., Jesionek, M. B., & Longo, L. L. (2015). *Sporolithon yoneshigueae* sp. nov. (Sporolithales, Corallinophycidae, Rhodophyta), a new rhodolith-forming coralline alga from the southwest Atlantic. *Phytotaxa, 224*, 140–158. https://doi.org/10.11646/phytotaxa.224.2.2

Bailey, J. C., & Chapman, R. L. (1998). A phylogenetic study of the Corallinales (Rhodophyta) based on nuclear small-subunit rRNA gene sequences. *Journal of Phycology, 34*, 692–705.

Bailey, J. C., Gabel, J. E., & Freshwater, D. W. (2004). Nuclear 18S rRNA gene sequence analyses indicate that the Mastophoroideae (Corallinaceae, Rhodophyta) is a polyphyletic taxon. *Phycologia, 43*, 3–12.

Bittner, L., Payri, C. E., Maneveldt, G. W., Couloux, A., Cruaud, C., Reviers, B., & Le Gall, L. (2011). Evolutionary history of the Corallinales (Corallinophycidae, Rhodophyta) inferred from nuclear, plastidial and mitochondrial genomes. *Molecular Phylogenetics and Evolution, 61, 697–713.*

Birkett, D. A., Maggs, C., & Dring, M. J. (1998). *MAERL: An overview of dynamics and sensitivity characteristics for conservation management of marine SACs*. UK Marine SACs Project.

Borges, V. P., Bastos, E., Batista, M. B., Bouzon, Z., Lhullier, C., Schmidt, E. C., Sissini, M. N., & Horta, P. A. (2014). The genus *Melobesia* (Corallinales, Rhodophyta) from the subtropical South Atlantic, with the addition of M. rosanoffii (Foslie) Lemoine. *Phytotaxa (Online), 190*, 268.

Broom, J. E. S., Hart, D. R., Farr, T. J., Nelson, W. A., Neill, K. F., Harvey, A. S., & Woelkerling, W. M. J. (2008). Utility of *psb*A and nSSU for phylogenetic reconstruction in the Corallinales based on New Zealand taxa. *Molecular Phylogenetics and Evolution, 46*, 958–973.

Burgos, D. C. (2011). Composicיִão e estrutura das comunidades de macroalgas do infralitoral do Arquipélago de Fernando de Noronha – Brasil, com ênfase nas calcárias incrustantes. Universidade Federal Rural de Pernambuco. Tese de Doutorado, 95 pp.

Cabioch, J., & Giraud, G. (1986). Structural aspects of biomineralization in the coralline algae (calcified Rhodophyceae). In B. S. C. Leadbeater & R. Riding (Eds.), *Biomineralization in lower plants and animals* (Vol. 30, pp. 141–156). Claredon Press.

Carro, B., Lopez, L., Peña, V., Bárbara, I., & Barreiro, R. (2014). DNA barcoding allows the accurate assessment of European maërl diversity: A proof-of-concept study. *Phytotaxa, 190*(1), 176–189.

Costa, I. O. (2013). Estudos taxonômicos em algas coralináceas não articuladas no litoral norte da Bahia, Brasil. Universidade do Estado da Bahia. Dissertacיִão de Mestrado, 99 pp. Farias, J. N. (2009) Aspectos taxonômicos de Lithothamnion superpositum e Mesophyllum engelhartii (Corallinales; Rhodophyta) – adicיִões para a flora brasileira. Universidade Federal de Santa Catarina. Dissertacיִão de Mestrado, 47 pp.

Costa, I. O., Horta, P. A., Bergstrom, E. R., & De Castro Nunes, J. M. (2014). Taxonomic study of crustose coralline algae off the northeastern Brazilian coast. *Phytotaxa, 190*(1), 130–161.

Costa, I. O., Jesus, P. B., Jesus, T. S., Souza, P. S., Horta, P. A., & Nunes, J. M. C. (2019). Reef-building coralline algae from the Southwest Atlantic: Filling gaps with the recognition of *Harveylithon* (Corallinaceae, Rhodophyta) on the Brazilian coast. *JPhycol, 55*(6), 1370–1385. https://doi.org/10.1111/jpy.12917

Coutinho, L. M. (2014). *Taxonomia de algas calcárias não geniculadas (Rhodophyta, Corallinales) e a influência do hidrodinamismo nas características morfológicas de rodolitos na Armação dos Búzios, RJ*. Dissertação de Mestrado. Museu Nacional/Universidade Federal do Rio de Janeiro. Departamento de Botânica, 98p.

Coutinho, L. M. (2019). *Aspectos moleculares e morfológicos da subfamília Melobesioideae (Rhodophyta, Hapalidiales) para o Brasil*. Tese de Doutorado. Museu Nacional/Universidade Federal do Rio de Janeiro. Departamento de Botânica, 168p.

Coutinho, L. M., Gomes, F. P., Sissini, M. N., Vieira-Pinto, T., de Oliveira Henriques, M. C. M., Oliveira, M. C., Horta, P. A., & de Barros Barreto, M. B. B. (2021). Cryptic diversity in non-geniculate coralline algae: A new genus *Roseolithon* (Hapalidiales, Rhodophyta) and seven new species from the Western Atlantic. *European Journal of Phycology, 57*, 227–250.

Crespo, T. M., da Gama Bahia, R., Maneveldt, G. W., & Amado Filho, G. M. (2014). Floristic composition of crustose coralline algae from the St. Peter and St. Paul Archipelago, a summit of the Mid-Atlantic Ridge. *Phytotaxa, 190*(1), 17–37.

FARIAS, J. (2009). Aspectos taxonômicos de Lithothamnion superpositum e Mesophyllum engelhartii (Corallinales; Rhodophyta) do Brasil. Dissertacmão de Mestrado. Universidade Federal de Santa Catarina, Florianópolis, Brasil.

Farias, J. N., Riosmena-Rodriguez, R., Bouzon, Z., Oliveira, E. C., & Horta, P. A. (2010). *Lithothamnion superpositum* (Corallinales; Rhodophyta): First description for the Western Atlantic or rediscovery of a species? *Phycological Research, 58*, 210–216.

Figueiredo, M. A. O. & Steneck, R. S. (2002). Floristic and ecological studies of crustose coralline algae on Brazil's Abrolhos reefs. *Proceedings of the 9th International Coral Reef Symposium, 1, 493–498.*

Figueiredo, M. A. O., Santos-de Menezes, K., Costa-Paiva, E. M., Paiva, P. C., & Ventura, C. R. R. (2007). Experimental evaluation of rhodoliths as living substrata for infauna at the Abrolhos Bank, Brazil. *Ciencias Marinas, 33*(4), 427–440.

Flora do Brasil. (2020). *Under construction*. Jardim Botânico do Rio de Janeiro. Available at: http://floradobrasil.jbrj.gov.br/. Accessed on: 16 Apr 2023.

Foslie, M. (1900). Revised systematical survey of the Melobesieae. *Kongelie Norske Vindenskabers Selskabs Skrifter, 5, 1–22.*

Foslie M (1905). Remarks on northern Lithothamnia. Det Kongelige Norske Videnskabers Selskabs Skrifter 1905, 1–138.

Foster, M. S. (2001). Rhodoliths: Between rocks and soft places. *Journal of Phycology, 37*, 659–667.

Fredericq, S., & Norris, J. N. (1995). A new order (Rhodogorgonales) and family (Rhodogorgonaceae) of red algae composed of two tropical calciferous genera, *Renouxia* gen. nov. and *Rhodogorgon*. *Cryptogamic Botany, 5*, 316–331.

Freiwald, A., & Henrich, R. (1994). Reefal coralline algal build-ups within the Arctic circle: Morphology and sedimentary dynamics under extreme environmental seasonality. *Sedimentology, 41*, 963–984.

Gabrielson, P. W., Hughey, J. R., & Diaz-Pulido, G. (2018). Genomics reveals abundant speciation in the coral reef building alga *Porolithon onkodes* (Corallinales, Rhodophyta). *Journal of Phycology, 54*, 429–434. https://doi.org/10.1111/jpy.12761

Gomes, F. O., Maggs, C. A., & Barros-Barreto, M. B. B. (2020). Integrative approach reveals four new cryptic species in the genus *Gayliella* (Ceramiaceae, Rhodophyta). *Journal of Phycology, 56*, 437.

Guindon, S., Dufayard, J. F., Lefort, V., Anisimova, M., Hordijk, W., & Gascuel, O. (2010). New algorithms and methods to estimate maximum-likelihood phylogenies: assessing the performance of PhyML 3.0. *Systematic Biology, 59*(3), 307–321.

Guiry, M. D., & Guiry, G. M. (2023). *AlgaeBase*. World-Wide Electronic Publication, National University of Ireland. http://www.algaebase.org. Searched on 23 Feb 2020.

Halfar, J., & Riegl, B. (2013). From coral framework to rhodolith bed: Sedimentary footprint of the 1982/1983 ENSO in the Galápagos. *Coral Reefs, 32*(4), 985.

Harvey, A. (2005). Coralline algae of central New Zealand: An identification guide to common'crustose'species. *NIWA Information Series, 57*, 1–145.

Harvey, A., & Woelkerling, W. (2007). A guide to nongeniculate coralline red algal (Corallinales, Rhodophyta) rhodolith identification. *Ciencias Marinas, 33*(4), 411–426.

Harvey, A. S., Broadwater, S., Woelkerling, W. J., & Mitrovski, Y. P. (2003). *Choreonema* (Corallinales, Rhodophyta): 18S rDNA phylogeny and resurrection of the Hapalidiaceae for the subffamilies Choreonematoidea, Austrolithoideae and Melobesioidae. *Journal of Phycology, 39*, 988–998.

Hebert, P. D., Cywinska, A., & Ball, S. L. (2003). Biological identifications through DNA barcodes. *Proceedings of the Royal Society of London B: Biological Sciences, 270*(1512), 313–321.

Henriques, M. C., Villas-Boas, A., Riosmena-Rodriguez, R. & Figueiredo, M. A. O. (2012). New records of rhodolith-forming species (Corallinales, Rhodophyta) from deep water in Espírito Santo State, Brazil. *Helgoland Marine Research, 66, 219–231.* https://doi.org/10.1007/s10152-011-0264-1

Henriques, M. C. (2016). *Taxonomia e filogenia molecular de algas calcárias incrustantes da Bacia de Campos e importância da gestão da sua biodiversidade no Brasil.* Tese de Doutorado. Universidade Federal do Rio de Janeiro, 222 pp.

Henriques, M. C., Riosmena-Rodriguez, R., Coutinho, L. M., & Figueiredo, M. A. (2014a). Lithophylloideae and Mastophoroideae (Corallinales, Rhodophyta) from the Brazilian continental shelf. *Phytotaxa, 190*(1), 112–129.

Henriques, M. C., Coutinho, L. M., Riosmena-Rodriguez, R., Barros-Barreto, M. B., Khader, S., & Figueiredo, M. A. (2014b). Three deep water species of *Sporolithon* (Sporolithales, Rhodophyta) from the Brazilian continental shelf, with the description of *Sporolithon elevatum* sp. *nov*. *Phytotaxa, 190*(1), 320–330.

Henriques, M. C. M. O., Nunes, D. S. P., Coutinho, L. M., Riosmena-Rodriguez, R., & Figueiredo, M. A. C. (2009). Deep water Sporolithon (Sporolithaceae, Corallinales, Rhodophyta) from Bahia and Espírito Santo states, Brazil. In: III INTERNATIONAL RODOLITH WORKSHOP, 2009, Búzios. III INTERNATIONAL RODOLITH WORKSHOP.

Hernández-Kantún, J. J., Riosmena-Rodriguez, R., Hall-Spencer, J. M., Peña, V., Maggs, C. A., & Rindi, F. (2015). Phylogenetic analysis of rhodolith formation in the Corallinales (Rhodophyta). *European Journal of Phycology, 50*(1), 46–61.

Holz, V. L., Bahia, R. G., Karez, C. S., Vieira, F. V., Moraes, F. C., Vale, N. F., Sudatti, D. B., Salgado, L. T., Moura, R. L., Amado-Filho, G. M., et al. (2020). Structure of rhodolith beds and surrounding habitats at the doce river Shelf (Brazil). *Diversity, 12, 75.* https://doi.org/10.3390/d12020075

Horta, P. A. (2000). *Macroalgas do infralitoral do sul e sudeste do Brasil: taxonomia e biogeografia.* Tese de Doutorado. Instituto de biociências da Universidade de São Paulo. Departamento de Botânica. 301p.

Horta, P. A. (2002). Bases para a identificação das coralináceas não articuladas do litoral brasileiro-uma síntese do conhecimento. *Biotemas, 15*(1), 7–44.

Horta, P. A., Scherner, F., Bouzon, Z. L., Riosmena-Rodríguez, R. & Oliveira, E. C. (2011). Morphology and reproduction of Mesophyllum erubescens (Foslie) Me. 170 Lemoine (Corallinales, Rhodophyta) from Southern Brazil. *Revista Brasileira de Botanica, 34(1), 125–134.*

Horta, P. A., Riul, P., Amado Filho, G. M., Gurgel, C. F. D., Berchez, F., Nunes, J. M. C., Scherner, F., Pereira, S., Lotufo, T., Letícia, L., Sissini, M., Bastos, E. O., Rosa, J., Munoz, P., Martins, C., Gouvêa, L., Carvalho, V., Bergstrom, E., Schubert, N., Bahia, R. G., Rodrigues, A. C., Rörig, L., Barufi, J. B., & Figueiredo, M. (2016). Rhodoliths in Brazil: Current knowledge and potential impacts of climate change. *Brazilian Journal of Oceanography, 64*, 117–136.

Huisman, J. M., Harper, J. T., & Saunders, G. W. (2004). Phylogenetic study of the Nemaliales (Rhodophyta) based on large-subunit ribosomal DNA sequences supports segregation of the Scinaiaceae fam. nov. and resurrection of *Dichotomaria* Lamarck. *Phycological Research, 52*, 224–234.

Irvine, L. M. & Chamberlain, Y. M. (1994). Seaweeds of the British Isles Vol 1: Rhodophyta; Part 2B Corallinales, Hildenbrandiales. Natural History Museum London. 276 pp.

Jeong, S. Y., Nelson, W. A., Sutherland, J. E., Peña, V., Le Gall, L., Diaz-Pulido, G., et al. (2021). Corallinapetrales and Corallinapetraceae: A new order and family of coralline red algae including Corallinapetra gabrielii comb. nov. *Journal of Phycology, 57*, 849.

Jesionek, M. B., Bahia, R. G., Hernández-Kantún, J. J., Adey, W. H., Yoneshigue-Valentin, Y., Longo, L. L., & Amado-Filho, G. M. (2016). A taxonomic account of non-geniculate coralline algae (Corallinophycidae, Rhodophyta) from shallow reefs of the Abrolhos Bank, Brazil. *Algae, 31*(4), 317–340.

Jesionek, M. B., Bahia, R. G., Lyra, M. B., Leão, L. A. B., Oliveira, M. C., & Amado-Filho, G. M. (2020). Newly discovered coralline algae in Southeast Brazil: *Tectolithon fluminense* gen. et sp. nov. and *Crustaphytum atlanticum* sp. nov. (Hapalidiales, Rhodophyta). *Phycologia, 59(2), 101–115.* https://doi.org/10.1080/00318884.2019.1702320

Jesus, P. B., Nauer, F., Lyra, G. M., Cassano, V., Oliveira, M. C., Nunes, J. M. C., & Schnadelbach, A. S. (2016). Species delimitation and phylogenetic analyses of some cosmopolitan species of *Hypnea* (Rhodophyta) reveal synonyms and misapplied names to H. cervicornis, including a new species from Brazil. *Journal of Phycology, 52*, 774–792.

Konar, B., Riosmena-Rodríguez, R. & Iken, K. (2006). Rhodolith bed: A newly discovered habitat in the North Pacific Ocean. *Botanica Marina, 49, 355–359.*

Kato, A., Baba, M., & Suda, S. (2011). Revision of the Mastophoroideae (Corallinales, Rhodophyta) and polyphyly in nongeniculate species widely distributed on Pacific coral reefs. *Journal of Phycology, 47*, 662–672.

Lamouroux, J. V. (1816). *Histoire des polypiers coralligenes flexibles*. IDC.

Le Gall, L., Payri, C. E., Bittner, L., & Saunders, G. W. (2010). Multigene phylogenetic analyses support recognition of the Sporolithales ord. nov. *Molecular Phylogenetics and Evolution, 54*(1), 302–305.

Leão, L. A. S., Bahia, R. G., Jesionek, M. B., Adey, W. H., Johnson, G., Salgado, L. T., & Pereira, R. C. (2020). *Sporolithon franciscanum sp. nov. (Sporolithales, Rhodophyta), a New Rhodolith-Forming Species from Northeast Brazil. Diversity, 12, 199.* https://doi.org/10.3390/d12050199

Linnaeus, C. (1767). *Systema naturae* (Vol. 2, 12th ed.). Salvi.

Littler, M. M., & Littler, D. S. (2013). The nature of crustose coralline algae and their interactions on reefs. *Smithsonian Contributions to the Marine Sciences, 39*, 199–212.

Littler, M. M., Littler, D. S., Blair, S. M., & Norris, J. N. (1986). Deep-water plant communities from an uncharted seamount off San Salvador Island, Bahamas: Distribution, abundance, and primary productivity. *Deep Sea Research Part A. Oceanographic Research Papers, 33*(7), 881–892.

Maddison, W. P., & Maddison, D. R. (2019). *Mesquite: A modular system for evolutionary analysis*. Version 3.2. Available at: http://mesquiteproject.org. Last accessed Apr 2020.

Mariath, R., Riosmena-Rodrigues, R., & Figueiredo, M. (2012). *Lithothamnion steneckii* sp. nov. and Pneophyllum conicum: new coralline red algae (Corallinales, Rhodophyta) for coral reefs of Brazil. *Algae, 27(4), 249–258.*

Marins, B. V. (2009). Aspectos biológicos de Laminaria spp.: taxonomia, filogenia molecular, parâmetros populacionais, composicmão química e flora associada. Instituto de Pesquisas Jardim Botânico do Rio de Janeiro. Tese de Doutorado, 122 pp.

Maria Carolina, Henriques Alexandre, Villas-Boas Rafael Riosmena, Rodriguez Marcia A. O., Figueiredo (2012) New records of rhodolith-forming species (Corallinales Rhodophyta) from deep water in Espírito Santo State Brazil. *Helgoland Marine Research* 66(2) 219-231 10.1007/s10152-011-0264-1

Moura, R. L., Secchin, N. A., Amado-Filho, G. M., Francini-Filho, R. B., Freitas, M. O., Minte-Vera, C. V., & Bastos, A. C. (2013). Spatial patterns of benthic megahabitats and conservation planning in the Abrolhos Bank. *Continental Shelf Research, 70*, 109–117.

Moura, R. L., Amado-Filho, G. M., Moraes, F. C., Brasileiro, P. S., Salomon, P. S., Mahiques, M. M., Bastos, A. C., Almeida, M. G., Silva, J. M., Araujo, B. F., Brito, F. P., Rangel, T. P., Oliveira, B. C. V., Bahia, R. G., Paranhos, R. P., Dias, R. J. S., Siegle, E., Figueiredo, A. G., Pereira, R. C., Leal, C. V., Hajdu, E., Asp, N. E., Gregoracci, G. B., Neumann-Leitao, S., Yager, P. L., Francini-Filho, R. B., Froes, A., Campeão, M., Silva, B. S., Moreira, A. P. B., Oliveira, L., Soares, A. C., Araujo, L., Oliveira, N. L., Teixeira, J. B., Valle, R. A. B., Rezende, C. E., & Thompson, C. C. (2016). An extensive reef system at the Amazon River mouth. *Science Advances, 2*(4), e1501252.

Nelson, W. A., Sutherland, J. E., Farr, T. J., Hart, D. R., Neill, K. F., Kim, H. J., & Yoon, H. S. (2015). Multi-gene phylogenetic analyses of New Zealand coralline algae: *Corallinapetra novaezelandiae* gen. et sp. nov. and recognition of the Hapalidiales ord. nov. *Journal of Phycology, 51*, 545–468.

Nunes, J. M. C., Guimarães, S. M. P. B., Donnangelo, A., Farias, J. & Horta, P. A. (2008). Aspectos taxonômicos de três espécies de coralínáceas não geniculadas do litoral do Estado da Bahia, Brasil. *Rodriguesia, 59(1), 75–86.*

Oliveira-Filho, E. C. (1977). Algas Marinhas Bentônicas do Brasil. Universidade de São Paulo. Tese de Livre Docência, 407 pp.

Pascelli, C., Riul, P., Riosmena-Rodríguez, R., Scherner, F., Nunes, M., Hall-Spencer, J. M., Oliveira, E. C., & Horta, P. A. (2013). Seasonal and depth-driven changes in rhodolith bed structure and associated macroalgae off Arvoredo Island (southeastern Brazil). *Aquatic Botany, 111*, 62–65.

Peña, V., De Clerck, O., Afonso-Carrillo, J., Ballesteros, E., Bárbara, I., Barreiro, R., & Le Gall, L. (2015). An integrative systematic approach to species diversity and distribution in the genus *Mesophyllum* (Corallinales, Rhodophyta) in Atlantic and Mediterranean Europe. *European Journal of Phycology, 50*(1), 20–36.

Peña, V., Vieira, C., Braga, J. C., Aguirre, J., Rösler, A., Baele, G., et al. (2020). Radiation of the coralline red algae (Corallinophycidae, Rhodophyta) crown group as inferred from a multilocus time-calibrated phylogeny. *Molecular Phylogenetics and Evolution, 150*, 106845.

Pereira-Filho, G. H., Amado-Filho, G. M., de Moura, R. L., Bastos, A. C., Guimaraes, S. M., Salgado, L. T., ... & Brasileiro, P. S. (2012). Extensive rhodolith beds cover the summits of southwestern Atlantic Ocean seamounts. *Journal of Coastal Research, 28*(1), 261–269.

Pereira-Filho, G. H., Shintate, S. G., Kitahara, M. V., Moura, R. L., Amado-Filho, G. M., Bahia, R. G., Moraes, F. C., Neves, L. M., Francini, C. L. B., Gibran, F. Z., & Motta, F. S. (2019). The southernmost Atlantic coral reef is off the subtropical Island of Queimada Grande (24°S), Brazil. *Bulletin of Marine Science, 95*(2), 277–287. https://doi.org/10.5343/bms.2018.0056

Philippi, R. A. (1837). Beweis, dass die Nulliporen Pflanzen sind. *Archiv für Naturgeschichte, 3*, 387–393. Figs 2–6, pl. IX.

Richards, J. L., Vieira-Pinto, T., Schmidt, W. E., Sauvage, T., Gabrielson, P. W., Oliveira, M. C., & Fredericq, S. (2016). Molecular and morphological diversity of *Lithothamnion* spp. (Hapalidiales, Rhodophyta) from deepwater rhodolith beds in the northwestern Gulf of Mexico. *Phytotaxa, 278*(2), 81–114.

Richards, J. L., Bahia, R. G., Jesionek, M. B., & Suzanne Fredericq, S. (2019). *Sporolithon amadoi* sp. nov. (Sporolithales, Rhodophyta), a new rhodolith-forming non-geniculate coralline alga from offshore the northwestern Gulf of Mexico and Brazil. *Phytotaxa, 423*(2), 049–067. https://doi.org/10.11646/phytotaxa.423.2.1

Richards, J. L., Kittle, R. P., III, Schmidt, W. E., Sauvage, T., Gurgel, C. F., Gabriel, D., & Fredericq, S. (2022). Assessment of rhodolith diversity in the northwestern Gulf of Mexico including the description of *Sporolithon gracile* sp. nov. (Sporolithales, Rhodophyta), and three new species of *Roseolithon* (Hapalidiales, Rhodophyta). *Frontiers in Marine Science, 9*, 906679.

Riosmena-Rodriguez, R., Woelkerling, W. J. & Foster, M. S. (1999). Taxonomic reassessment of rhodolith forming species of *Lithophyllum* (Corallinales, Rhodophyta) in the Gulf of California, Mexico. *Phycologia, 38(5), 401–417.*

Riosmena-Rodriguez, R. R., Maneveldt, G. W., Horta, P., & Figueiredo, M. (2014). Systematics and biogeography of the subclass Corallinophycideae (Rhodophyta) from the Atlantic Ocean. *Phytotaxa, 190*(1), 3–4.

Rocha, M. R., Metri, R. & Omuro, J. Y. (2006). Spatial distribution and abundance of ascidians in a bank of coralline algae at Porto Norte, Arvoredo Island, Santa Catarina. *Journal of Coastal Research, 40, 1676–1679.*

Rösler, A., Perfectti, F., Peña, V., & Braga, J. C. (2016). Phylogenetic relationships of corallinaceae (Corallinales, Rhodophyta): Taxonomic implications for reef-building corallines. *Journal of Phycology, 52*, 412–431.

Saunders, G. W. (2005). Applying DNA barcoding to red macroalgae: A preliminary appraisal holds promise for future applications. *Philosophical Transactions of the Royal Society of London B: Biological Sciences, 360*(1462), 1879–1888.

Sissini, M.N., Oliveira, M.C., Gabrielson, P.W., Robinson, N.R., Okolodkov, Y.B., Riosmena-Rodríguez, R. & Horta, P.A. (2014). Mesophyllum erubescens (Corallinales, Rhodophyta) – So many species in one epithet. Phytotaxa, v. 190, n 1, p. 299–319.

Sissini, M. N., Koerich, G., de Barros-Barreto, M. B., Coutinho, L. M., Gomes, F. P., Oliveira, W., Costa, I. O., de Castro Nunes, J. M., Henriques, M. C., Vieira-Pinto, T., Torrano-Silva, B. N., Oliveira, M. C., Le Gall, L., Horta, P. A. (2022). Diversity, distribution, and environmental drivers of coralline red algae: the major reef builders in the Southwestern Atlantic. *Coral Reefs, 41, 711–725.* https://doi.org/10.1007/s00338-021-02171-1

Steneck, R. S. (1986). The ecology of coralline algal crusts: Convergent patterns and adaptative strategies. *Annual Review of Ecology, Evolution and Systematics, 17*, 273–303.

Tamega, F. T. S., Oliveira, O. S., Figueiredo, M. A. O. (2013). Catalogue of the benthic marine life from Peregrino Oil Field, Campos Basin, Brazil. Report Instituto Biodiversidade Marinha, Rio de Janeiro.

Tâmega, F. T. S., & Figueiredo, M. A. O. (2005). Distribuicmão das algas calcárias incrustantes (Corallinales, Rhodophyta) em diferentes habitats na Praia do Forno, Armacmão dos Búzios, Rio deJaneiro. *Rodriguésia, 56*(87), 123–132.

Tâmega, F. T. D. S., Bassi, D., Figueiredo, M. A. D. O., & Cherkinsky, A. (2014). Deep-water rhodolith bed from central Brazilian continental shelf, Campos Basin: Coralline algal and faunal taxonomic composition. *Galaxea, Journal of Coral Reef Studies, 16*(1), 21–31.

Taylor, W. R. (1960). Marine algae of the eastern tropical and subtropical coasts of the America. The University of Michigan Press, Ann Arbor, 870 pp.

Tomita, N. Y. (1976). Contribuicmão ao conhecimento do gênero Sporolithon (Corallinaceae, Cryptonemiales) no Brasil. Universidade de São Paulo. Tese de Doutorado, 138 pp

Torrano-Silva, B. N. (2015). Taxonomia e Filogenia de Lithophylloideae no Brasil. Universidade de São Paulo. Tese de doutorado. 250pp.

Torrano-Silva, B. N., Vieira, B. R., Riosmena-Rodríguez, R., & Oliveira, M. C. (2018). Guidelines for DNA barcoding of coralline algae, focusing on Lithophylloideae (Corallinales) from Brazil. *Botanica Marina, 61*(2), 1–14.

Vidal, R., Meneses, I., & Smith, M. (2003). Molecular genetic identification of crustose representatives of the order Corallinales (Rhodophyta) in Chile. *Molecular Phylogenetics and Evolution, 28*, 404–419.

Vieira-Pinto, T. V. (2011). *Aspectos morfoanatômicos, reprodutivos e moleculares do gênero Lithophyllum (Lithophylloideae, Corallinales, Rhodophyta) do sul do Brasil.* Dissertação de mestrado. Universidade Federal de Santa Catarina—Departamento de Botânica, 88p.

Vieira-Pinto, T. (2016). *Diversidade das algas calcárias crostosas do Brasil baseada em marcadores moleculares e morfologia.* Tese de Doutorado. Universidade de São Paulo, 312p.

Vieira-Pinto, T., Oliveira, M. C., Bouzon, J., Sissini, M., Richards, J. L., Riosmena-Rodríguez, R., & Horta, P. A. (2014). *Lithophyllum* species from Brazilian coast: Range extension of *Lithophyllum margaritae* and description of *Lithophyllum atlanticum* sp. nov. *Phytotaxa, 190*(1), 355–369.

Villas-Bôas, A. B., Figueiredo, M. A. O., & Villaça, R. C. (2005). Colonization and growth of crustose coralline algae (Corallinales, Rhodophyta) on the Rocas Atoll. *Brazilian Journal of Oceanography, 53*(3–4), 147–156.

Villas-Boas, A. B., Riosmena-Rodriguez, R., Amado-Filho, G. M., Maneveldt, G. W., & Figueiredo, M. A. O. (2009). Rhodolith-forming species of *Lithophyllum (Corallinales; Rhodophyta) from Espírito Santo State, Brazil, including the description of L. depressum sp. nov. Phycologia, 48*(4), 237–248.

Villas-Boas, A. B. (2008). Comunidades de organismos incrustantes e identificacmão de algas calcárias em bancos de rodolitos no Estado do Espírito Santo. Universidade Federal do Rio de Janeiro. Tese de Doutorado, 123 pp.

Villas-Boas, A. B., Riosmena-Rodriguez, R., Tamega, F. T. S., Amado-Filho, G. M., Maneveldt, G. W. & Figueiredo, M. A. O. (2015). Rhodolith-forming species of the subfamilies Neogoniolithoideae, and Hydrolithoideae (Rhodophyta, Corallinales) from Espírito Santo State, Brazil. *Phytotaxa, 222(3), 169–184.*

Woelkerling, W. M. J. (1996). Subfamily Melobesioideae. In H. B. S. W. Womersley (Ed.), *The marine benthic flora of southern Australia – Part IIIB. Gracilariales, Rhodymeniales, Corallinales and Bonnemaisoniales* (164–210p.). Australian Biological Resources Study.

Woelkerling, W. J., Irvine, L. M., & Harvey, A. S. (1993). Growth forms in non-geniculate coralline red algae (Corallinales, Rhodophyta). *Australian Systematic Botany, 6*, 277–293.

Woelkerling, W. J., Granier, B., & Dias-Brito, D. (2014). Heydrichia (?) poignantii, sp. nov. (Sporolithaceae, Sporolithales, Rhodophyta), a 100 million year old fossil coralline red alga from North-Eastern Brazil, and a new Hauterivian record of Sporolithon from Switzerland. *Carnets de Geologie-Notebooks on Geology*. https://doi.org/10.4267/2042/53736

Žuljević, A., Kaleb, S., Peña, V., Despalatović, M., Cvitković, I., De Clerck, O., Le Gall, L., Falace, A., Vita, F., Braga, J. C., & Antolić, B. (2016). First freshwater coralline alga and the role of local features in a major biome transition. *Scientific Reports, 6*, 19642.

Chapter 4
Biodiversity Associated with Rhodolith Beds in Brazil

Frederico T. S. Tâmega, Guido Grimaldi, André Macedo, Sulamita O. Barbosa, Antônio B. Anderson, Eduardo Bastos, Vanessa Freire, Anderson Moreira, Celso Fernandes, Marcia A. O. Figueiredo, Pedro Bastos Carneiro, Lidiane P. Gouvêa, and Paulo A. Horta

Abstract Brazilian rhodolith-associated flora and fauna are abundant and diverse, providing a scenario of multicolored marine forests. The Brazilian coast has a high diversity of seaweeds, with 381 taxa associated with rhodolith beds. This includes the Phylum Chlorophyta (65), Ochrophyta (46), and Rhodophyta (270), comprising ~30% of seaweed *taxa* known to Brazil. In this context, the cryptofauna associated with Brazilian rhodoliths is significantly more abundant, with eight times higher density and greater taxonomic richness than the unconsolidated sediment. The rhodolith beds are an oasis of diversity in the Brazilian continental shelf with many

F. T. S. Tâmega (✉)
Departamento de Biotecnologia Marinha, Instituto de Estudos do Mar Almirante Paulo Moreira,
Arraial do Cabo, RJ, Brazil

G. Grimaldi · A. Macedo · E. Bastos · V. Freire · P. A. Horta
Departamento de Botânica, Centro de Ciências Biológicas, Universidade Federal de Santa Catarina, Florianópolis, SC, Brazil

S. O. Barbosa
Programa de Pós-Graduação em Botânica, Museu Nacional – UFRJ, Rio de Janeiro, RJ, Brazil

A. B. Anderson
Laboratório de Ichthyologia, Departamento de Oceanografia, Universidade Federal do Espírito Santo, Vitória, Brazil

A. Moreira · C. Fernandes
Laboratório de Meios Porosos e Propriedades Termofísicas—EMC, CTC, Universidade Federal de Santa Catarina, Florianópolis, SC, Brazil

M. A. O. Figueiredo
Instituto de Pesquisas Jardim Botânico do Rio de Janeiro, Rio de Janeiro, RJ, Brazil

P. B. Carneiro
Universidade Federal do Delta do Parnaíba (UFDPar), Parnaíba, Brazil

L. P. Gouvêa
Centre of Marine Sciences, University of Algarve, Faro, Portugal

P. A. Horta, M. N. Sissini (eds.), *Brazilian Rhodolith Beds*, Brazilian Marine Biodiversity, https://doi.org/10.1007/978-3-031-61449-1_4

singularities, interacting with different coastal and ocean ecosystems, with countless relationships and species intertwined by the evolutionary history and environments built by coralline algae. All these biodiversity and ecological functions are threatened due to the combined effects of global and local stressors, mainly represented by oil exploitation, trawling, ocean warming, and acidification.

Keywords Marine ecology · Ecological cascade · Biodiversity · Fauna · Flora

4.1 Introduction

Rhodolith beds are one of the main benthic communities dominated and constructed by marine macrophytes, considered the world's largest coralline red algae carbonate deposits (Foster, 2001). These structurally complex benthic environments (Fig. 4.1.) are distributed across a wide range of latitudes and depths, spanning from tropical to polar regions deepening to 268 meters (Bosence, 1983; Littler et al., 1986; Foster, 2001; Henriques et al., 2014a; McCoy & Kamenos, 2015; Riosmena-Rodríguez et al., 2017). In Brazil, rhodolith beds are widespread, covering extensive areas of the north, northeastern, and southeastern Brazilian continental shelf (Kempf, 1970; Milliman & Amaral, 1974), reaching the southern coast in isolated formations beds in the Marine Protected Area (MPA) of Reserva Biológica Marinha do Arvoredo (Carvalho et al., 2020a). The range of the latitudinal high niche suitability is between 5°N and 27°S covering an area of more than 200,000 Km^2 (Carvalho et al., 2020b).

Regarding their morphological complexity (Schubert et al., 2022), which increases the heterogeneity of the surrounding environment (Jones et al., 1994), rhodoliths can be defined as foundation species, considering the coralline red algae as ecosystems engineers (Dayton, 1972). According to Schubert et al. (2022), the complex surface and related physiological variability, coupled with the interstitial space formed inside rhodoliths, results in a highly complex tridimensional living

Fig. 4.1 Different aspects of rhodolith bed presenting different dominant species and composition. (**a**) Rhodolith bed dominated by *Melyvonnea erubescens.* (**b**) Rhodolith bed dominated by *Lithothamnion crispatum*

structure. This unique configuration enables the production of oxygen and absorption of CO_2, offering favorable conditions for the settlement and growth of species within these microhabitats (Jackson et al., 2004). Consequently, this ecological cascade increases the abundance and richness of associated *taxa* (Peña et al., 2014a, b; Ordines et al., 2015).

To standardize the different terminologies used for non-geniculated coralline algae growth forms, Woelkerling et al. (1993) defined ten focal points in the literature that characterizes the external morphology (unconsolidated, encrusting, warty, lumpy, fruticose, discoid, layered, foliose, ribbonlike, arborescent). Each of these forms is distinctly characterized; "warty," "lumpy," and "fructose" plants can grow attached to a substrate or grow freely; and the nodules composed by plants with these characteristics can be entirely constituted by protuberances (Woelkerling et al., 1993). According to the same author, "protuberances" can be defined as "a cylindrical to compressed structure, or with a more irregular shape that grows outward or branched, which normally has a radial organization" (Woelkerling, 1988). Although it seems clear the importance of the tridimensional structure (Fig. 4.2) in driving associated species patterns, there are just a few studies aiming at the

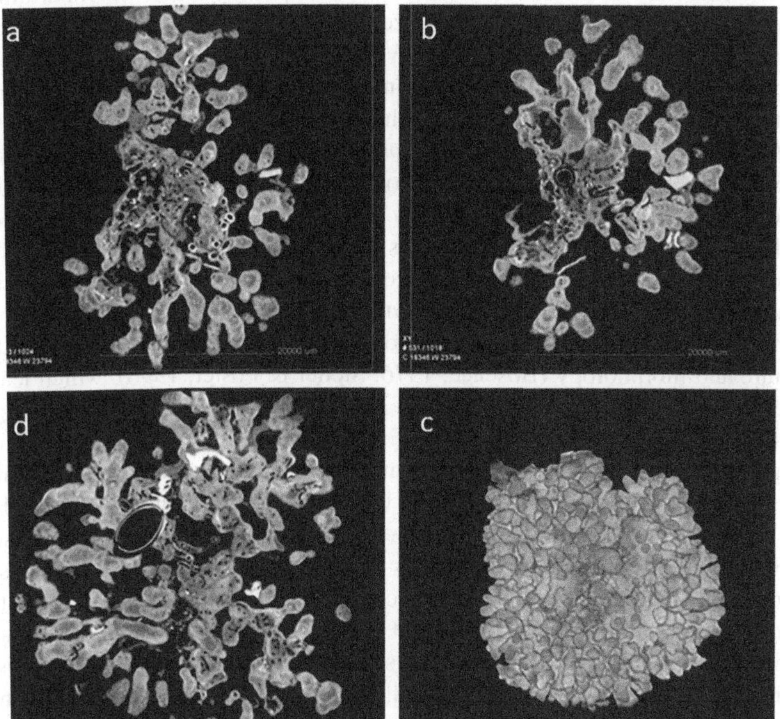

Fig. 4.2 Microtomography of *Melyvonnea erubescens* highlighting different optical sections with abundant internal spaces (**a**–**c**) and surface with expressive rugosity (**d**)

settlement of cryptofaunal communities in rhodoliths, leading to conclusions about what drivers are most important to these patterns (Steller & Cáceres-Martínez, 2009).

Regarded to the structural complexity of the nodules, De Grave (1999) hypothesized that the shape of interstitial space supports the associated crustacean's assemblage, and attesting the importance of the tridimensional structure of rhodoliths, Otero-Ferrer et al. (2019) concluded that early faunal colonization is rather driven by heterogeneity of the nodules (defined by concavity, convexity, epibionts, etc.), than the size of experimental rhodoliths. In agreement, Steller et al. (2003) found that thallus complexity is an important driver for an associated community settlement, although space available (thallus volume) appears to have great importance too. Availability of space, refuge, and resources provided by the complexity of the thalli seems to be a good predictor for abundance and richness of cryptofauna; this supports the hypothesis that an increase in complexity structure drives species diversity (Weber-Van Bosse & Foslie, 1904).

In another study, Méndez Trejo et al. (2020), yet considering two species with different growth forms (*Lithothamnion crispatum* Hauck and *Sporolithon* sp.), verified that cryptofaunal diversity associated to *L. crispatum*, which has a higher branching density, was greater related to *Sporolithon* sp., inferring that species and growth forms may drive the community structure patterns. Also corroborating the influence of the tridimensional structure in cryptofauna, Abrecht et al. (2022) observed a higher abundance and diversity of organisms in branching and larger size rhodoliths, compared to laminar and columnar morphotypes. Evaluating the macroinvertebrate composition within two rhodolith forming species (*Mesophyllum engelhartii* (Foslie) Adey; *Neogoniolithon brassica-florida* (Harvey) Setchell and Mason), with different growth forms (foliose and fruticose, respectively), in the temperate waters from Australia. Harvey and Bird (2008) observed Phylum combined and individually, and no differences in cryptofauna and abundance was found between the two rhodolith species, suggesting that there is no interspecific divergence, despite considering different growth forms.

Physical factor variations in the environment are thought to influence rhodolith structure and morphology (Bosence, 1976; Steller & Foster, 1996), which leads to altered community-associated structure. Méndez Trejo et al. (2020) investigated two different rhodolith species at three different rhodolith beds, each one differing in regard to water motion, light level, and sedimentation characteristic. Authors observed that, for *Sporolithon*, at the bed with more light availability and much less mud, free space available in the nodules was proportionally higher in comparison to the other bed, presumably because they were capable of reaching bigger sizes. Thought, running counter the prevailing expectation, that sedimentation has strong negative impact in rhodoliths (Bernard et al., 2019), thus harboring less organisms associated, when both rhodolith-forming species were combined to the analysis, the sedimented site had larger nodules regarding to volume and free space, with greater cryptofaunal species richness. Abrecht et al. (2022) also investigated the influence of different wave energy habitats in rhodolith abundance, morphotype, volume, and shape, with more energetic environments indirectly driving rhodolith cryptofauna as it harbors larger, rounder, and more branching nodules.

Seasonality can influence cryptofauna in rhodoliths (Foster et al., 2007; Riosmena-Rodríguez & Medina-López, 2010; Otero-Ferrer et al., 2020), and depth also may be an important driver, as it directly modules growth rates (Bahia et al., 2010). Recently, Veras et al. (2020) found that euphotic zones can harbor twice the abundance of organisms and number of *taxa* than mesophotic zones. Besides that, porosity was found to increase biodiversity in an Arctic environment (Teichert, 2014). Furthermore, the presence of bioengineers, as *Malacanthus plumieri* (Bloch, 1786) (Pereira-Filho et al., 2015), and rhodolith age (McConnico et al., 2014) can indirectly affect the structure of a rhodolith bed or individually.

4.2 Associated Flora

Regarding Brazilian shallow water rhodolith beds (Testa & Bosence, 1999; Gherardi, 2004; Amado-Filho et al., 2007, 2010; Riul et al., 2008, 2009; Bahia et al., 2010; Dias & Villaça, 2012; Dominguez et al., 2013; Pascelli et al., 2013; Villas-Boas et al., 2009, 2014) and some formation oceanic islands (Villas-Boas et al., 2005; Pereira-Filho et al., 2012; Amado-Filho et al., 2012b), associated flora are abundant and diverse, providing a cenarium of multicolored marine forest. With increasing depth, mesophotic rhodolith beds (Della Giustina, 2006; Amado Filho et al., 2012a, 2016; Figueiredo et al., 2012; Henriques et al., 2012, 2014a, b; Pereira-Filho et al., 2012; Moura et al., 2013, 2016; Tâmega et al., 2013, 2014) have marine animal as dominant epibionts.

Associated with these foundation species in the world, and mainly in Brazil, are observed an abundant and diverse benthic flora (Fig. 4.3), including endemic (Anderson et al., 2021) and economically important species (Solano-Barquero et al., 2022). In a general way, the biomass of associated flora increases with the latitude reduction (Carvalho et al., 2020b). At intermediary latitudes, at depths of 45–120 meters, the rhodolith beds provide substrate for the endemic kelp-forming species *Laminaria abyssalis* A.B.Joly and E.C.Oliveira (Amado-Filho et al., 2007; Rothman et al., 2017). In the southernmost latitudes, these species become rare or are replaced by species more tolerant to colder temperatures (Pascelli et al., 2013). Although tropical species dominate these environments, *L. glaciale* has already been found at depths of 60 meters, where temperatures below 15 °C allow its development (Henriques et al., 2012).

Wide occurrence of rhodolith beds at a wide latitudinal gradient exposes these organisms and associated flora to a high amplitude of environmental factors, which seems to be determinant of the community structure and diversity (Hernandez-Kantun et al., 2017). Current velocity, waves, and depth are among the local factors that most influence the formation of rhodoliths and abundance of epibiotic flora (Hinojosa-Arango et al., 2009; Sañé et al., 2016). Under very weak currents, the coralline algae would not come off and would not roll increasing the probability of canopy formation. Under very strong currents, the thallus will be more frequently cleaned or eroded with the contact with the substrate reducing canopy cover (Foster,

Fig. 4.3 Different aspects of foundation species and associated flora of Brazilian rhodolith beds, where (**a**) and (**b**) represent foundation species in a unispecific and multispecific rhodolith, constructed by *Lithothamnion crispatum* and by *Lithophyllum atlanticum* (left) with *Melyvonnea erubescens* (right), respectively; (**c**) *Asparagopsis taxiformis*; (**d**) *Dictyota* sp. (**e**) Detail of *Cryptonemia* sp.; (**f**) Different species with brown and green seaweeds, with high cover of *Caulerpa* sp

2001; Hinojosa-Arango et al., 2009). In addition, regions with greater movement tend to have less abundance of mobile organisms and an increase of organisms with stronger fixation structures (Maughan & Barnes, 2000; Hinojosa-Arango et al., 2009). Other authors have reported a seasonal difference in epiphyte coverage, with its decrease in winter due to storms occurring in this season (Amado-Filho et al., 2007; Pascelli et al., 2013). It is important to point out that local stressors, as those related with rhodolith mining, or impacts of trawling, can also threat their beds and associated biodiversity, composing together with ocean warming and acidification reasons of major concern of the academic community (Fragkopoulou et al., 2021).

However, despite their abundance and the ubiquitous vertical and latitudinal distribution, the majority of Brazilian rhodolith beds are poorly known due to mostly beds occurring in deeper habitats (~30–120 m depth), with difficult access (Amado-Filho et al., 2017). In this chapter, we describe the associate flora related to the rhodolith beds, which encompass the main Brazilian marine forests.

The Brazilian coast encompasses 48 *taxa* of non-geniculate coralline algae (NGCA) including crustose coralline algae (CCA) and rhodolith-forming species (9 Sporolithales, 25 Corallinales, and 14 Hapalidiales; Horta et al., 2016). The NGCA species richness found in Brazilian waters is more than twice than recorded worldwide: Europe and Azores (12; Peña and Barbara (2008); Peña et al. (2011)), Australia and New Zealand (7; Harvey and Bird (2008); Nelson et al. (2012)), Gulf of California (4; Ávila and Riosmena-Rodríguez (2010)), Alaska (1; Konar et al. (2006)), and Canada (1; Gagnon et al. (2012)).

Coralline red algae were subject of integrative systematic studies, encompassing molecular and morpho-anatomical data in Brazil (Bahia et al., 2014a, b; Sissini et al., 2014; Torrano-Silva et al., 2014, 2018; Vieira-Pinto et al., 2014; Tâmega et al. (2021)) and worldwide (Peña et al., 2011, 2014a, b; Harvey et al., 2013; Kato et al., 2013; Mateo-Cid et al., 2014; Adey et al., 2015; Hernandez-Kantun et al., 2015; Pardo et al., 2015; Wolf et al., 2016), which have resulted into reassess species diversity and detect cryptic species. To more discussion, see Chap. 2, Rhodolith Beds: Occurrence, Diversity, and Evolution.

A recent survey along Arraial do Cabo Bay, southeastern Brazil, has shown a new type of coralline red algae bed that are dominated by free-living geniculate coralline algae (GCA; Tâmega et al., 2017), instead of rhodoliths (Riosmena-Rodríguez et al., 2017). Morphological based studies have identified *Amphiroa beauvoisii* (J.V. Lamouroux), *Jania adhaerens* (J.V.Lamouroux), and *Jania capillacea* (Harvey). Rhodolith beds are well known (Foster, 2001). However, little is known about the species and its interactions found in GCA beds, creating this unique structural benthic formation. These unknown GCA beds are of significance because of the branched structure thalli that influences the abundance and richness of associated organisms by providing available hard substrate, space, refuge, and resources (Chapman et al., 2005; Asnaghi et al., 2015).

The literature reported a high seaweed diversity to the Brazilian coast, and 381 *taxa* are associated with rhodolith beds, corresponding to the Phylum Chlorophyta (65), Ochrophyta (46), and Rhodophyta (270; Table 4.1). New seaweed records have been reported associated with the Brazilian rhodolith beds, as Amado-Filho et al. (2010), Guimarães and Amado-Filho (2009), Riul et al. (2009), and Carneiro et al. (2021). Rhodolith beds were highly diverse holding nearly 30% of seaweed *taxa* known to Brazil (Forzza et al., 2010).

The abundance of associated flora is related to site characteristics, such as depth, turbidity, and current velocity, and factors that change with latitude as temperature and available irradiance. These environmental drivers, boosted by biological interactions, provide a variable and multicolored cenarium. Despite the dominance of brown seaweed in most of the tropical littoral, with *taxa* like *Dictyopteris* spp., *Canistrocarpus* spp., *Sargassum* spp., and *Padina* spp., some of these areas will

Table 4.1 Seaweeds taxa associated with rhodolith beds in Brazil. Abbreviation: epiphytic (E), rhodolith forming species (R)

Taxon	Habit
Phylum Chlorophyta	
Anadyomene lacerata D.S.Littler and Littler	E
Anadyomene linkiana D.S.Littler and M.M.Littler	E
Anadyomene saldanhae A.B.Joly and E.C.Oliveira	E
Anadyomene stellata (Wulfen in Jacq.) J.Agardh	E
Avrainvillea nigricans Decaisne	E
Bolbocoleon sp.	E
Boodleopsis pusilla (Collins) W.R.Taylor, A.B.Joly and Bernatowicz	E
Bryopsis hypnoides J.V.Lamouroux	E
Bryopsis pennata J.V.Lamouroux	E
Bryopsis plumosa (Hudson) C.Agardh	E
Caulerpa ambigua Okamura	E
Caulerpa cupressoides (Vahl) C.Agardh	E
Caulerpa cupressoides var. *lycopodium* Weber Bosse	E
Caulerpa lanuginosa C.Agardh	E
Caulerpa mexicana Sonder ex Kützing	E
Caulerpa paspaloides var. *compressa* (W/eber Boose) M.Howe	E
Caulerpa prolifera (Forsskal) J.V.Lamouroux	E
Caulerpa pusilla (Kützing) J.Agardh	E
Caulerpa racemosa (Firssjak) J.Agardh	E
Caulerpa scalpelliformis (R. Brown ex Turner) C.Agardh	E
Caulerpa sertularioides (S.G.Gmelin) M.Howe	E
Caulerpa verticillata C.Agardh	E
Caulerpa verticillata f. *charoides* Weber-van Bosse	E
Caulerpa webbiana Montagne	E
Caulerpa webbiana f. *tomentella* (Harvey ex J.Agardh) Weber Bosse	E
Caulerpa sp.	E
Chaetomorpha sp.	E
Chamaedoris peniculum (J.Ellis and Solander) Kuntze	E
Cladophora brasiliana G.Martens	E
Cladophora catenata (Linnaeus) Kützing	E
Cladophora corallicola Borgesen	E
Cladophora dalmatica Kützing	E
Cladophora ordinata (Børgesen) C. Hoek	E
Cladophora pygmaea Reinke	E
Cladophora rupestris (Linnaeus) Kützing	E
Cladophora vagabunda (Linnaeus) Hoek	E
Cladophora sp.	E
Codium intertextum Collins and Hervey	E
Codium isthmocladum Vickers	E
Codium sp.	E

(continued)

Table 4.1 (continued)

Taxon	Habit
Dasycladus vermicularis (Scop.) Krasser	E
Dictyosphaeria cavernosa (Forsskal) Børgesen	E
Entocladia viridis Reinke	E
Ernodesmis verticillata (Kützing) Børgesen	E
Halimeda cuneata Hering	E
Halimeda discoidea Decaisne	E
Halimeda gracilis Harvey ex J.Agardh	E
Halimeda incrassata (J.Ellis) J.V.Lamouroux	E
Halimeda opuntia (Linnaeus) J.V.Lamouroux	E
Halimeda sp.	E
Microdictyon marinum (Bory de Saint-Vincent) P.C. Silva	E
Neomeris annulata Dickie	E
Penicillus capitatus Lamouroux	E
Rhipilia tomentosa Kützing	E
Udotea abbottiorum D.Litter and M.Litter	E
Udotea cyathiformis Decaisne	E
Udotea flabellum (J.Ellis and Solander) M.Howe	E
Udotea unistratea D.S.Littler and Littler	E
Ulva fasciata Delile	E
Ulva flexuosa Wulfen	E
Ulva lactuca Linnaeus	E
Valonia utricularis (Roth) C. Agardh	E
Ventricaria ventricosa (J. Agardh) J.L.Olsen and J.A.West	E
Verdigellas fimbriata D.L.Ballantine and J.N. Norris	E
Verdigellas peltata D.L.Ballantine and J.N. Norris	E
Phylum Ochrophyta	
Canistrocarpus cervicornis (Kützing) DePaula and DeClerck	E
Chnoospora minima (K.Hering) Papenf.	E
Colpomenia sinuosa (Mertens ex Roth) Derbès and Solier	E
Dictyopteris delicatula J.V.Lamouroux	E
Dictyopteris jolyana E.C.Oliveira and R.P.Furtado	E
Dictyopteris justii J.V.Lamouroux	E
Dictyopteris plagiogramma (Montagne) Vickers	E
Dictyopteris polypodioides (DC. in Lam. and DC.) J.V.Lamouroux	E
Dictyota bartayresiana J.V.Lamouroux	E
Dictyota ciliolata Sonder ex Kützing	E
Dictyota menstrualis (Hoyt) Schnetter, Hornig, and Weber-Peukert	E
Dictyota mertensii (C.Martius) Kützing	E
Dictyota pinnatifida Kützing	E
Dictyota puchella Hornig and Schnetter	E
Dictyota sp.	E
Feldmannia irregularis (Kützing) Hamel	E

(continued)

Table 4.1 (continued)

Taxon	Habit
Feldmannia simplex (H.Crouan and P.Crouan) Hamel	E
Hincksia mitchelliae (Harvey) P.C.Silva	E
Hincksia onslowensis (Amsler and Kapraum) P.C.Silva in P.C.Silva et al.	E
Laminaria abyssalis A.B.Joly and E.C.Oliveira	E
Lobophora variegata (J.V.Lamouroux) Womersley ex E.C.Oliveira	E
Nereia tropica (W.R.Taylor) W.R.Taylor	E
Padina boergesenii Allender and Kraft	E
Padina gymnospora (Kützing) Sonder	E
Padina pavonica (Linnaeus) Thivy	E
Padina profunda S.A.Earle	E
Padina sanctae-crucis Børgesen	E
Rosenvingea intricata (J.Agardh) Børgesen	E
Rosenvingea orientalis (J.Agardh) Børgesen	E
Rosenvingea sanctae-crucis Børgesen	E
Sargassum filipendula C.Agardh	E
Sargassum furcatum Kützing	E
Sargassum hystrix J.Agardh	E
Sargassum hystrix var. *spinulosum* (Kützing) Grunow	E
Sargassum platycarpum Montagne	E
Sargassum stenophyllum C. Martius	E
Sargassum vulgare C.Agardh	E
Sargassum sp.	E
Spatoglossum schroederi (C.Agardh) Kützing	E
Sphacelaria novae-hollandiae Sonder	E
Sphacelaria rigidula Kützing	E
Sphacelaria sp.	E
Sporochnus bolleanus Mont	E
Sporochnus pedunculatus (Hudson) C.Agardh	E
Stypopodium zonale (J.V.Lamouroux) Papenfuss	E
Zonaria tournefortii (J.V.Lamouroux) Montagne	E
Phylum Rhodophyta	
Acanthophora spicifera (Vahl) Børgesen	E
Acrochaetium antillarum W.R.Taylor	E
Acrochaetium globosum Børgesen	E
Acrochaetium infestans M.A.Howe and Hoyt	E
Acrochaetium liagorae Børgesen	E
Acrochaetium microscopicum (Nägeli ex Kützing) Nägeli	E
Acrochaetium yamadae Y.Lee and I.K.Lee	E
Acrothamnion butlerae (Collins) Kylin	E
Acrosymphyton caribaeum (J.Agardh) Sjöstedt	E
Acrosymphyton sp.	E
Agardhiella floridana (Kylin) P.W.Gabrielsen ex Guimarães and Oliveira	E

(continued)

Table 4.1 (continued)

Taxon	Habit
Agardhiella ramosissima (Harvey) Kylin	E
Agardhiella subulata (C.Agardh) Kraft and M.J.Wynne	E
Aglaothamnion boergesenii (Aponte and D.L.Ballant.) L'Hardy-Halos and Rueness	E
Aglaothamnion felipponei (Howe) Aponte, Ballantine and J.N.Norris	E
Aglaothamnion cf. *gallicum* (Nägeli) L'Hardy-Halos and Ardré	E
Aglaothamnion halliae (Collins) Aponte, D.L. Ballant and J.N.Norris	E
Aglaothamnion sp.	E
Amansia multifida J.V.Lamouroux	E
Amansia sp. 1	E
Amansia sp. 2	E
Amphiroa beauvoisii J.V.Lamouroux	E
Amphiroa fragilissima (Linnaeus) J.V.Lamouroux	E
Amphiroa rigida J.V.Lamouroux	E
Amphiroa van-bosseae Me.Lemoine	E
Amphiroa sp.	E
Anotrichium tenue (C.Agardh) Nägeli	E
Antithamnion antillanum Børgesen	E
Antithamnion cf. *decipiens* (J.Agardh) Athanasiadis	E
Antithamnionella atlantica (E.C.Oliveira) C.W.Schneid.	E
Antithamnionella breviramosa (Dawson) E.M.Woll.	E
Asparagopsis taxiformis (Delile) Trevisan	E
Asteromenia peltata (W.R.Taylor) Huisman and A.J.K.Millar	E
Bostrychia sp.	E
Botryocladia bahamensis Ballantine and Aponte	E
Botryocladia occidentalis (Børgesen) Kylin	E
Botryocladia pyriformis (Børgesen) Kylin	E
Bryothamnion seaforthii (Turner) Kützing	E
Bryothamnion triquetrum (S.G.Gmelin) M.Howe	E
Caloglossa leprieurii (Montagne) G.Martens	E
Centroceras clavulatum (C.Agardh) Montagne	E
Carradoriella denudata (Dillwyn) A.M.Savoie and G.W.Saunders	E
Ceramium brasiliense A.B.Joly	E
Ceramium cf. *comptum* Børgesen	E
Ceramium dawsonii A.B.Joly	E
Ceramium flaccidum (Harvey ex Kützing) Ardissone	E
Ceramium luetzelburgii O.C. Schmidt	E
Ceramium vagans P.C.Silva	E
Ceramium sp.	E
Ceratodictyon intricatum (C.Agardh) R.E.Norris	E
Champia parvula (C.Agardh) Harvey	E
Champia salicornioides Harvey	E
Champia sp.	E

(continued)

Table 4.1 (continued)

Taxon	Habit
Chondria dasyphylla (Woodward) C.Agardh	E
Chondria platyramea A.B.Joly and Ugadim	E
Chondria sp.	E
Chondracanthus acicularis (Roth) Fredericq	E
Chondrophycus furcatus (Cordeiro-Marino and M.T. Fujii) M.T.Fujii and Sentíes	E
Chondrophycus translucidus (M.T.Fujii and M.Cordeiro-Marino) D.J.Garbary and J.T.Harper	E
Chrysymenia enteromorpha Harvey	E
Chrysymenia ventricosa (J.V.Lamouroux) J.Agardh	E
Chrysymenia sp.	E
Claudea elegans J.V.Lamour.	E
Coelarthrum cliftonii (Harv.) Kylin	E
Colaconema ophioglossa (C.W.Schneid.) Afonso-Carrillo, Sanson and Sangil	E
Corallina panizzoi Schneter and U.Richt.	E
Corallina sp.	E
Corynomorpha clavata (Harvey) J.Agardh	E
Cottoniella filamentosa (M. Howe) Borgesen	E
Craspedocarpus jolyi (E.C.Oliveira) C.W.Schneider	E
Crouania attenuata (C.Agardh) J.Agardh	E
Cryptonemia bengryi W.R.Taylor	E
Cryptonemia crenulata (J.Agardh) J.Agardh	E
Cryptonemia delicatula Joly and Cordeiro	E
Cryptonemia seminervis (C.Agardh) J.Agardh	E
Cryptonemia sp.	E
Cryptopleura crispa Kylin	E
Dasya baillouviana (S.G.Gmelin) Montagne	E
Dasya brasiliensis E.C.Oliveira-Filho and Y.Y.Braga	E
Dasya collinsiana M.A.Howe	E
Dasya elongata Sonder	E
Dasya ramosissima Harvey	E
Dasya rigidula (Kützing) Ardissone	E
Dichotomaria marginata (J.Ellis and Solander) Lamarck	E
Dichotomaria obtusata (J.Ellis and Solander) Lamarck	E
Dictyurus occidentalis J.Agardh	E
Digenia simplex (Wulfen) C.Agardh	E
Dipterosiphonia dendritica (C.Agardh) F.Schmitz	E
Dipterosiphonia reversa C.W.Schneider	E
Dorhniella antillarum (W.R.Taylor) Feldmann-Mazoyer	E
Dotyophycus pacificum I.A.Abbott	E
Dudresnaya crassa M.Howe	E
Erythrotrichia carnea (Dillwyn) J.Agardh	E
Falkenbergia hillebrandii (Bornet) Falkenberg	E

(continued)

Table 4.1 (continued)

Taxon	Habit
Flahaultia tegetiformans W.R.Taylor	E
Galaxaura rugosa (J.Ellis and Sol.) J.V.Lamour.	E
Gayliella flaccida (Harvey ex Kützing) T.O.Cho and L.J. McIvor.	E
Gelidiopsis variabilis (Greville ex J.Agardh) F.Schmitz	E
Gelidiopsis sp.	E
Gelidium crinale (Hare ex Turner) Gaillon	E
Gelidium cf. *pusillum* (Stackhouse) LeJolis	E
Gelidium sp.	E
Gloiocladia iyoensis (Okamura) R.E.Norris	E
Gracilaria birdiae Palastino and E.C.Oliveira	E
Gracilaria caudata J.Agardh	E
Gracilaria cervicornis (Turner) J.Agardh	E
Gracilaria cuneata Areschoug	E
Gracilaria curtissiae J.Agardh	E
Gracilaria domingensis (Kützing) Sonder ex Dickie	E
Gracilaria mammillaris (Montagne) M.Howe	E
Gracilaria sp.	E
Grateloupia filicina (J.V.Lamouroux) C.Agardh	E
Griffithsia globulifera Harvey ex Kütz.	E
Griffithsia schousboei Montagne	E
Gymnogongrus griffithsiae (Turner) C.Martius	E
Gymnothamnion elegans (Schousboe ex C.Agardh) J.Agardh	E
Haliptilon cubense (Montagne ex Kützing) Garbary and H.W.Johansen	E
Haliptilon subulatum (J.Ellis and Solander) H.W.Johansen	E
Halopithys schottii (W.R.Taylor) L.E.Phillips and DeClerck	E
Haloplegma duperreyi Montagne	E
Halymenia brasiliana S.M.P.B.Guimarães and M.T.Fujii	E
Halymenia elongata C.Agardh	E
Halymenia floresii (Clemente) C.Agardh	E
Halymenia floridana J.Agardh	E
Halymenia vinacea M.A.Howe and W.R.Taylor	E
Herposiphonia bipinnata M.A.Howe	E
Heterosiphonia crispella (C.Agardh) M.J.Wynne	E
Heterosiphonia gibesii (Harvey) Falkenberg	E
Herposiphonia secunda (C.Agardh) Ambronn	E
Herposiphonia tenella (C.Agardh) Ambronn	E
Herposiphonia sp.	E
Heterodasya mucronata (Harvey) M.J.Wynne	E
Heterosiphonia crispella (C. Agardh) M.J.Wynne	E
Heterosiphonia gibesii (Harvey) Falkenberg	E
Hydrolithon boergesenii (Foslie) Foslie	E
Hydrolithon breviclavium (Foslie) Foslie	R

(continued)

Table 4.1 (continued)

Taxon	Habit
Hydrolithon farinosum (J.V.Lamouroux) D.Penrose and Y.M.Chamberlain	R
Hydrolithon onkodes (Heydrich) Penrose and Woelkerling	R
Hydrolithon reinboldii (Weber-van Bosse and Foslie) Foslie	R
Hydrolithon samoense (Foslie) D.Keats and Y.Chamberlain	R
Hydrolithon sp.	R
Harveylithon rupestre (Foslie) A.Rösler, Perfectti, V.Peña & J.C.Braga (Foslie) Penrose	R
Hydropuntia caudata (J.Agardh) Gurgel and Fredericq	E
Hydropuntia cornea (J.Agardh) M.J.Wynne	E
Hypnea cenomyce J.Agardh	E
Hypnea musciformis (Wulfen) J.V.Lamouroux	E
Hypnea spinella (C.Agardh) Kützing	E
Hypnea valentiae (Turner) Mont.	E
Hypnea sp.	E
Hypoglossum hypoglossoides (Stackhouse) F.S.Collins and Hervey	E
Hypoglossum tenuifolium (Harvey) J.Agardh	E
Jania adhaerens J.V.Lamouroux	E
Jania capillacea Harvey	E
Jania crassa J.V.Lamouroux	E
Jania cubensis Montagne ex Kützing	E
Jania pumila J.V.Lamouroux	E
Jania subulata (Ellis and Solander) Sonder	E
Jania ungulata (Yendo) Yendo	E
Kallymenia limminghei Montagne	E
Kallymenia sp.	E
Laurencia sp.	E
Lejolisia mediterranea Bornet	E
Leptofauchea brasiliensis A.B.Joly	E
Liagora ceranoides J.V.Lamouroux	E
Lithophyllum atlanticum Vieira-Pinto, Oliveira et Horta	R
Lithophyllum corallinae (P.Crouan and H.Crouan) Heydrich	R
Lithophyllum depressum Villas-Boas, Figueiredo and Riosmena-Rodriguez	R
Lithophyllum johansenii Woelkerling and S.J.Campbell	R
Lithophyllum margaritae (Hariot) Heydrich	R
Lithophyllum stictiforme (Areschoug) Hauck	R
Lithophyllum sp.	R
Lithothamnion brasiliense Foslie	R
Lithothamnion crispatum Hauck	R
Lithothamnion glaciale Kjellman	R
Lithothamnion muelleri Lenormand ex Rosanoff	R
Lithothamnion superpositum Foslie	R
Lithothamnion sp.	R
Lejolisia mediterranea Bornet	E

(continued)

Table 4.1 (continued)

Taxon	Habit
Leptofauchea brasiliensis A.B.Joly	E
Lomentaria rawitscheri A.B.Joly	E
Melanothamnus sphaerocarpus (Børgesen) Díaz-Tapia and Maggs	E
Melobesia membranacea (Esper) J.V.Lamouroux	R
Melobesia rosanoffi (Esper) J.V.Lamouroux	R
Melyvonnea erubescens (Foslie) Athanasiadis and D. L.Ballantine	R
Meristiella gelidium (J.Agardh) D.P.Cheney and P.W.Gabrielson	E
Meristotheca echinocarpa (Areshoug) Faye and Masuda	E
Mesophyllum engelhartii (Foslie) W.H.Adey	R
Mesophyllum erubescens (Foslie) Lemoine	R
Mesophyllum sp.	R
Murrayella periclados (C.Agardh) F.Schmitz	E
Naccaria corymbosa J.Agardh	E
Neogoniolithon brassica-florida (Harvey) Setchell et Mason	R
Neogoniolithon fosliei (Heydrich) Setchell and L.R.Mason	R
Neogoniolithon mamillare (Harvey) Setchell and Mason	R
Neogoniolithon sp.	R
Neosiphonia ferulacea (Suhr ex. J.Agardh) S.M. Guimarães and M.T. Fujii.	E
Neosiphonia gorgoniae (Harvey) S.M.Guimarães and M.T.Fujii	E
Neosiphonia sphaerocarpa (Børgesen) M.S.Kim and I.K.Lee	E
Nitophyllum adhaerens M.J.Wynne	E
Nitophyllum sp.	E
Ochtodes secundiramea (Montagne) M.Howe	E
Ophidocladus simpliciusculus (P.Crouan and H.Crouan) Falkenberg	E
Osmundaria obtusiloba (C.Agardh) R.E.Norris	E
Osmundea lata (M.Howe and W.R.Taylor) Y.Yoneshigue-Valentin, M.T.Fujii and C.F.Gurgel	E
Palisada furcata (Cordeiro-Marino and M.T.Fujii) Cassano and M.T.Fujii	E
Palisada flagellifera (J. Agardh) K.W.Nam	E
Periphykon delesserioides A.B.Joly, Ugadim and E.C.Oliveira	E
Petroglossum undulatum C.W.Schneider	E
Peyssonnelia inamoena Pilger	E
Peyssonnelia simulans Weber Bosse	E
Peyssonnelia sp.	E
Phymatolithon sp.	R
Platoma gelatinosum (M.A.Howe) C.W.Schneider et al.	E
Platoma sp.	E
Pleonosporium caribaeum (Børgesen) R.E.Norris	E
Plocamium brasiliense (Greville) M.Howe and W.R.Taylor	E
Pneophyllum sp.	R
Polysiphonia howei Hollenberg	E
Polysiphonia subtilissima Mont	E

(continued)

Table 4.1 (continued)

Taxon	Habit
Polysiphonia sp.	E
Predaea feldmannii Børgesen	E
Predaea sp.	E
Protokuetzingia schottii W.R.Taylor	E
Pterocladiella capillacea (S.G.Gmelin) Santelices and Hommersand	E
Pterosiphonia pennata (C.Agardh) Sauvageau	E
Reticulocaulis mucosissimus I.A.Abbott	E
Rhodymenia divaricata E.Y.Dawson	E
Rhodymenia pseudopalmata (J.V.Lamouroux) P.C.Silva	E
Titanoderma pustulatum (J.V.Lamouroux) Nägeli	R
Tricleocarpa cylindrica (J.Ellis and Solander) Huisman and Borowitzka	E
Tsengia sp.	E
Sahlingia subintegra (Rosenvinge) Kornmann	E
Scinaia aborealis Huisman	E
Scinaia sp.	E
Sebdenia flabellata (J.Agardh) P.G.Parkinson	E
Solieria filiformis (Kützing) P.W.Gabrielson	E
Spermothamnion investiens (P.Crouan and H.Crouan) Vickers	E
Spongites fruticulosus Kützing	R
Spongites yendoi (Foslie) Chamberlain	R
Sporolithon australasicum (Foslie) N.Yamaguishi-Tomita ex M.J.Wynne	R
Sporolithon elevatum Henriques et Riosmena-Rodriguez	R
Sporolithon episoredion (Adey, Townsend and Boykins) Verheij	R
Sporolithon episporum (M.Howe) E.Y.Dawson	R
Sporolithon erythraeum (Rothpletz) Kylin	R
Sporolithon howei (Lemoine) N.Yamaguishi-Tomita ex M.J.Wynne	R
Sporolithon molle (Heydrich) Heydrich	R
Sporolithon pacificum E.Y.Dawson	R
Sporolithon ptychoides Heydrich	R
Sporolithon tenue Bahia et al.	R
Sporolithon yoneshigueae Bahia, Amado-Filho, Maneveldt and W.H.Adey	R
Sporolithon sp.	R
Spyridia clavata Kützing	E
Spyridia filamentosa (Wulfen) Harvey	E
Spyridia hypnoides (Bory) Papenfuss	E
Spyridiocolax capixaba AB Joly and EC Oliveira	E
Stylonema alsidii (Zanardini) K.M.Drew	E
Tiffaniella gorgonea (Mont.) Doty and Meñez	E
Titanoderma prototypum (Foslie) Woelkerling et al.	R
Titanoderma pustulatum (J.V.Lamouroux) Nägeli	R
Titanophora submarina K.E.Bucher and J.N.Norris	R
Thuretia bornetii Vickers	E

(continued)

Table 4.1 (continued)

Taxon	Habit
Tricleocarpa cylindrica (J.Ellis and Solander) Huisman and Borowitzka	E
Tricleocarpa fragilis (Linnaeus) Huisman and R.A.Townsend	E
Tsengia sp.	E
Vertebrata foetidissima (Cocks ex Bornet) Díaz-Tapia and Maggs in Díaz-Tapia	E
Vidalia obtusiloba (C.Agardh) J.Agardh	E
Wrangelia argus (Mont.) Mont.	E
Wrightiella tumanowiczii (Gatty ex Harvey) F.Schmitz	E
Wurdemannia miniata (Sprengel) Feldmann and G.Hamel	E

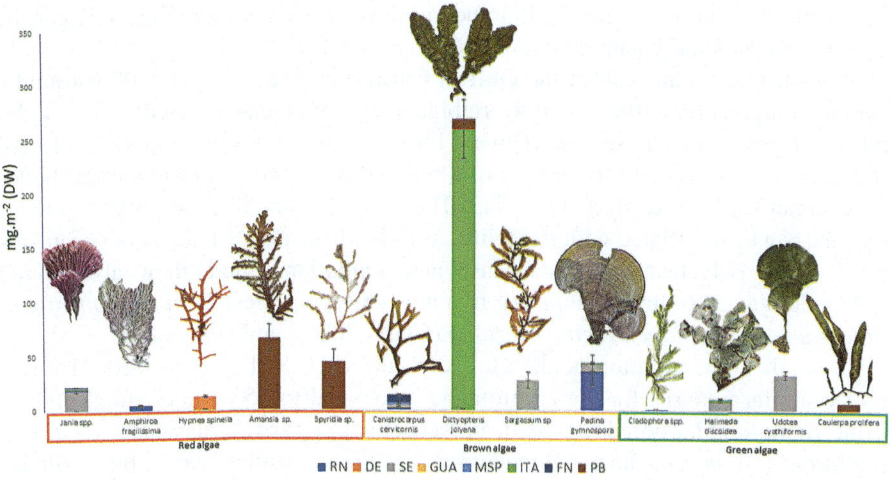

Fig. 4.4 Mean (±SD bars) maximum biomass (dry weight) of main dominant seaweed associated with rhodolith beds from Brazilian coast. RN: Rancho Norte/Florianópolis, SC; DE: Deserta Island/Florianópolis, SC; SE: Sequim/Itaipava, ES; GUA: Rasas Island/Guarapari, ES; MSP: Morro de São Paulo/Cairu, BA; ITA: Itaparica Island/Itaparica, BA; FN: Rata Island/Fernando de Noronha, PE; PB: João Pessoa/Paraíba, PB. (Data from Riul et al., 2009; Carvalho et al., 2020a, b)

present the dominance of red seaweed as *Amansia* spp. and *Spyridia* spp. Green seaweed despite being less abundant can be also conspicuous (Fig. 4.4).

Overall, the epiphytic seaweed nearshore was twice as high in species richness than offshore, showing deep-water depauperate reefs and soft-bottoms (Figueiredo et al., 2008). There were NGCA *taxa* exclusive to offshore sites and a few *taxa* widespread along four up to nine coastline states of Brazil (20% epiphytics and 15% corallines). Nevertheless, rhodolith beds shared over 25% epiphytic seaweed and 30% NGCA *taxa* with nearby coral reefs and rocky shores. Although rhodoliths are known to hold canopy seaweed on top, the associated flora beneath is not easily distinguished from surrounding habitats. There is a trend of species richness increasing in intermediate productive sites where seaweed is neither too high nor too low (Figueiredo et al., 2008). This survey indicated a need to work more on identifying seaweed *taxa* from rhodolith beds and other hard-bottom habitats at same sampling time.

4.3 Associated Fauna

Rhodoliths host a wide range of benthic organisms, serving as nursery substrate, providing microhabitats for diverse *taxa* to refuge and settle, including some commercially important species, such as scallops, crabs, and fishes. Due to the difficulty of processing and counting the *taxon* that lives in rhodoliths, the majority of the studies focusing on rhodolith cryptofauna investigate macrobenthic elements (> 0.5 mm). Identifying these *taxa* in the smaller possible taxonomic unit consists of hard work and demand specialized labor.

The main groups observed within rhodoliths are representative of the Phylum Arthropoda, Annelida, Echinodermata, and Mollusca (in order of magnitude), but Sipuncula, Cnidaria, Nemertea, Platyhelminthes, and Chordata (Fig. 4.5) can be present and eventually abundant (Abrecht et al., 2022).

In subtropical waters from the Gulf of California, Steller et al. (2003) found a similar composition in the macroinvertebrates cryptofauna associated to the rhodoliths analyzes, with a 52 species richness for only eight rhodolith nodules collected (richness was estimated for a subset of three nodules), and from individual rhodoliths, organisms ranged from 11 to 542. These numbers confirm the high biodiversity that can be associated with rhodoliths as individuals also. Majority of organisms were from polychaetes, crustaceans, and cnidarians, and they also found Echinodermata; Mollusca, including two new chiton species: *Ischnochiton rhodolithophilus* and *Lepidochitona corteziana* (Clark, 2000); and nematode.

In another investigation at the Gulf of California, Steller et al. (2007) found pretty similar patterns for the cryptofauna associated to 15 nodules (total of 114, some rhodoliths harbored 40 *taxa*), polychaetes (52 *taxa*), crustacean (29 *taxa*), molluscs (14 *taxa*), and echinoderms (9 *taxa*), which were the primary contributors.

Harvey and Bird (2008) found domain from polychaetes (89%) regarded to cryptofaunal community from rhodolith forming species at the cold-temperate waters from South Australia, followed by Mollusca, Arthropoda, and Echinodermata, and those results are similar to some studies that observed crustaceans, polychaetes, and molluscs as being the most recurrent elements to occur in rhodoliths (De Grave et al., 2000; Hinojosa-Arango & Riosmena-Rodríguez, 2004; Figueiredo et al., 2007).

At the Gulf of Mexico again, Riosmena-Rodríguez and Medina-López (2010) inferred that most of cryptofaunal organism founded were juveniles, although varying in relation to bed type and season.

One of the few studies in which the authors also investigated organisms smaller than 0.5 mm is the one conducted by Méndez Trejo et al. (2020). In this study, conducted at three different beds in New Zealand, 78 rhodoliths were collected, processed, and analyzed, and they counted a total of 4342 individuals, which were separated into 112 *taxa* across 16 phyla. Approximately 57% of the organisms counted were ciliophorans, protists organisms with ~1 mm. Polychaetes had 1022 individuals counted, divided into 18 different families.

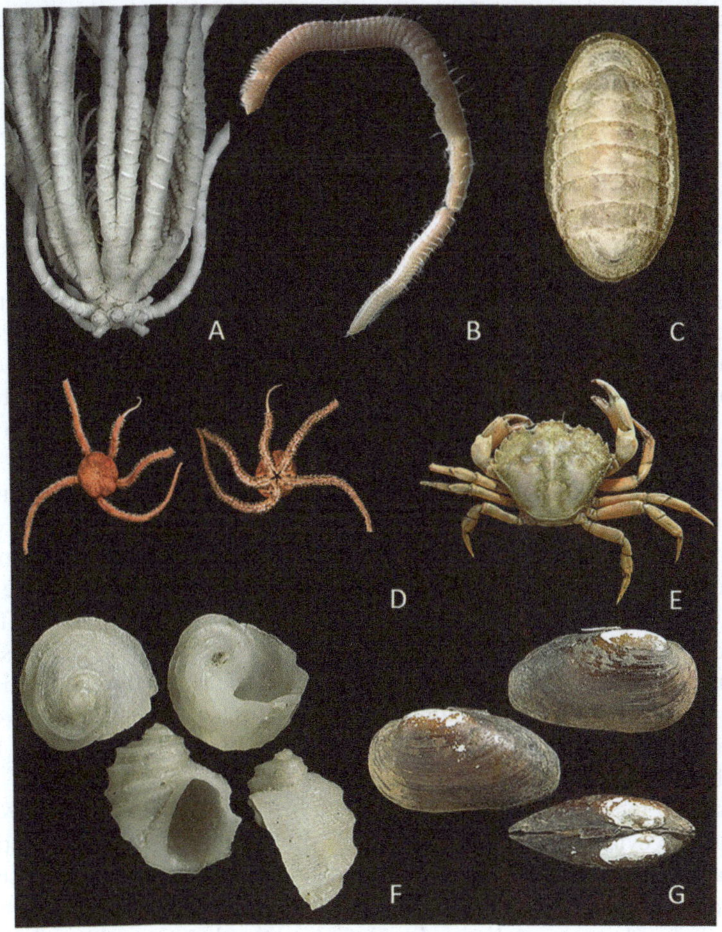

Fig. 4.5 Representative organisms of some of the common classes associated with rhodolith cryptofauna; (**a**) Crinoidea, by Charles Messing; (**b**) Polychaeta by Adrian G. Glover; (**c**) Polyplacophora, by Joop Trausel and Frans Slieker; (**d**) Ophiuroidea, by Kajetan Deja; (**e**) Malacostraca, by Hans Hillewaert; (**f**) Gastropoda, by Rudolf Kapeller; (**g**) Bivalvia, by Rudolf Kapeller. (All images from marinespecies.org)

Peaks of recruitment may occur in rhodoliths regarded to ontogenetic stages and sizes for different organisms, as its rhodolith structure probably allows a better place for organisms to accommodate specially in comparison with adjacent habitats (Otero-Ferrer et al., 2018).

In Brazil, a study conducted by Neves and Costa (2022) on the southeast coast revealed that the abundance of cryptofauna associated with rhodoliths is eight times higher than that found in the unconsolidated sediment. This resulted in a higher density of cryptofauna (1790 ind./100 cm^3) and a greater richness of *taxa* in rhodoliths compared to the unconsolidated sediment (100 ind./100 cm^3). Regarding the

Table 4.2 Permanova evaluation of the marine flora assemblage associated with the Brazilian rhodolith bed, considering the species observed in coastal and offshore environments (where Diagonal in bold, similarity; below diagonal, dissimilarity, above diagonal, *p* value of Monte Carlo test)

Presence	Coastal and offshore	Offshore	Coastal
Coastal and offshore	**64.12**	0.068	0.841
Offshore	60.23	**58.82**	0.076
Coastal	64.83	60.56	**64.27**

macrofauna (> 0.5 mm), the abundance of invertebrates in the unconsolidated sediment accounted for 65% of the sample; however, the *taxon* richness remained higher in the rhodoliths with 104 (52 exclusive) *taxa* compared to 74 (22 exclusive) in the unconsolidated sediment. The rhodolith cryptofauna (infauna) composition was dominated by Copepoda (56.4%), Nematoda (25.1%), Polychaeta (11.3%), Ostracoda (2.7%), Acari (2.1%), and others (2.8%).

Besides, the well-known latitudinal patterns shaping a tropical and a warm temperate province (Carvalho et al., 2020b) should be highlighted for floristic changes considering the coastal and offshore environments (Table 4.2). The similarities observed considering the coast are higher than that observed in offshore assemblages. On the other hand, despite the absence of significance, the differences observed among assemblages vary around 60%. These vertical gradients should receive further efforts to clarify expected patterns considering the changes imposed by the irradiance abundance and quality and the intrinsic ecological and evolutionary filter imposed by these chances that are also followed by changes in nutrient availability and temperature.

The role of these epibionts to the rhodolith bed function are still poorly known. The available evidence indicates a major positive impact on its calcification and reproduction (Carvalho et al., 2020b). The calcification facilitation is a consequence of CO_2 absorption by the associated flora during photosynthesis, which increases saturation of carbonate, as observed in the interaction between other calcifiers and primary producers (Bergstrom et al., 2019). On the other hand, the presence of these epiphytes provides an alternative source of food, reducing grazing pressure on vulnerable initial stages of development of spores. All these hypotheses need further evaluation to map at least the interspecific interaction considering different groups and regions.

Considering the impacts of global stressors and their interactions with pollution and overfishing, it can have direct destructive impacts, when considered bottom trawling, which act synergistically in the deterioration of the rhodolith ecosystem health and function (Fragkopoulou et al., 2021). Therefore, management and conservation of these key ecosystems should consider the observed latitudinal and vertical particularities, and the creation of MPAs considering priority regions should receive further attention (Magris et al. 2021).

4.4 Associated Ichthyofauna

Rhodolith beds present a high richness of reef fishes functioning as source of refuge and feeding (Moura et al., 2021; Anderson et al., 2022). Although these habitats are still much less sampled than the iconic shallow water reefs, the available information reinforces their importance hosting threatened and economically important species (Fig. 4.6).

Despite variations in the diversity of communities of marine organisms in relation to the biogeographical peculiarities of the Brazilian coast, the vast majority of nektonic are shared among the eight ecoregions that make up the Brazilian Biogeographic Province. Even species with tropical affinities that thrive in northeastern Brazil [e.g., *Sparisoma axillare* (parrotfish), *Holacanthus ciliaris* (angelfish), *Malacanthus plumieri* (sand tilefish)] can be found at their southern limit of distribution, the coast of Santa Catarina (Anderson et al., 2022). Recently, scientists indicate that population connectivity (e.g., larval dissemination, clonal dissemination) can be enhanced or facilitated by the presence of extensive rhodolith beds along the entire coast of Brazil (Moura et al., 2021).

The central portion of the Brazilian coast, a region that encompasses the south of Bahia, the coast of Espírito Santo, and the north of Rio de Janeiro, is considered a marine ecotone (i.e., transition zone between different biomes), with nektonic and benthic composition, encompassing species belonging to the tropical waters of northern Brazil and warm temperate waters to the south. This intersection gives the region the greatest ichthyofaunal diversity on the Brazilian coast (Aued et al., 2018; Pinheiro et al., 2018; Anderson et al., 2022), including the only refuge of the kelp species endemic to the Brazilian coast *Laminaria abyssalis* (Anderson et al., 2021). Along these ecosystems of the central portion of the Brazilian coast, species that play important roles in the ecosystem, for example, groupers: *Epinephelus adscensionis, E. morio, Mycteroperca bonaci, M. interstitialis*) and are also important economic resources, found a suitable niche in rhodolith beds. Rhodolith beds also play a key role regarding the nursery of these marine species. Many economically important marine fish species inhabit the rhodolith beds in the early stages of their life cycles (Costa et al., 2020; Anderson et al., 2023).

Snappers (e.g., *Pagrus pagrus, Calamus penna, C. pennatula*) belonging to the family Sparidae are examples of species associated with the rhodolith beds of the central portion of the Brazilian coast. The Sparidae *P. pagrus* is the dominant species, exponentially surpassing in biomass (~45% of the total) all other species associated with the rhodolith beds of the central Brazilian region. As carnivores/mesoconsumers, these species play a key role (Papaconstantinou & Caragitsou, 1989; Anderson et al., 2023), feeding on marine vertebrates and invertebrates influencing the balance between primary and secondary productivity. Subsequently, they are preyed upon (e.g., groupers and sea basses), completing their role as a connector between the lower and higher trophic levels (Simon et al., 2016; Pinheiro et al., 2018; Anderson et al., 2023).

Fig. 4.6 Different aspects of ichthyofauna associated with different physiognomies of rhodolith beds. (**a**) Nurse shark (*Ginglymostoma cirratum*) lying down on a rhodolith bed inside a shallow cave; (**b**) a scrawled cowfish (*Acanthostracion quadricornis*) swimming in a rhodolith bed with abundant associated seaweed community

Further, south rhodolith beds are found at their southern limit in the Brazilian coast, in the Arvoredo Marine Biological Reserve (Carvalho et al., 2020a, b). This region is also the biogeographic limit of distribution for 96.4% of all Brazilian marine ichthyofauna with tropical affinity (Anderson et al., 2015). According to these authors, the marine fish associated with rhodolith beds along the Brazilian Biogeographic Province (BBP) is composed of 150 of *taxa*, belonging to 31 families: 20.46% of all marine reef fish inhabiting the Brazilian coast (Anderson et al., 2022). Thirty-six species (24%) are endemic to the BBP, 13 species (8.3%) are included in the IUCN red list of endangered species (IUCN, 2020), and 59 species (39.3%) are targeted for some type of artisanal, subsistence, or commercial/industrial fishing (Table 4.3).

Table 4.3 List of marine fish associated with the rhodolith beds that inhabit the Brazilian Biogeographic Province

Type	Order	Family	*Species*	Trophic	IUCN	Targeted	Endemic
OSTEICH	Perciformes	Acanthuridae	*Acanthurus bahianus*	HERB	LC	1	1
OSTEICH	Perciformes	Acanthuridae	*Acanthurus chirurgus*	HERB	LC	1	0
OSTEICH	Perciformes	Acanthuridae	*Acanthurus coeruleus*	HERB	LC	1	0
OSTEICH	Perciformes	Apogonidae	*Astrapogon puncticulatus*	PLANK	LC	0	0
OSTEICH	Tetraodontiformes	Balistidae	*Balistes capriscus*	MINV	VU	1	0
OSTEICH	Tetraodontiformes	Balistidae	*Balistes vetula*	MINV	VU	1	0
OSTEICH	Tetraodontiformes	Balistidae	*Melichthys niger*	OMNI	LC	1	0
OSTEICH	Tetraodontiformes	Balistidae	*Xanthichthys ringens*	MINV	LC	0	0
OSTEICH	Perciformes	Blenniidae	*Parablennius marmoreus*	OMNI	LC	0	0
OSTEICH	Perciformes	Blenniidae	*Parablennius pilicornis*	OMNI	LC	0	0
OSTEICH	Perciformes	Blenniidae	*Hypleurochilus fissicornis*	OMNI	LC	0	0
OSTEICH	Pleuronectiformes	Bothidae	*Bothus lunatus*	MINV	LC	1	0
OSTEICH	Pleuronectiformes	Bothidae	*Bothus ocellatus*	MINV	LC	1	0
OSTEICH	Perciformes	Callionymidae	*Callionymus bairdi*	MINV	LC	0	0
OSTEICH	Perciformes	Carangidae	*Caranx crysos*	MCAR	LC	1	0
OSTEICH	Perciformes	Carangidae	*Pseudocaranx dentex*	PLANK	LC	1	0
CONDR	Carcharhiniformes	Carcharhinidae	*Carcharhinus obscurus*	MCAR	VU	1	0
CONDR	Carcharhiniformes	Carcharhinidae	*Galeocerdo cuvier*	MCAR	NT	1	0
OSTEICH	Perciformes	Chaetodontidae	*Chaetodon sedentarius*	SINV	LC	0	0
OSTEICH	Perciformes	Chaetodontidae	*Chaetodon striatus*	SINV	LC	0	0
OSTEICH	Perciformes	Chaetodontidae	*Prognathodes guyanensis*	SINV	LC	0	0
OSTEICH	Anguilliformes	Chlopsidae	*Chlopsis bicolor*	MINV	LC	0	0
OSTEICH	Anguilliformes	Chlopsidae	*Chlopsis dentatus*	MINV	DD	0	0
OSTEICH	Anguilliformes	Chlopsidae	*Kaupichthys hyoproroides*	MINV	LC	0	0
OSTEICH	Perciformes	Cirrhitidae	*Amblycirrhitus pinos*	MINV	LC	0	0

(continued)

Table 4.3 (continued)

Type	Order	Family	*Species*	Trophic	IUCN	Targeted	Endemic
OSTEICH	Anguilliformes	Congridae	*Ariosoma opistophthalmum*	MCAR	LC	0	0
OSTEICH	Anguilliformes	Congridae	*Heteroconger camelopardalis*	PLANK	LC	0	1
OSTEICH	Pleuronectiformes	Cynoglossidae	*Symphurus rhytisma*	MINV	LC	0	0
OSTEICH	Scorpaeniformes	Dactylopteridae	*Dactylopterus volitans*	MINV	LC	1	0
OSTEICH	Perciformes	Dactyloscopidae	*Dactyloscopus crossotus*	MINV	LC	0	0
CONDR	Myliobatiformes	Dasyatidae	*Dasyatis centroura*	MINV	LC	1	0
CONDR	Myliobatiformes	Dasyatidae	*Dasyatis guttata*	MINV	DD	1	0
CONDR	Myliobatiformes	Dasyatidae	*Hypanus marianae*	MINV	DD	1	1
OSTEICH	Perciformes	Epinephelidae	*Cephalopholis fulva*	MCAR	LC	1	0
OSTEICH	Perciformes	Epinephelidae	*Paranthias furcifer*	PLANK	LC	1	0
OSTEICH	Perciformes	Epinephelidae	*Dermatolepis inermis*	MCAR	NT	1	0
OSTEICH	Perciformes	Epinephelidae	*Epinephelus morio*	MCAR	NT	1	0
OSTEICH	Perciformes	Epinephelidae	*Hyporthodus mystacinus*	MCAR	NT	1	0
OSTEICH	Perciformes	Epinephelidae	*Hyporthodus nigritus*	MCAR	CR	1	0
OSTEICH	Perciformes	Epinephelidae	*Hyporthodus niveatus*	MCAR	VU	1	0
OSTEICH	Perciformes	Liopropomatidae	*Liopropoma carmabi*	MINV	LC	0	0
OSTEICH	Perciformes	Epinephelidae	*Mycteroperca venenosa*	MCAR	NT	1	0
OSTEICH	Perciformes	Epinephelidae	*Rypticus saponaceus*	MINV	LC	0	0
CONDR	Orectolobiformes	Ginglymostomatidae	*Ginglymostoma cirratum*	MCAR	VU	1	0
OSTEICH	Perciformes	Gobiidae	*Chriolepis fisheri*	MINV	LC	0	0
OSTEICH	Perciformes	Gobiidae	*Coryphopterus dicrus*	MINV	LC	0	0
OSTEICH	Perciformes	Gobiidae	*Coryphopterus thrix*	MINV	VU	0	0
OSTEICH	Perciformes	Gobiidae	*Elacatinus figaro*	MINV	VU	1	1
OSTEICH	Perciformes	Gobiidae	*Elacatinus pridisi*	MINV	NE	0	1
OSTEICH	Perciformes	Gobiidae	*Gnatholepis thompsoni*	MINV	LC	0	0

OSTEICH	Perciformes	Gobiidae	*Pinnichthys aimoriensis*	MINV	LC	0	1
OSTEICH	Perciformes	Grammatidae	*Gramma brasiliensis*	MINV	NE	1	1
OSTEICH	Perciformes	Haemulidae	*Anisotremus moricandi*	MINV	END	1	1
OSTEICH	Perciformes	Haemulidae	*Anisotremus surinamensis*	MINV	LC	1	0
OSTEICH	Perciformes	Haemulidae	*Anisotremus virginicus*	MINV	LC	1	0
OSTEICH	Perciformes	Haemulidae	*Conodon nobilis*	MINV	LC	1	0
OSTEICH	Perciformes	Haemulidae	*Haemulon aurolineatum*	MINV	LC	1	0
OSTEICH	Perciformes	Haemulidae	*Haemulon parra*	MINV	LC	1	0
OSTEICH	Perciformes	Haemulidae	*Haemulon plumieri*	MINV	LC	1	0
OSTEICH	Perciformes	Haemulidae	*Haemulon steindachneri*	MINV	LC	1	0
OSTEICH	Perciformes	Haemulidae	*Orthopristis ruber*	MINV	LC	1	0
OSTEICH	Beryciformes	Holocentridae	*Holocentrus adscensionis*	MINV	LC	1	0
OSTEICH	Beryciformes	Holocentridae	*Myripristis jacobus*	MINV	LC	0	0
OSTEICH	Perciformes	Labridae	*Bodianus pulchellus*	MINV	LC	1	0
OSTEICH	Perciformes	Labridae	*Bodianus rufus*	MINV	LC	1	0
OSTEICH	Perciformes	Labridae	*Doratonotus megalepis*	MINV	LC	0	0
OSTEICH	Perciformes	Labridae	*Halichoeres brasiliensis*	MINV	DD	0	1
OSTEICH	Perciformes	Labridae	*Halichoeres dimidiatus*	MINV	LC	0	1
OSTEICH	Perciformes	Labridae	*Halichoeres penrosei*	MINV	LC	0	1
OSTEICH	Perciformes	Labridae	*Halichoeres poeyi*	MINV	LC	0	0
OSTEICH	Perciformes	Labridae	*Halichoeres rubrovirens*	MINV	NE	0	1
OSTEICH	Perciformes	Labridae	*Halichoeres sazimai*	MINV	NE	0	1
OSTEICH	Perciformes	Labridae	*Thalassoma noronhanum*	PLANK	LC	0	1
OSTEICH	Perciformes	Labridae	*Xyrichtys splendens*	MINV	LC	0	0
OSTEICH	Perciformes	Labridae	*Xyrichtys novacula*	MINV	LC	0	0
OSTEICH	Perciformes	Labridae-Scarinae	*Cryptotomus roseus*	HERB	LC	0	0

(continued)

Table 4.3 (continued)

Type	Order	Family	*Species*	Trophic	IUCN	Targeted	Endemic
OSTEICH	Perciformes	Labridae-Scarinae	*Scarus zelindae*	HERB	DD	1	1
OSTEICH	Perciformes	Labridae-Scarinae	*Sparisoma amplum*	HERB	LC	1	1
OSTEICH	Perciformes	Labridae-Scarinae	*Sparisoma axillare*	HERB	DD	1	1
OSTEICH	Perciformes	Labridae-Scarinae	*Sparisoma frondosum*	HERB	DD	1	1
OSTEICH	Perciformes	Labridae-Scarinae	*Sparisoma rocha*	HERB	NE	0	1
OSTEICH	Perciformes	Labridae-Scarinae	*Sparisoma tuiupiranga*	HERB	LC	0	1
OSTEICH	Perciformes	Lutjanidae	*Etelis oculatus*	MCAR	DD	1	0
OSTEICH	Perciformes	Lutjanidae	*Lutjanus analis*	MCAR	VU	1	0
OSTEICH	Perciformes	Lutjanidae	*Lutjanus vivanus*	MCAR	LC	1	0
OSTEICH	Perciformes	Lutjanidae	*Ocyurus chrysurus*	MCAR	DD	1	0
OSTEICH	Perciformes	Lutjanidae	*Rhomboplites aurorubens*	MCAR	VU	1	0
OSTEICH	Perciformes	Malacanthidae	*Malacanthus plumieri*	MCAR	LC	1	0
OSTEICH	Tetraodontiformes	Monacanthidae	*Aluterus schoepfii*	OMNI	LC	0	0
OSTEICH	Tetraodontiformes	Monacanthidae	*Aluterus scriptus*	OMNI	LC	1	0
OSTEICH	Tetraodontiformes	Monacanthidae	*Cantherhines macrocerus*	OMNI	LC	0	0
OSTEICH	Tetraodontiformes	Monacanthidae	*Cantherhines pullus*	OMNI	LC	0	0
OSTEICH	Tetraodontiformes	Monacanthidae	*Monacanthus ciliatus*	OMNI	LC	0	0
OSTEICH	Tetraodontiformes	Monacanthidae	*Stephanolepis hispidus*	OMNI	LC	0	0
OSTEICH	Perciformes	Mullidae	*Mulloidichthys martinicus*	MINV	LC	1	0
OSTEICH	Perciformes	Mullidae	*Mullus argentinae*	MINV	LC	1	0
OSTEICH	Perciformes	Mullidae	*Pseudupeneus maculatus*	MINV	LC	1	0
OSTEICH	Anguilliformes	Muraenesocidae	*Cynoponticus savanna*	MCAR	LC	0	0
OSTEICH	Anguilliformes	Muraenidae	*Channomuraena vittata*	MCAR	LC	0	0
OSTEICH	Anguilliformes	Muraenidae	*Gymnothorax funebris*	MCAR	LC	1	0
OSTEICH	Anguilliformes	Muraenidae	*Gymnothorax maderensis*	MCAR	LC	0	0

OSTEICH	Anguilliformes	Muraenidae	*Gymnothorax miliaris*	MCAR	LC	1	0
OSTEICH	Anguilliformes	Muraenidae	*Gymnothorax moringa*	MCAR	LC	1	0
OSTEICH	Anguilliformes	Muraenidae	*Gymnothorax ocellatus*	MCAR	LC	0	0
OSTEICH	Anguilliformes	Muraenidae	*Gymnothorax polygonius*	MCAR	LC	0	0
OSTEICH	Anguilliformes	Muraenidae	*Gymnothorax vicinus*	MCAR	LC	1	0
OSTEICH	Anguilliformes	Muraenidae	*Monopenchelys acuta*	MINV	LC	0	0
CONDR	Myliobatiformes	Myliobatidae	*Aetobatus narinari*	MCAR	NT	1	0
OSTEICH	Lophiiformes	Ogcocephalidae	*Ogcocephalus vespertilio*	MINV	LC	0	0
OSTEICH	Anguilliformes	Ophichthidae	*Myrichthys breviceps*	MINV	LC	0	0
OSTEICH	Anguilliformes	Ophichthidae	*Myrichthys ocellatus*	MINV	LC	0	0
OSTEICH	Anguilliformes	Ophichthidae	*Ophichthus gomesii*	MINV	LC	0	0
OSTEICH	Anguilliformes	Ophichthidae	*Ophichthus ophis*	MINV	LC	1	0
OSTEICH	Anguilliformes	Ophichthidae	*Pseudomyrophis frio*	MINV	LC	0	0
OSTEICH	Anguilliformes	Ophichthidae	*Quassiremus ascensionis*	MCAR	LC	0	0
OSTEICH	Perciformes	Opistognathidae	*Opistognathus aurifrons*	PLANK	LC	0	1
OSTEICH	Perciformes	Opistognathidae	*Opistognathus brasiliensis*	PLANK	LC	0	1
OSTEICH	Perciformes	Opistognathidae	*Opistognathus cuvierii*	PLANK	LC	0	1
OSTEICH	Perciformes	Opistognathidae	*Opistognathus lonchurus*	PLANK	LC	0	0
OSTEICH	Tetraodontiformes	Ostraciidae	*Acanthostracion polygonius*	OMNI	LC	0	0
OSTEICH	Tetraodontiformes	Ostraciidae	*Acanthostracion quadricornis*	OMNI	LC	0	0
OSTEICH	Pleuronectiformes	Paralichthyidae	*Etropus crossotus*	MINV	LC	0	0
OSTEICH	Pleuronectiformes	Paralichthyidae	*Syacium micrurum*	MINV	LC	0	0
OSTEICH	Perciformes	Pomacanthidae	*Centropyge aurantanotus*	HERB	LC	1	1
OSTEICH	Perciformes	Pomacanthidae	*Holacanthus ciliaris*	SINV	LC	1	0
OSTEICH	Perciformes	Pomacanthidae	*Holacanthus tricolor*	SINV	LC	1	0

(continued)

Table 4.3 (continued)

Type	Order	Family	*Species*	Trophic	IUCN	Targeted	Endemic
OSTEICH	Perciformes	Pomacanthidae	*Pomacanthus arcuatus*	SINV	LC	1	0
OSTEICH	Perciformes	Pomacanthidae	*Pomacanthus paru*	SINV	LC	1	0
OSTEICH	Perciformes	Pomacentridae	*Chromis enchrysura*	PLANK	LC	0	0
OSTEICH	Perciformes	Pomacentridae	*Chromis flavicauda*	PLANK	DD	0	1
OSTEICH	Perciformes	Pomacentridae	*Chromis jubauna*	PLANK	LC	0	1
OSTEICH	Perciformes	Pomacentridae	*Microspathodon chrysurus*	HERB	LC	0	0
OSTEICH	Perciformes	Pomacentridae	*Stegastes pictus*	HERB	LC	0	1
OSTEICH	Perciformes	Pomacentridae	*Stegastes variabilis*	HERB	LC	0	1
OSTEICH	Perciformes	Priacanthidae	*Priacanthus arenatus*	MINV	LC	1	0
OSTEICH	Perciformes	Microdesmidae	*Ptereleotris randalli*	PLANK	LC	0	1
OSTEICH	Perciformes	Sciaenidae	*Pareques acuminatus*	MINV	LC	0	0
CONDR	Rhinopristiformes	Rhinobatidae	*Pseudobatos horkelii*	MCAR	CR	1	1
OSTEICH	Scorpaeniformes	Scorpaenidae	*Pontinus longispinis*	MCAR	LC	0	0
OSTEICH	Scorpaeniformes	Scorpaenidae	*Pontinus rathbuni*	MCAR	LC	0	0
OSTEICH	Scorpaeniformes	Scorpaenidae	*Scorpaena agassizii*	MCAR	LC	0	0
OSTEICH	Scorpaeniformes	Scorpaenidae	*Scorpaena albifimbria*	MCAR	LC	0	0
OSTEICH	Scorpaeniformes	Scorpaenidae	*Scorpaena brasiliensis*	MCAR	LC	1	0
OSTEICH	Scorpaeniformes	Scorpaenidae	*Scorpaena dispar*	MCAR	LC	0	0
OSTEICH	Scorpaeniformes	Scorpaenidae	*Scorpaena isthmensis*	MCAR	LC	0	0
OSTEICH	Scorpaeniformes	Scorpaenidae	*Scorpaena melasma*	MCAR	LC	0	0
OSTEICH	Scorpaeniformes	Scorpaenidae	*Scorpaena petricola*	MCAR	LC	0	1
OSTEICH	Scorpaeniformes	Scorpaenidae	*Scorpaena plumieri*	MCAR	LC	0	0
OSTEICH	Scorpaeniformes	Sebastidae	*Helicolenus lahillei*	MCAR	NE	0	0
OSTEICH	Perciformes	Anthiadidae	*Anthias menezesi*	PLANK	LC	0	1
OSTEICH	Perciformes	Serranidae	*Diplectrum formosum*	MCAR	LC	0	0

OSTEICH	Perciformes	Serranidae	*Diplectrum radiale*	MCAR	LC	0	0
OSTEICH	Perciformes	Serranidae	*Pronotogrammus martinicensis*	PLANK	LC	0	0
OSTEICH	Perciformes	Serranidae	*Pseudogramma gregoryi*	MINV	LC	1	0
OSTEICH	Perciformes	Serranidae	*Serranus annularis*	MINV	LC	0	0
OSTEICH	Perciformes	Serranidae	*Serranus atrobranchus*	MINV	LC	0	0
OSTEICH	Perciformes	Serranidae	*Serranus baldwini*	MINV	LC	0	0
OSTEICH	Perciformes	Serranidae	*Serranus flaviventris*	MINV	LC	0	0
OSTEICH	Perciformes	Serranidae	*Serranus phoebe*	MINV	LC	0	0
OSTEICH	Perciformes	Sparidae	*Calamus penna*	MINV	LC	1	0
OSTEICH	Perciformes	Sparidae	*Calamus pennatula*	MINV	LC	1	0
OSTEICH	Perciformes	Sparidae	*Pagrus pagrus*	MINV	LC	1	0
OSTEICH	Aulopiformes	Synodontidae	*Synodus intermedius*	MCAR	LC	0	0
OSTEICH	Aulopiformes	Synodontidae	*Synodus synodus*	MCAR	LC	0	0
OSTEICH	Tetraodontiformes	Tetraodontidae	*Canthigaster figueiredoi*	OMNI	LC	0	1
OSTEICH	Tetraodontiformes	Tetraodontidae	*Sphoeroides spengleri*	MINV	LC	0	0
CONDR	Carcharhiniformes	Triakidae	*Mustelus canis*	MINV	NT	0	0
OSTEICH	Scorpaeniformes	Triglidae	*Prionotus punctatus*	MINV	NE	0	0

IUCN Status: *CR* critically endangered (extreme risk of extinction in the wild), *EN* threatened (high risk of extinction in the wild), *VU* vulnerable (risk of becoming endangered), *NT* nearly at risk (high probability of becoming endangered in the near future), *LC* low concern (low risk), *DD* deficient data (little information or insufficient information to draw a risk profile), *NE* not evaluated. Trophic categories: *MCAR* carnivore, *MINV* movable invertebrate eaters, *SINV* sessile invertebrate eaters, *OMNI* omnivores, *PLANK* planktivores, *HERB* herbivores

4.5 Final Remarks

The rhodolith beds are an oasis of diversity in the Brazilian continental shelf with many singularities, interacting with different coastal and ocean ecosystems, with countless relationships and species intertwined by the evolutionary history and environments built by coralline algae. All these diversities and ecological functions are threatened by the interaction of global and local stressor, mainly represented by oil exploitation, trawling and ocean warming, and acidification.

Considering that "the devil lives in the details," in them, we must find many of the answers that explain the resilience of these algae and the ecosystems constituted by them, throughout their millions of years of evolutionary history. Accessing this universe requires an attentive look at the different dimensions and their subtleties to appreciate these fantastic evolutionary and ecological relationships that can be explored from thousands of kilometers to a scale of a few millimeters. It takes a poetic look to take us on the mirific expeditions into the hardened world of limestone forests of rhodoliths.

References

Abrecht, M., Peinemann, V. N., Yazaryan, A. K., Kestler, M., De Mattei, B. C., Hà, B. A., Ryznar, E., & Jacobs, D. K. (2022). Cryptofaunal diversity in fringing reef rhodoliths. *Coral Reefs, 41*, 199–212.

Adey, W. H., Hernandez-Kantun, J. J., Johnson, G., & Gabrielson, P. W. (2015). DNA sequencing, anatomy, and calcification patterns support a monophyletic, subarctic, carbonate reef-forming *Clathromorphum* (Hapalidiaceae, Corallinales, Rhodophyta). *Journal of Phycology, 51*, 189–203. https://doi.org/10.1111/jpy.12266

Amado-Filho, G. M., Maneveldt, G., Manso, R. C. C., Marins-Rosa, B. V., Pacheco, M. R., & Guimarães, S. M. P. B. (2007). Structure of rhodolith beds from 4 to 55 meters deep along the southern coast of Espírito Santo state, Brazil. *Ciencias Marinas, 33*, 399–410.

Amado-Filho, G. M., Maneveldt, G. W., Pereira-Filho, G. H., Manso, R. C. C., Bahia, R. G., Barros-Barreto, M. B., & Guimarães, S. M. P. B. (2010). Seaweed diversity associated with a Brazilian tropical rhodolith bed. *Ciencias Marinas, 36*, 371–391.

Amado-Filho, G. M., Pereira-Filho, G. H., Bahia, R. G., Abrantes, D. P., Veras, P. C., & Matheus, Z. (2012a). Occurence and distribution of rhodolith beds on the Fernando de Noronha archipelago of Brazil. *Aquatic Botany, 101*, 41–45.

Amado-Filho, G. M., Moura, R. L., Bastos, A. C., Salgado, L. T., Sumida, P. Y., Guth, A. Z., Francini-Filho, R. B., Pereira-Filho, G. H., Abrantes, D. P., Brasileiro, P. S., Bahia, R. G., Leal, R. N., Kaufman, L., Kleypas, J. A., Farina, M., & Thompson, F. L. (2012b). Rhodolith beds are major $CaCO_3$ bio-factories in the tropical south West Atlantic. *PLoS One, 7*(4), e35171. https://doi.org/10.1371/journal.pone.0035171

Amado-Filho, G. M., Moura, R. L., Bastos, A. C., Francini-Filho, R. B., Pereira-Filho, G. H., Bahia, R. G., Moraes, F. C., & Motta, F. S. (2016). Mesophotic ecosystems of the unique South Atlantic atoll are composed by rhodolith beds and scattered consolidated reefs. *Marine Biodiversity, 46*, 933–936. https://doi.org/10.1007/s12526-015-0441-6

Amado-Filho, G. M., Bahia, R. G., Pereira-Filho, G. H., & Longo, L. L. (2017). South Atlantic rhodolith beds: Latitudinal distribution, species composition, structure and ecosystem functions, threats and conservation status. In R. Riosmena-Rodríguez, W. Nelson, & J. Aguirre

(Eds.), *Rhodolith/maërl beds: A global perspective* (pp. 299–317). Springer, Series, Coastal Research Library. https://doi.org/10.1007/978-3-319-29315-8_7

Anderson, A. B., Carvalho-Filho, A., Morais, R. A., Nunes, L. T., Quimbayo, J. P., & Floeter, S. R. (2015). Brazilian tropical fishes in their southern limit of distribution: Checklist of Santa Catarina's rocky reef ichthyofauna, remarks and new records. *Check List, 11*, 1688.

Anderson, A. B., Assis, J., Batista, M. B., Serrão, E. A., Guabiroba, H. C., Delfino, S. D. T., Pinheiro, H. T., Pimentel, C. R., Gomes, L. E. O., Vilar, C. C., Bernardino, A. F., Horta, P., Ghisolfi, R. D., & Joyeux, J. C. (2021). Global warming assessment suggests the endemic Brazilian kelp beds to be an endangered ecosystem. *Marine Environmental Research, 168*, 105307.

Anderson, A. B., Bernardes, M. B., Pinheiro, H. T., Guabiroba, H. C., Pimentel, C. R., Vilar, C. C., Gomes, L. E. O., Bernardino, A. F., Delfino, S. D. T., Giarrizzo, T., Ferreira, C. E. L., & Joyeux, J. C. (2022). Niche availability and habitat affinities of the red porgy *Pagrus pagrus* (Linnaeus, 1758): An important ecological player on the world's largest rhodolith beds. *Journal of Fish Biology, 101*, 179–189.

Anderson, A. B., Pinheiro, H. T., Batista, M. B., Francini-Filho, R. B., Gomes, L. E. O., et al. (2023). Biogeographic patterns of marine fishes associated with rhodolith beds in the Southwestern Atlantic reveal an ecotone of biodiversity. *Biodiversity and Conservation, 32*, 821–837. https://doi.org/10.1007/s10531-022-02528-0

Asnaghi, V., Thrush, S. F., Hewitt, J. E., Mangialajo, L., Cattaneo-Vietti, R., & Chiantore, M. (2015). Colonisation processes and the role of coralline algae in rocky shore community dynamics. *Journal of Sea Research, 95*, 132–138. https://doi.org/10.1016/j.seares.2014.07.012

Aued, A. W., Smith, F., Quimbayo, J. P., Candido, D. V., Longo, G. O., Ferreira, C. E., et al. (2018). Large-scale patterns of benthic marine communities in the Brazilian province. *PLoS One, 13*(6), e0198452.

Ávila, E., & Riosmena-Rodríguez, R. (2010). Rhodoliths beds as critical habitats for monitoring in marine protected areas in the Gulf of California. In G. Polisciano & O. Farina (Eds.), *National parks, vegetation, wildlife and threats* (pp. 207–214). Nova Science Publishers.

Bahia, R. G., Abrantes, D. P., Brasileiro, P. S., Pereira Filho, G. H., & Amado Filho, G. M. (2010). Rhodolith bed structure along a depth gradient on the northern coast of Bahia state, Brazil. *Brazilian Journal of Oceanography, 58*, 323–337.

Bahia, R. G., Amado-Filho, G. M., Maneveldt, G. W., Adey, W. A., Johnson, G., Marins, B. V., & Longo, L. L. (2014a). *Sporolithon tenue* sp. nov. (Sporolithales, Corallinophycidae, Rhodophyta): A new rhodolith-forming species from the tropical southwestern Atlantic. *Phycological Research, 62*, 44–54.

Bahia, R. G., Amado-Filho, G. M., & Maneveldt, G. W. (2014b). *Sporolithon molle* (Heydrich) Heydrich (Sporolithales, Corallinophycidae, Rhodophyta): An addition to the Atlantic an addition to the Atlantic flora found on a remote oceanic Island. *Cryptogamie Algologie, 35*, 7–14.

Bergstrom, E., Silva, J., Martins, C., & Horta, P. (2019). Seagrass can mitigate negative ocean acidification effects on calcifying algae. *Scientific Reports, 9*, 1932. https://doi.org/10.1038/s41598-018-35670-3

Bernard, G., Romero-Ramirez, A., Tauran, A., Pantalos, M., Deflandre, B., Grall, J., & Grémare, A. (2019). Declining maerl vitality and habitat complexity across a dredging gradient: Insights from in situ sediment profile imagery (SPI). *Scientific Reports, 9*, 16463.

Bosence, D. W. J. (1976). Ecological studies on two unattached coralline algae from western Ireland. *Palaeontology, 19*, 365–395.

Bosence, D. W. J. (1983). The occurrence and ecology of recent rhodoliths: A review. In T. M. Peryt (Ed.), *Coated grains* (pp. 225–242). Springer-Verlag.

Carneiro, P. B. M., Lima, J. P., Bandeira, E. V. P., Neto, A. R. X., Barreira, C. A. R., Tâmega, F. T. S., Matthews-Cascon, H., Franklin Junior, W., & Morais, J. O. (2021). Structure, growth and $CaCO_3$ production in a shallow rhodolith bed from a highly energetic siliciclastic carbonate coast in the equatorial SW Atlantic Ocean. *Marine Environmental Research, 166*, 105280.

Carvalho, V. F., Assis, J., Serrão, E. A., Nunes, J. M., Anderson, A. B., Batista, M. B., Barufi, J. B., Silva, J., Pereira, S. M. B., & Horta, P. A. (2020a). Environmental drivers of rhodolith beds and epiphytes community along the South Western Atlantic coast. *Marine Environmental Research, 154*, 104827.

Carvalho, V. F., Silva, J., Kerr, R., Anderson, A. B., Bastos, E. O., Cabral, D., Gouvêa, L. P., Peres, L., Martins, C. D. L., Silveira-Andrade, V. M., Sissini, M. N., & Horta, P. H. (2020b). When descriptive ecology meets physiology: A study in a South Atlantic rhodolith bed. *Journal of the Marine Biological Association of the United Kingdom, 100*, 347–360.

Chapman, M. G., People, J., & Blockley, D. (2005). Intertidal assemblages associated with natural *corallina* turf and invasive mussel beds. *Biodiversity and Conservation, 14*, 1761–1776. https://doi.org/10.1007/s10531-004-0698-8

Clark, R. N. (2000). The chiton fauna of the Gulf of California rhodolith beds (with descriptions of four new species). *Nemouria, 43*, 1–20.

Costa, A. C. P., Garcia, T. M., Paiva, B. P., Neto, A. R. X., & de Oliveira Soares, M. (2020). Seagrass and rhodolith beds are important seascapes for the development of fish eggs and larvae in tropical coastal areas. *Marine Environmental Research, 161*, 105064.

Dayton, P. K. (1972). Toward an understanding of community resilience and the potential effects of enrichments to the benthos at McMurdo Sound, Antarctica. In B. C. Parker (Ed.), *Proceedings of the colloquium on conservation problems in Antarctica* (pp. 81–96).

De Grave, S. (1999). The influence of sedimentary heterogeneity on within maërl bed differences in infaunal crustacean community. *Estuarine, Coastal and Shelf Science, 49*, 153–163.

De Grave, S., Fazakerley, H., Kelly, L., Guiry, M. D., Ryan, M., & Walshe, J. (2000). *A study of selected maerl beds in Irish waters and their potential for sustainable extraction*. Marine Institute.

Della Giustina, I. D. (2006). *Sedimentação carbonática de algas vermelhas coralináceas da plataforma continental da bacia de Campos: um modelo carbonático análogo para o Terciário*. Master thesis. Universidade Federal do Rio Grande do Sul.

Dias, G. T. M., & Villaça, R. C. (2012). Coralline algae depositional environments on the Brazilian central southeastern shelf. *Journal of Coastal Research, 28*, 270–279.

Dominguez, J. M. L., Silva, R. P., Nunes, A. S., & Freire, A. F. M. (2013). The narrow, shallow, low accommodation shelf of Central Brazil: Sedimentology, evolution, and human uses. *Geomorphology, 203*, 46–59.

Figueiredo, M. A. O., de Menezes, K. S., Costa Paiva, E. M., Paiva, P. C., & Ventura, C. R. R. (2007). Experimental evaluation of rhodoliths as living substrata for infauna at the Abrolhos Bank Brazil. *Ciencias Marinas, 33*, 427–440.

Figueiredo, M. A. O., Horta, P. A., Pedrini, A. G., & Nunes, J. M. C. (2008). Benthic marine algae of the coral reefs of Brazil: A literature revision. *Oecologia Brasiliensis, 12*, 258–269.

Figueiredo, M. A., Coutinho, R., Villas-Bôas, A. B., Tâmega, F. T. S., & Mariath, R. (2012). Deep-water rhodolith productivity and growth in the southwestern Atlantic. *Journal of Applied Phycology, 24*, 487–493. https://doi.org/10.1007/s10811-012-9802-8

Forzza, R. C., et al. (2010). *Catálogo de plantas e fungos do Brasil*. Instituto de Pesquisas Jardim Botânico do Rio de Janeiro.

Foster, M. S. (2001). Rhodoliths: Between rocks and soft places. *Journal of Phycology, 37*, 659–667.

Foster, M. S., McConnico, M. S., Lundsten, L., Wadsworth, T., Kimball, T., Brooks, L. B., Medina-López, M., Riosmena-Rodríguez, R., Hernández-Carmona, G., Vásquez-Elizondo, R. M., Johnson, D., & Steller, D. S. (2007). Diversity and natural history of *Lithothamnion muelleri-Sargassum horridum* community in the Gulf of California. *Ciencias Marinas, 33*, 367–384.

Fragkopoulou, E., Serrão, E. A., Horta, P. A., Koerich, G., & Assis, J. (2021). Bottom trawling threatens future climate refugia of rhodoliths globally. *Frontiers in Marine Science, 7*, 1246.

Gagnon, P., Matheson, K., & Stapleton, M. (2012). Variation in rhodolith morphology and biogenic potential of newly discovered rhodolith beds in Newfoundland and Labrador (Canada). *Botanica Marina, 55*, 85–99. https://doi.org/10.1515/bot-2011-0064

Gherardi, D. F. M. (2004). Community structure and carbonate production of a temperate rhodolith bank from Arvoredo Island, southern Brazil. *Brazilian Journal of Oceanography, 52*, 207–224.
Guimarães, S. M. P. B., & Amado-Filho, G. M. (2009). First record of *Reticulocaulis mucosissimus* I. A. Abbott (Naccariaceae, Rhodophyta) for the western Atlantic Ocean. *Revista Brasileira de Botânica, 32*, 671–675.
Harvey, A. S., & Bird, F. L. (2008). Community structure of a rhodolith bed from cold temperate waters (southern Australia). *Australian Journal of Botany, 56*, 437–450.
Harvey, A. S., Woelkerling, W. J., Huisman, J. M., & Gurgel, C. F. D. (2013). A monographic account of Australian species of *Amphiroa* (Corallinaceae, Rhodophyta). *Australian Systematic Botany, 26*, 81–144. https://doi.org/10.1071/SB13010
Henriques, M. C., Villas-Boas, A., Riosmena-Rodriguez, R., & Figueiredo, M. A. O. (2012). New records of rhodolith-forming species (Corallinales, Rhodophyta) from deep water in Espírito Santo state, Brazil. *Helgoland Marine Research, 66*, 219–231.
Henriques, M. C., Riosmena-Rodríguez, R., Coutinho, L. M., & Figueiredo, M. A. O. (2014a). Lithophylloideae and Mastophoroideae (Corallinales, Rhodophyta) from the Brazilian continental shelf. *Phytotaxa, 190*, 112–129.
Henriques, M. C., Coutinho, L. M., Riosmena-Rodríguez, R., Barros-Barreto, M. B., Khader, S., & Figueiredo, M. A. O. (2014b). Three deep water species of *Sporolithon* (Sporolithales, Rhodophyta) from the Brazilian continental shelf, with the description of *Sporolithon elevatum* sp. nov. *Phytotaxa, 190*, 320–330.
Hernandez-Kantun, J. J., Rindi, F., Adey, W. H., Heesch, S., Peña, V., Le Gall, L., & Gabrielson, P. W. (2015). Sequencing type material resolves the identity and distribution of the generitype *Lithophyllum incrustans*, and related european species *L. hibernicum* and *L. bathyporum* (Corallinales, Rhodophyta). *Journal of Phycology, 51*, 791–807. https://doi.org/10.1111/jpy.12319
Hernandez-Kantun, J. J., Hall-Spencer, J. M., Grall, J., Adey, W., Rindi, F., Maggs, C. A., Bárbara, I., & Peña, V. (2017). North Atlantic rhodolith beds. In R. Riosmena-Rodríguez, W. Nelson, & J. Aguirre (Eds.), *Rhodolith/maërl beds: A global perspective* (pp. 265–279). Springer, Series, Coastal Research Library.
Hinojosa-Arango, G., & Riosmena-Rodríguez, R. (2004). Influence of rhodolith forming species and growth forms on associated fauna of rhodolith beds in the central-west Gulf of California, Mexico. *Marine Ecology, 25*, 109–127.
Hinojosa-Arango, G., Maggs, C. A., & Johnson, M. P. (2009). Like a rolling stone: The mobility of maerl (Corallinaceae) and the neutrality of the associated assemblages. *Ecology, 90*, 517–528.
Horta, P. A., Riul, P., Amado Filho, G. M., Gurgel, C. F. D., Berchez, F., Nunes, J. M. C., Scherner, F., Pereira, S., Lotufo, T., Peres, L., Sissini, M., Bastos, E. O., Rosa, J., Munoz, P., Martins, C., Gouvêa, L., Carvalho, V., Bergstrom, E., Schubert, N., Bahia, R. G., Rodrigues, A. C., Rörig, L., Barufi, J. B., & Figueiredo, M. (2016). Rhodoliths in Brazil: Current knowledge and potential impacts of climate change. *Brazilian Journal of Oceanography, 64*(spe2), 117–136.
IUCN. (2020). *The IUCN red list of threatened species. Version 2020-1*. https://www.iucnredlist.org. Acessed on June, 9th, 2020. https://www.iucnredlist.org. Accessed 9 May 2023.
Jackson, C. M., Kamenos, N. A., Moore, P. G., & Young, M. (2004). Meiofaunal bivalves in maerl and other substrata; their diversity and community structure. *Ophelia, 58*, 48–60.
Jones, C. G., Lawton, J. H., & Shachak, M. (1994). Organisms as ecosystem engineers. *Oikos, 69*, 373–386.
Kato, A., Baba, M., & Suda, S. (2013). Taxonomic circumscription of heterogeneous species *Neogoniolithon brassica-florida* (Corallinales, Rhodophyta) in Japan. *Phycological Research, 61*, 15–26.
Kempf, M. (1970). Notes on the benthic bionomy of the N–NE Brazilian shelf. *Marine Biology, 5*, 213–224.
Konar, B., Riosmena-Rodríguez, R., & Iken, K. (2006). Rhodolith bed: A newly discovered habitat in the North Pacific Ocean. *Botanica Marina, 49*, 355–359.

Littler, M. M., Littler, D. S., Blair, S. M., & Norris, J. N. (1986). Deep-water plant communities from an uncharted seamount off San Salvador Island, Bahamas: Distribution, abundance, and primary productivity. *Deep Sea Research Part A. Oceanographic Research Papers, 33*, 881–892.

Magris, R. A., Costa, M. D., Ferreira, C. E., Vilar, C. C., Joyeux, J. C., Creed, J. C., Copertino, M. S., Horta, P. A., Sumida, P. Y. G., Francini-Filho, R. B., & Floeter, S. R. (2021). A blueprint for securing Brazil's marine biodiversity and supporting the achievement of global conservation goals. *Diversity and Distributions, 27*, 198–215.

Mateo-Cid, L. E., Mendoza-González, A. C., & García-López, D. Y. (2014). Systematic survey of *Lithothamnion, Melobesia* and *Mesophyllum* species (Hapalidiaceae, Corallinales, Rhodophyta) recorded along the Atlantic coast of Mexico. *Phytotaxa, 164*, 226–238.

Maughan, B. C., & Barnes, D. K. A. (2000). Epilithic boulder communities of Lough Hyne, Ireland: The influences of water movement and sediment. *Journal of the Marine Biological Association of the United Kingdom, 80*, 767–776. https://doi.org/10.1017/S0025315400002733

McConnico, L. A., Foster, M. S., Steller, D. L., & Riosmena-Rodríguez, R. (2014). Population biology of a long-lived rhodolith: The consequences of becoming old and large. *Marine Ecology Progress Series, 504*, 109–118.

McCoy, S. J., & Kamenos, N. A. (2015). Coralline algae (Rhodophyta) in a changing world: Integrating ecological, physiological, and geochemical responses to global change. *Journal of Phycology, 51*, 6–24.

Mendez Trejo, M. D. C., Neill, K. F., Twist, B. A., & Nelson, W. A. (2020). Cryptofauna associated with rhodoliths: Diversity is species-specific and influenced by habitat. *Marine Ecology, 42*, 12647.

Millimann, J. D., & Amaral, C. A. B. (1974). Economic potential of Brazilian continental margin sediments. *Annals of 28th Brazilian Congress of Geology, 28*, 335–344.

Moura, R. L., Secchin, N. A., Amado-Filho, G. M., Francini Filho, R. B., Freitas, M. O., Minte-Vera, C. V., Teixeira, J. B., Thompson, F. L., Dutra, G. F., Sumida, P. Y. G., Guth, A. Z., Lopes, R. M., & Bastos, A. C. (2013). Spatial patterns of benthic megahabitats and conservation planning in the Abrolhos Bank. *Continental Shelf Research, 70*, 109–117.

Moura, R. M., Amado-Filho, G. M., Moraes, F. C., Brasileiro, P. S., Salomon, P. S., Mahiques, M. M., Bastos, A. C., Almeida, M. G., Silva-Jr, J. M., Araujo, B. F., Brito, F. P., Rangel, T. P., Oliveira, B. C. V., Bahia, R. G., Paranhos, R. P., Dias, R. J. S., Siegle, E., Figueiredo-Jr, A. G., Pereira, R. C., Leal, C. V., Hajdu, E., Asp, N. E., Gregoracci, G. B., Neumann-Leitão, S., Yager, P. L., Francini-Filho, R. B., Fróes, A., Campeão, M., Silva, B. S., Moreira, A. P. B., Oliveira, L., Soares, A. C., Araujo, L., Oliveira, N. L., Teixeira, J. B., Valle, R. A. B., Thompson, C. C., Rezende, C. E., & Thompson, F. L. (2016). An extensive reef system at the Amazon River mouth. *Science Advances, 2*, 1–11.

Moura, R. L., Abieri, M. L., Castro, G. M., Carlos-Júnior, L. A., Chiroque-Solano, P. M., Fernandes, N. C., Teixeira, C. D., Ribeiro, F. V., Salomon, P. S., Freitas, M. O., Gonçalves, J. T., Neves, L. M., Hackradt, C. W., Felix-Hackradt, F., Rolim, F. A., Motta, F. S., Gadig, O. B. F., Pereira-Filho, G. H., & Bastos, A. C. (2021). Tropical rhodolith beds are a major and belittled reef fish habitat. *Scientific Reports, 11*(1), 794–794.

Nelson, W., Neill, K., Farr, T., Barr, N., D'archino, R., Miller, S., & Stewart, R. (2012). Rhodolith beds in northern New Zealand: Characterisation of associated biodiversity and vulnerability to environmental stressors. *New Zealand Aquatic Environment and Biodiversity Report.*

Neves, S. B., & Costa, K. G. (2022). Diversity of benthic fauna of rhodoliths and sediments deposited on sandstone reefs in Southeast Brazil. *Ocean and Coastal Research, 70*, e22010. https://doi.org/10.1590/2675-2824070.21029sbn

Ordines, F., Bauzá, M., Sbert, M., Roca, P., Gianotti, M., & Massutí, E. (2015). Red algal beds increase the condition of nekto-benthic fish. *Journal of Sea Research, 95*, 115–123.

Otero-Ferrer, F., Mannarà, E., Cosme, M., Falace, A., Montiel-Nelson, J. A., Espino, F., Haroun, R., & Tuya, F. (2018). Heterogeneity rather than size determines faunal colonization of discrete habitat units: a case study with rhodolith-associated macrofauna. In *Abstract of the VI international Rhodolith workshop,* Brittany, 25–29 June 2018.

Otero-Ferrer, F., Mannarà, E., Cosme, M., Falace, A., Montiel-Nelson, J. A., Espino, F., Haroun, R., & Tuya, F. (2019). Early-faunal colonization patterns of discrete habitat units: A case study with rhodolith-associated vagile macrofauna. *Estuarine, Coastal and Shelf Science, 218*, 9–22.

Otero-Ferrer, F., Cosme, M., Tuya, F., Espino, F., & Haroun, R. (2020). Effect of depth and seasonality on the functioning of rhodolith seabeds. *Estuarine, Coastal and Shelf Science, 235*, 106579.

Papaconstantinou, C., & Caragitsou, E. (1989). Feeding interaction between two sympatric species Pagrus pagrus and Phycis phycis around Kastellorizo Island (Dodecanese, Greece). *Fisheries Research, 7*(4), 329–342. https://doi.org/10.1016/0165-7836(89)90065-9

Pardo, C., Peña, V., Barreiro, R., & Bárbara, I. (2015). A molecular and morphological study of *Corallina sensu lato* (Corallinales, Rhodophyta) in the Atlantic Iberian Peninsula. *Cryptogamie Algologie, 36*, 31–54. https://doi.org/10.7872/crya.v36.iss1.2015.31

Pascelli, C., Riul, P., Riosmena-Rodríguez, R., Scherner, F., Nunes, M., Hall-Spencer, J. M., Oliveira, E. C., & Horta, P. (2013). Seasonal and depth-driven changes in rhodolith bed structure and associated macroalgae off Arvoredo Island (southeastern Brazil). *Aquatic Botany, 111*, 62–65.

Peña, V., & Bárbara, I. (2008). Maërl community in the northwestern Iberian Peninsula: A review of floristic studies and long-term changes. *Aquatic Conservation, 18*, 339–366.

Peña, V., Adey, W. H., Riosmena-Rodríguez, R., Jung, M. Y., Afonso-Carrillo, J., Choi, H. G., & Bárbara, I. (2011). *Mesophyllum sphaericum* sp. nov. (Corallinales, Rhodophyta): A new maerl-forming species from the Northeast Atlantic. *Journal of Phycology, 47*, 911–927.

Peña, V., Bárbara, I., Grall, J., Maggs, C. A., & Hall-Spencer, J. M. (2014a). The diversity of seaweeds on maerl in the NE Atlantic. *Marine Biodiversity, 44*, 533–551.

Peña, V., Hernandez-Kantun, J. J., Grall, J., Pardo, C., Lopez, L., Bárbara, I., Le Gall, L., & Barreiro, R. (2014b). Detection of gametophytes in the maerl-forming species *Phymatolithon calcareum* (Melobesioideae, Corallinales) assessed by DNA barcoding. *Cryptogamie Algologie, 35*, 15–25.

Pereira-Filho, G. H., Amado-Filho, G. M., Moura, R. L., Bastos, A. C., Guimarães, S. M. P. B., Salgado, L. T., Francini-Filho, R. B., Bahia, R. G., Abrantes, D. P., Guth, A. Z., & Brasileiro, P. S. (2012). Extensive Rhodolith beds cover the summits of Southwestern Atlantic Ocean seamounts. *Journal of Coastal Research, 28*, 261–269.

Pereira-Filho, G. H., Veras, P. D. C., Francini-Filho, R. B., Moura, R. L., Pinheiro, H. T., Gibran, F. Z., Matheus, Z., Neves, L. M., & Amado-Filho, G. M. (2015). Effects of the sand tilefish *Malacanthus plumieri* on the structure and dynamics of a rhodolith bed in the Fernando de Noronha Archipelago, tropical West Atlantic. *Marine Ecology Progress Series, 541*, 65–73.

Pinheiro, H. T., Rocha, L. A., Macieira, R. M., Carvalho-Filho, A., Anderson, A. B., Bender, M. G., Di Dario, F., Ferreira, C. E. L., Figueiredo-Filho, J., Francini-Filho, R., Gasparini, J. L., Joyeux, J.-C., Luiz, O. J., Mincarone, M. M., Moura, R. L., Nunes, J. A. C. C., Quimbayo, J. P., Rosa, R. S., Sampaio, C. L. S., Sazima, I., Simon, T., Vila-Nova, D. A., & Floeter, S. R. (2018). South-Western Atlantic reef fishes: Zoogeographical patterns and ecological drivers reveal a secondary biodiversity Centre in the Atlantic Ocean. *Diversity and Distributions, 24*(7), 951–965. https://doi.org/10.1111/ddi.12729

Riosmena-Rodriguez, R., & Medina-López, M. A. (2010). The role of rhodolith beds in the recruitment of invertebrate species from the southwestern Gulf of California, México. In A. Israel (Ed.), *Seaweeds and their role in global changing environments: Cellular origin, life in extreme habitats, astrobiology* (pp. 127–138). Springer.

Riosmena-Rodríguez, R., Nelson, W., & Aguirre, J. (2017). *Rhodolith/Maërl beds: A global perspective*. Springer, Series, Coastal Research Library.

Riul, P., Targino, C. H., Farias, J. N., Visscher, P. T., & Horta, P. A. (2008). Decrease in *Lithothamnion* sp. (Rhodophyta) primary production due to the deposition of a thin sediment layer. *Journal of the Marine Biological Association of the United Kingdom, 88*, 17–19.

Riul, P., Lacouth, P., Pagliosa, P. R., Christoffersen, M. L., & Horta, P. A. (2009). Rhodolith beds at the easternmost extreme of South America: Community structure of an endangered environment. *Aquatic Botany, 90*, 315–320.

Rothman, M. D., Mattio, L., Anderson, R. J., & Bolton, J. J. (2017). A phylogeographic investigation of the kelp genus *Laminaria* (Laminariales, Phaeophyceae), with emphasis on the South Atlantic Ocean. *Journal of Phycology, 53*, 778–789. https://doi.org/10.1111/jpy.12544

Sañé, E., Chiocci, F. L., Basso, D., & Martorelli, E. (2016). Environmental factors controlling the distribution of rhodoliths: An integrated study based on seafloor sampling, ROV and side scan sonar data, offshore the W-Pontine Archipelago. *Continental Shelf Research, 129*, 10–22. https://doi.org/10.1016/j.csr.2016.09.003

Schubert, N., Peña, V., Salazar, V. W., et al. (2022). Rhodolith physiology across the Atlantic: Towards a better mechanistic understanding of intra- and interspecific differences. *Frontiers in Marine Science, 9*. https://doi.org/10.3389/fmars.2022.921639

Simon, T., Pinheiro, H. T., Moura, R. L., Carvalho-Filho, A., Rocha, L. A., Martins, A. S., Mazzei, E., Francini-Filho, R. B., Amado-Filho, G. M., & Joyeux, J.-C. (2016). Mesophotic fishes of the Abrolhos Shelf, the largest reef ecosystem in the South Atlantic. *Journal of Fish Biology, 89*(1), 990–1001.

Sissini, M. N., Oliveira, M. C., Gabrielson, P. W., Robinson, N. M., Okolodkov, Y. B., Riosmena-Rodriguez, R., & Horta, P. A. (2014). *Mesophyllum erubescens* (Corallinales, Rhodophyta)—So many species in one epithet. *Phytotaxa, 190*, 299–319. https://doi.org/10.11646/phytotaxa.190.1.18

Solano-Barquero, A., Sibaja-Cordero, J. A., & Cortés, J. (2022). Macrofauna associated with a rhodolith bed at an oceanic Island in the eastern tropical pacific (Isla del coco National Park, Costa Rica). *Frontiers in Marine Science, 9*, 1–16. https://doi.org/10.3389/fmars.2022.858416

Steller, D. L., & Cáceres-Martínez, C. (2009). Coralline algal rhodoliths enhance larval settlement and early growth of the Pacific calico scallop *Argopecten ventricosus*. *Marine Ecology Progress Series, 396*, 49–60.

Steller, D. L., & Foster, M. S. (1996). Environmental factors influencing distribution and morphology of rhodolits in Bahia Concepcion, B.C.S., Mexico. *Oceanographic Literature Review, 43*, 818–819.

Steller, D. L., Riosmena-Rodríguez, R., Foster, M. S., & Roberts, C. A. (2003). Rhodolith bed diversity in the Gulf of California: The importance of rhodolith structure and consequences of disturbance. *Aquatic Conservation, 13*, S5–S20.

Steller, D. L., Hernández-Ayón, J. M., Riosmena Rodríguez, R., & Cabello-Pasini, A. (2007). Effect of temperature on photosynthesis, growth and calcification rates of the free-living coralline algae *Lithophyllum margaritae*. *Ciencias Marinas, 33*, 441–456.

Tâmega, F. T. S., Spotorno-Oliveira, P., & Figueiredo, M. A. O. (2013). *Catalogue of the benthic marine life from Peregrino oil field, Campos basin, Brazil*. Instituto Biodiversidade Marinha.

Tâmega, F. T. S., Bassi, D., Figueiredo, M. A. O., & Cherkinsky, A. (2014). Deep-water rhodolith bed from central Brazilian continental shelf, Campos basin: Coralline algal and faunal taxonomic composition. *Galaxea, 16*, 21–31.

Tâmega, F. T. S., Perna, G. H. H., Spotorno-Oliveira, P., Riosmena-Rodriguez, R., & Gonçalves, J. E. A. (2017). A unique free-living geniculate coralline algal bed formation. *Marine Biodiversity, 47*, 373–374. https://doi.org/10.1007/s12526-016-0487-0

Tâmega, F. T. S., Torrano-Silva, B. N., Oliveira, M. C., Spotorno-Oliveira, P., Calazans, S. H., Rosas-Alquicira, E. F., Coutinho, R., & Peña, V. (2021). Identification of 'articuliths' in a unique algal bed formation from Brazil and description of *Jania cabista* sp. nov. (Corallinales, Rhodophyta). *Phycologia, 60*, 283–302. https://doi.org/10.1080/00318884.2021.1916279

Teichert, S. (2014). Hollow rhodoliths increase Svalbard's shelf biodiversity. *Scientific Reports, 4*, 6972.

Testa, V., & Bosence, D. W. J. (1999). Physical and biological controls on the formation of carbonate and siliciclastic bedforms on the northeast Brazilian shelf. *Sedimentology, 46*, 279–301.

Torrano-Silva, B. N., Riosmena-Rodrígues, R., & Oliveira, M. C. (2014). Systematic position of *Paulsilvella* in the Lithophylloideae (Corallinaceae, Rhodophyta) confirmed by molecular data. *Phytotaxa, 190*, 94–111.

Torrano-Silva, B. N., Vieira, B. R., Riosmena-Rodríguez, R., & Oliveira, M. C. (2018). Guidelines for DNA barcoding of coralline algae, focusing on Lithophylloideae (Corallinales) from Brazil. *Botanica Marina, 61*, 127–140. https://doi.org/10.1515/bot-2017-0040

Veras, P. C., Pierozzi, I., Lino, J. B., Amado-Filho, G. M., de Senna, A. R., Santos, C. S. G., & Pereira-Filho, G. H. (2020). Drivers of biodiversity associated with rhodolith beds from euphotic and mesophotic zones: Insights for management and conservation. *Perspectives in Ecology and Conservation, 18*, 37–43.

Vieira-Pinto, T., Oliveira, M. C., Bouzon, J., Sissini, M., Richards, J. L., Riosmena-Rodríguez, R., & Horta, P. A. (2014). *Lithophyllum* species from Brazilian coast: Range extension of *Lithophyllum margaritae* and description of *Lithophyllum atlanticum* sp. nov. (Corallinales, Corallinophycidae, Rhodophyta). *Phytotaxa, 190*, 355–369.

Villas-Bôas, A. B., Figueiredo, M. A. O., & Villaça, R. C. (2005). Colonization and growth of crustose coralline algae (Corallinales Rhodophyta) on the Rocas Atoll. *Brazilian Journal of Oceanography, 53*, 147–156. https://doi.org/10.1590/S1679-87592005000200005

Villas-Bôas, A. B., Riosmena-Rodriguez, R., Amado-Filho, G. M., Maneveldt, G., & Figueiredo, M. A. (2009). Rhodolith-forming species of *Lithophyllum* (Corallinales; Rhodophyta) from Espírito Santo state, Brazil, including the description of *L. depressum* sp. nov. *Phycologia, 48*, 237–248.

Villas-Bôas, A. B., Riosmena-Rodríguez, R., & Figueiredo, M. A. O. (2014). Community structure of rhodolith-forming beds on the central Brazilian continental shelf. *Helgoland Marine Research, 68*, 27–35.

Weber-van Bosse, A., & Foslie, M. (1904). *The Corallinaceae of the Siboga-expedition*. Brill Publishers and Printers.

Woelkerling WJ (1988) *The coralline red algae: an analysis of the genera and sub-families of nongeniculate Corallinaceae*. British Museum (Natural History) and Oxford University Press.

Woelkerling, W. J., Irvine, L. M., & Harvey, A. S. (1993). Growth-forms in non-geniculate coralline red algae (Corallinales, Rhodophyta). *Australian Systematic Botany, 6*, 277–293.

Wolf, M. A., Falace, A., Kaleb, S., & Moro, I. (2016). Molecular data confirm the existence of attached crustose tetrasporangial thalli in *Phymatolithon calcareum* (Melobesioideae, Hapalidiaceae, Rhodophyta) from the Mediterranean Sea. *Aquatic Botany, 134*, 75–81.

Chapter 5
Mollusks Associated with Rhodolith Beds from the Southwestern Atlantic

Aramys Rocchetti de Mello Cesar, Izabela Felice Paes, Ana Clara Becker, Flávio Dias Passos, and Guilherme Henrique Pereira-Filho

Abstract As presented in Chaps. 1 and 2, rhodolith beds are found worldwide from polar to tropical regions, but their greatest latitudinal concentration is found in the Southwestern Atlantic (SWA), along the Brazilian Economic Exclusive Zone. In Chaps. 3 and 4, the authors showed how rhodolith beds are of great importance to marine biodiversity. However, many are the gaps on this matter when it comes to the beds from the SWA—habitats imminently threatened by a series of global and local anthropogenic stressors, including oil and gas industry (shown in Chap. 8). Here, to represent the importance of rhodolith bed-associated diversity, we deeply detail the diversity of mollusks associated with rhodolith beds and reveal the importance of these habitats to the SWA marine biodiversity. For instance, >10% of the gastropods occurring in this ocean were registered associated with rhodoliths from only a few small sampling sites at Fernando de Noronha Archipelago (3°S), Abrolhos Bank (17°S), and Queimada Grande Island (24°S). Besides logistical difficulties in sampling extensive areas covered by the SWA rhodoliths, taxonomic investments and human resource training have not been attractive to the young generations of biologists and scientists, triggering a scarcity of high-quality knowledge and material of reference. In consequence, technical staff of environmental agencies, consultant bodies, and committed companies face the lack of technical material to guide the identification of invertebrates associated with rhodolith beds. Thus, most technical reports seem to underestimate the huge diversity of life forms inhabiting the SWA rhodolith beds. In this context, this chapter aims to provide a general panorama of the mollusks associated with rhodolith beds. Also, we provide a photographic guide that will help a large body of technical workers in the identification of the high biodiversity associated with the SWA and thus improve this discussion during environmental licensing on rhodolith beds.

A. R. de Mello Cesar · I. F. Paes · F. D. Passos
Departamento de Biologia Animal, Instituto de Biologia, Universidade Estadual de Campinas (UNICAMP), São Paulo, Brazil

A. C. Becker · G. H. Pereira-Filho (✉)
Instituto do Mar, Laboratório de Ecologia e Conservação Marinha, Universidade Federal de São Paulo, São Paulo, Brazil
e-mail: pereira.filho@unifesp.br

P. A. Horta, M. N. Sissini (eds.), *Brazilian Rhodolith Beds*, Brazilian Marine Biodiversity, https://doi.org/10.1007/978-3-031-61449-1_5

Keywords Marine biodiversity · Macrofauna · Benthic communities

5.1 South Atlantic Rhodolith Beds Overview

Rhodolith beds extend globally from tropical to polar regions, but it is in the South Atlantic (SA) where they are most latitudinally concentrated. More precisely, in terms of SA, such structures have only been identified in its western portion, along the Brazilian coast that bear 34 rhodolith-forming species (see Chap. 3 for more details). Despite the great variability of environmental conditions across the Brazil's coast, rhodolith beds span from close to the Amazon river mouth (5°N) (Lavagnino et al., 2020) to the north of Rio de Janeiro State (23°S) (Villas-Bôas et al., 2014) with sparse beds occurrence in Queimada Grande and Arvoredo Islands (24°S and 27°S, respectively) (Pereira-Filho et al., 2019; Gherardi, 2004)—that is, rhodolith beds occur from tropical to subtropical waters, including oceanic islands, seamounts, and the single SA atoll (Rocas Atoll) (Pereira-Filho et al., 2011, 2012; Amado-Filho et al., 2012a, 2016). Not surprisingly, it is in the Brazilian Exclusive Economic Zone (EEZ) that lies the rhodolith bed with the largest continuous distribution, located in the Abrolhos Bank, covering an area of 20,904 km^2 (Amado-Filho et al., 2012b; Brasileiro et al., 2015).

Rhodolith beds develop over sedimentary bottoms and play a bioengineering role by altering physicochemical properties of the surrounding environment and by forming three-dimensional substrata that provide important habitats for a diversity of organisms (e.g., Otero-Ferrer et al., 2019; de Cerqueira Veras et al., 2020). Although the importance of rhodolith beds has become a worldwide consensus by the day, much there is to understand about their physicochemical and ecological roles in the South Atlantic. It is known, for instance, that most beds from the Brazilian EEZ occur in mesophotic conditions (i.e., between ~30 and 120 m), which represents mapping and sampling difficulties that have only been overcome in the past few years with the aid of remotely operated vehicles (ROVs) observations, mixed-gas diving, and geomorphometry techniques (e.g., Pereira-Filho et al., 2012; Moura et al., 2013; Lavagnino et al., 2020). Even though it is estimated that Brazil's coast detains an extension of rhodolith beds comparable to that of global coral reef in terms of biological importance (Jones et al., 2015; Amado-Filho et al., 2012a, b), studies regarding Southwestern Atlantic (SWA) rhodolith beds and its associated biodiversity are scarce in comparison with the large extensions of these habitats (see Amado-Filho & Pereira-Filho, 2012 for details). On the other hand, they cover marine bottoms of large economic interest such as oil and gas exploration (Santos et al., 2016; Marques & Araújo, 2019) and fishing industry and are threatened by ocean acidification (Schubert et al., 2019), as well as by unprecedented environmental accidents (Sissini et al., 2020; Holz et al., 2020). In this context, basic taxonomic studies become increasingly urgent. How many species are yet to be discovered in regions of the world that for any reason are not well sampled and studied? This seems to be the case of the SWA rhodolith beds. For instance, the last years were marked by intense scientific efforts to register the high biodiversity associated with rhodolith beds and the discovery of new occurrences of benthic marine algae, as well as some new invertebrates (e.g., Amado-Filho & Pereira-Filho, 2012; Santos et al., 2016; Senna et al. 2021). However, taxonomic investments and human resource training have not been attractive to the young generations of biologists and scientists, triggering a scarcity of high-quality knowledge and material of reference (Buckley, 2015). In consequence, technical staff of environmental agencies, consultant bodies, and committed companies face the scarcity of technical material to guide identification of the invertebrates associated with rhodolith beds. Thus, most technical reports seem to underestimate the huge diversity of life forms inhabiting rhodolith beds, one of the most prominent SWA marine benthic habitats. In this context, this chapter aims to provide a general panorama of the mollusks associated with SWA rhodolith beds. We also present photographic guides that we hope will help a large body of technical workers to identify this biological group and improve their reports concerning biodiversity associated with rhodoliths.

5.2 Mollusks and Rhodolith Beds

In areas where purely sedimentary soft bottoms prevail with limited life-supporting conditions, the presence of rhodolith beds increases the diversity of macroalgae and invertebrates by enhancing viable settling substrata, primary production, and habitat heterogeneity (Steller & Foster, 1995). The latter is highly related to the shape of the nodules, branching density, and presence of cavities, all of which make rhodolith beds suitable habitats for a range of different organisms (Steller et al., 2003). Mollusks also benefit from association with rhodoliths, and they are the third most abundant group of organisms associated with algal beds, just behind polychaetes and crustaceans (e.g., Steller et al., 2003; Tâmega et al., 2013; de Cerqueira Veras et al., 2020). Known as being primary consumers, mollusks not only take advantage of the foraging conditions rhodoliths provide but also use rhodolith beds as refuges from predators and as important nursery areas (Kamenos et al., 2003; Kamenos et al., 2004; Steller & Cáceres-Martinez, 2009).

Mollusca—the second-largest phylum of invertebrates—comprises seven classes (Aplacophora, Monoplacophora, Scaphopoda, Cephalopoda, Gastropoda, Polyplacophora, and Bivalvia). Aplacophora is a small class represented by <300 species, most of which inhabit deep-sea benthos from the edge continental shelves to thousands of meters deep and are characterized by their vermiform body covered by a cuticle in which are embedded innumerable spicules of aragonite (Scheltema, 1996). Monoplacophora is a group morphologically simple, limpet-like mollusks often considered as living fossils (Stöger et al., 2016). Scaphopoda is among the smallest classes; they are infaunal microcarnivores that manipulate their prey with specialized tentacles, or captacula. Their tusk-like shell is open at both ends where a burrowing foot and the captacula protrude anteriorly, while respiratory currents pass through the posterior opening (Reynolds, 2002). Cephalopoda includes symmetrical mollusks with a well-developed head that contains a circumoral crown of appendages that bear suckers and/or hooks. The mouth has chitinous beak-like jaws and chitinous tonguelike radula (Haimovici et al., 2009). Gastropods are the most diverse class of Mollusca, comprising almost 75% of the ~110,000 species described. They have a univalve shell and possess a file-like radula used in feeding the biofilm covering rocks, plants, and algae thallus (Brown & Lydeard, 2010). Polyplacophora are recognizable by their distinctive eight overlapping shell plates called valves connected to a girdle. All valves articulate to form an oval to elongate dorsal or ventral profile (Kaas & Van Belle, 1985). Bivalvia is the second most diverse class of the Phylum Mollusca, with ~10,000 extant species (Lemer et al., 2019). They are characterized by a single shell divided in two parts (valves), a muscular foot and the absence of head and radula. They are mostly filter feeders (suspensivores) or depositivorous, but there are also carnivorous species that have modified gills that help the feeding process (Reid & Reid, 1975; Giribet, 2008).

Here, we focused on the most abundant Classes inhabiting the SWA rhodolith beds (i.e., Gastropoda, Bivalvia, and Polyplacophora). The presented material was obtained during several expeditions between the years 2015 and 2019 and came from Fernando de Noronha Archipelago (FNA), Abrolhos Bank (AB), and Queimada Grande Island (QGI). A total of 1254 organisms were identified belonging to 26 families of gastropods, 14 families of bivalves, and two families of polyplacophorans, totaling ~100 infrageneric taxa (Table 4.1). Only specimens in which the soft parts were present were included in the present guide. Despite the limited spatial sampling area, our efforts add new evidence of the high biodiversity associated with the SWA rhodolith beds. For instance, ~5% of the prosobranch gastropods that occur throughout >15,000 km of coast from Caribbean to the Argentinian southern coast and ~ 10% of those occurring in Brazilian marine habitats were found in only few square meters of rhodolith beds from only three punctual sampled areas (see Barroso et al., 2016) (Table 5.1).

The most common gastropods found in FNA are *Pseudodiala puncturina*, *Zafrona idalina*, *Eulithidium affine*, and *Simulamerelina caribaea*. For the AB, the main species are *Caecum ryssotitum*, *Turbonilla pupoides*, *Caecum brasilicum*, *Alvania auberiana*, and *Meioceras nitidum*. As for QGI, most common gastropod species are *Bostrycapulus aculeatus*, *Astyris lunata*, *Cerithiopsis flava*, and *Sinum perspectivum*. A photo-guide of gastropods associated with rhodolith beds is provided in Figs. 5.1, 5.2, 5.3, 5.4, 5.5, and 5.6.

In FNA, the most common bivalves associated with rhodoliths are *Lima caribaea*, *Barbatia cancellaria*, *B. domingensis*, and *Leiosolenus bisulcatus*. In the AB, the most common bivalves associated with rhodoliths are *Lamychaena hians*, a non-identified species from the subfamily Chioninae, *Musculus lateralis*, *Papyridea semisulcata* and *Crassinella lunulata*. For QGI most common bivalves are *Gouldia cerina*, *Leiosolenus bisulcatus*, *Pinctada imbricata*, *Musculus lateralis*, and *Barbatia domingensis*. A photo-guide of bivalves associated with rhodolith beds is provided in Figs. 5.7 and 5.8.

Finally, among Polyplacophora, the family Acanthochitonidae is the most frequent in FNA and AB, with representatives of *Acanthochitona terezae* and

Table 5.1 Mollusks associated with the Southwest Atlantic rhodolith beds

Class	Family	Species	Series number
Gastropoda	**Rissoidae**	*Simulamerelina caribaea* (d'Orbigny, 1842)	ZUEC-GAS 7272-74
		Alvania auberiana (d'Orbigny, 1842)	ZUEC-GAS 7275-76
	Barleeiidae	*Pseudodiala puncturina* (Dos Santos & Absalão, 2006)	ZUEC-GAS 7277-99
		Caelatura spirocordata (Absalão & Rios, 1995)	ZUEC-GAS 7639-44
		Caelatura barcelosii (Absalão & Rios, 1995)	ZUEC-GAS 7653-55
	Eulimidae	*Hemiliostraca auricincta* (Abbott, 1958)	ZUEC-GAS 7300-01
		Eulima fulvocincta C. B. (Adams, 1850)	ZUEC-GAS 7585
		Vitreolina sp. 1	ZUEC-GAS 7911
		Vitreolina sp. 2	ZUEC-GAS 7912
	Litiopidae	*Alaba incerta* (d'Orbigny, 1841)	ZUEC-GAS 7302-04
	Marginellidae	*Volvarina albolineata* (d'Orbigny, 1842)	ZUEC-GAS 7887-89
		Marginellidae 1	ZUEC-GAS 7305
		Marginellidae 2–3	ZUEC-GAS 7307-16;7891-903
	Columbellidae	*Astyris lunata* (Say, 1826)	ZUEC-GAS 7436-53
		Zafrona idalina (Duclos, 1840)	ZUEC-GAS 7317-35
		Columbella mercatoria (Linnaeus, 1758)	ZUEC-GAS 7337
		Anachis fenneli (Radwin, 1968)	ZUEC-GAS 7582
	Pyramidellidae	*Turbonilla pupoides* (d'Orbigny, 1841)	ZUEC-GAS 7504-25
		Turbonilla multicostata (C. B. Adams, 1850)	Not deposited yet
		Eulimastoma didymum (Verrill & Bush, 1900)	ZUEC-GAS 7348;7537;7814
	Muricidae	*Aspella* sp.	ZUEC-GAS 7593
		Favartia sp.	ZUEC-GAS 7596
		Coralliophila aberrans (C. B. Adams, 1850)	ZUEC-GAS 7590-92
		Coralliophila sp.	ZUEC-GAS 7349
		Trachypollia turricula (Maltzan, 1884)	ZUEC-GAS 7594
	Triviidae	*Pusula pediculus* (Linnaeus, 1758)	ZUEC-GAS 7360; 7806
		Niveria suffusa (J.E. Gray, 1827)	ZUEC-GAS 7498
	Nassariidae	*Phrontis alba* (Say, 1826)	ZUEC-GAS 7599

(continued)

Table 5.1 (continued)

Class	Family	Species	Series number
	Phasianellidae	*Eulithidium affine* (C. B. Adams, 1850)	ZUEC-GAS 7354-59; 7463-72; 7794
	Triphoridae	*Metaxia excelsa* Faber & Moolenbeek, 1991	ZUEC-GAS 7591-92
		Cosmotriphora melanura (C. B. Adams, 1850)	ZUEC-GAS 7351; 7803
		Iniforis pseudothomae (Rolán & Fernández-Garcés, 1993)	ZUEC-GAS 7494
		Nanaphora verbenei (Moolenbeek & Faber, 1989)	ZUEC-GAS 7497
		Marshallora sp.	ZUEC-GAS 7496
	Fissurellidae	*Emarginula* sp. 1	ZUEC-GAS 7612
		Emarginula sp. 2	ZUEC-GAS 7613
	Calyptraeidae	*Bostrycapulus aculeatus* (Gmelin, 1791)	ZUEC-GAS 7852-65
	Caecidae	*Meioceras nitidum* (Stimpson, 1851)	ZUEC-GAS 7545-56
		Caecum brasilicum (Folin, 1874)	ZUEC-GAS 7727-44
		Caecum ryssotitum (Folin, 1867)	ZUEC-GAS 7569-76
	Anabathridae	*Amphithalamus glabrus* (Simone, 1996)	ZUEC-GAS 7662-63; 7677
	Skeneidae	*Parviturbo weberi* (Pilsbry & McGinty, 1945)	ZUEC-GAS 7678-81
		Cyclostrema sp.	ZUEC-GAS 7869-70
	Rissoellidae	*Rissoella ornata* (Simone, 1995)	Not deposited yet
	Scissurellidae	*Scissurella* sp.	ZUEC-GAS 7769;7773
	Turbinidae	*Astralium latispina* (Philippi, 1844)	ZUEC-GAS 7775
	Pisaniidae	*Engina turbinella* (Kiener, 1836)	ZUEC-GAS 7600; 7602
	Naticidae	*Sinum perspectivum* (Say, 1831)	ZUEC-GAS 7761-62
	Olividae	*Olivella nivea* (Gmelin, 1791)	ZUEC-GAS 7759-60
	Zebinidae	*Zebina* sp.	ZUEC-GAS 7836-37
	Cerithiopsidae	*Cerithiopsis flava* (C. B. Adams, 1850)	ZUEC-GAS 7362-64;7382-84
		Seila adamsii (H. C. Lea, 1845)	ZUEC-GAS 7368
		Cerithiopsis cf. *gemmulosa* (C. B. Adams, 1850)	ZUEC-GAS 7371
Bivalvia			
	Arcidae	*Barbatia domingensis* (Lamarck, 1819)	ZUEC-BIV 7701-11; 8249-53
		Barbatia cancellaria (Lamarck, 1819)	ZUEC-BIV 7712-24; 8248-01
		Arcidae 1	ZUEC-BIV 8305

(continued)

Table 5.1 (continued)

Class	Family	Species	Series number
		Arcidae. 2	ZUEC-BIV 8306
	Chamidae	*Chama sinuosa* (Broderip, 1835)	ZUEC-BIV 8332-43
	Limidae	*Lima caribaea* (d'Orbigny, 1853)	ZUEC-BIV 7725-35; ZUEC-BIV 8243-47
	Crassatellidae	*Crassinella lunulata* (Conrad, 1834)	ZUEC-BIV 8190-8263
	Cardiidae	*Americardia media* (Linnaeus, 1758)	ZUEC-BIV 8212
		Papyridea semisulcata (Gray, 1825)	ZUEC-BIV 8203-69
	Philobryidae	Philobryidae	ZUEC-BIV 8344-45
	Veneridae	Chioninae	ZUEC-BIV 8256-61; 8309-14
		Gouldia cerina (C. B. Adams, 1845)	ZUEC-BIV 8315-31
	Pteriidae	*Pinctada imbricata* (Röding, 1798)	ZUEC-BIV 8200-02
	Mytilidae	*Musculus lateralis* (Say, 1822)	ZUEC-BIV 8240-42; 8284-95
		Leiosolenus bisulcatus (d'Orbigny, 1853)	ZUEC-BIV 7736; 8235-38
		Dacrydium hendersoni (Salas & Gofas, 1997)	ZUEC-BIV 8230-34
		Crenella decussata (Montagu, 1808)	ZUEC-BIV 8225-29;8280-82
		Gregariella sp.	ZUEC-BIV 8296-99
	Pectinidae	*Leptopecten bavayi* (Dautzenberg, 1900)	ZUEC-BIV 8308
		Pectinidae 1	ZUEC-BIV 8255
	Condylocardiidae	*Carditopsis smithii* (Dall, 1896)	ZUEC-BIV 8220-24;8278-79
	Gastrochaenidae	*Lamychaena hians* (Gmelin, 1791)	ZUEC-BIV 8213-19; ZUEC-BIV 8270-77
Polyplacophora			
	Acanthochitonidae	*Acanthochitona terezae* (Guerra Júnior, 1983)	ZUEC-POP 05-21;42-73
		Acanthochitona sp.	ZUEC-POP 22-24;92-93
	Ischnochitonidae	*Ischnochiton striolatus* (Gray, 1828)	ZUEC-POP 25-41;74-91
		Ischnochiton sp.	Not deposited yet

Acanthochitona sp. being the most common. For QGI, only one species of chiton is observed (*Ischnochiton striolatus*). A photo-guide of polyplacophorans associated with rhodolith beds is provided in Fig. 5.9.

Fig. 5.1 Photomicrographs of the gastropods species associated with rhodoliths of Brazilian coast islands. Rissoidae: (**a**) *Alvania auberiana*, (**b**) *Simulamerelina caribaea*; Columbellidae: (**c**) *Columbella mercatoria*, (**d**) *Zafrona idalina*, (**e**) *Zafrona idalina* (variation), (**f**) *Astyris lunata*, (**g**) *Anachis fenneli*; Phasianellidae: (**h**) *Eulithidium affine*; Cerithiopsidae: (**i**) *Seila adamsii*, (**j**) *Cerithiopsis* cf. *gemmulosa*, (**k**) *Cerithiopsis flava*. Scale bars: **a** and **b** = 500 μm; **c** = 2000 μm; **d–h** = 1000 μm; **i** = 2000 μm; **j** = 500 μm; **k** = 1000 μm

Fig. 5.2 Photomicrographs of the gastropods species associated with rhodoliths of Brazilian coast islands. Muricidae: (**a**) *Aspella* sp., (**b**) *Trachypolia turricula*, (**c**) *Favartia* sp., (**d**) *Coralliophila* sp., (**e**) *Stramonita brasiliensis*; Nassariidae: (**f**) *Phrontis alba*; Triviidae: (**g**) *Niveria suffusa*, (**h**) *Pusula pediculus;* Caecidae: (**i**) *Meioceras nitidum*, (**j**) *Caecum ryssotitum*, (**k**) *Caecum brasilicum*. Scale bars: **a** and **b** = 500 μm; **c** and **d** = 1000 μm; **e** = 2000 μm; **f** = 500 μm; **g** and **h** = 2000 μm; **i**–**k** = 500 μm

Fig. 5.3 Photomicrographs of the gastropods species associated with rhodoliths of Brazilian coast islands. Eulimidae: (**a**) *Hemiliostraca auricincta*, (**b**) *Melanella* sp., (**c**) *Vitreollina* sp. 1, (**d**) *Vitreollina* sp. 2, (**e**) *Eulima fulvocincta*; Zebinidae: (**f**) *Zebina* sp.; Litiopidae: (**g**) *Alaba incerta*; Marginellidae: (**h**) Marginellidae 1, (**i**) Marginellidae 2, (**j**) Marginellidae 3, (**k**) *Volvarina albolineata*. Scale bars: **a** = 500 μm; **b** = 1000 μm; **c**–**k** = 500 μm

Fig. 5.4 Photomicrographs of the gastropod species associated with rhodoliths of Brazilian coast islands. Triphoridae: (**a**) *Metaxia excelsa*, (**b**) *Cosmotriphora melanura*, (**c**) *Marshallora* sp., (**d**) *Nanaphora verbenei*, (**e**) *Isotriphora* sp., (**f**) *Iniforis pseudothomae*, (**g**) *Latitriphora albida*; Pyramidellidae: (**h**) *Turbonilla pupoides*, (**i**) *Turbonilla multicostata*, (**j**) *Eulimastoma didymum*, (**k**) *Fargoa bushiana*. Scale bars: **a** = 1000 μm; **b–k** = 500 μm

5.3 Promising Future for Discoveries on the Biodiversity Associated with Rhodolith Beds

Large portions of the Brazilian EEZ are covered by rhodolith beds, and it is only natural that the more beds are sampled, the greater will be the knowledge on their associated biodiversity (Foster et al., 2013). This becomes even more likely to be true when we assume that the majority of rhodolith beds occur in Brazil's mesophotic zone and many of them remain unmapped. A model developed by Carvalho et al. (2019) predicted that, along the Brazilian EEZ, there are around 230,000 km^2 of suitable areas for the occurrence of rhodolith beds.

Fig. 5.5 Photomicrographs of the gastropods species associated with rhodoliths of Brazilian coast islands. Fissurellidae: (**a**) *Emarginula* sp. 1, (**b**) *Emarginula* sp. 2; Calyptraeidae: (**c**) *Bostrycapulus aculeatus*; Naticidae: (**d**) *Sinum perspectivum*; Olividae: (**e**) *Olivella nivea*; Pisaniidae: (**f**) *Engina turbinella*. Scale bars: **a** = 500 μm; **b** = 1000 μm; **c** = 2000 μm; **d** = 500 μm; **e** = 1000 μm; **f** = 2000 μm

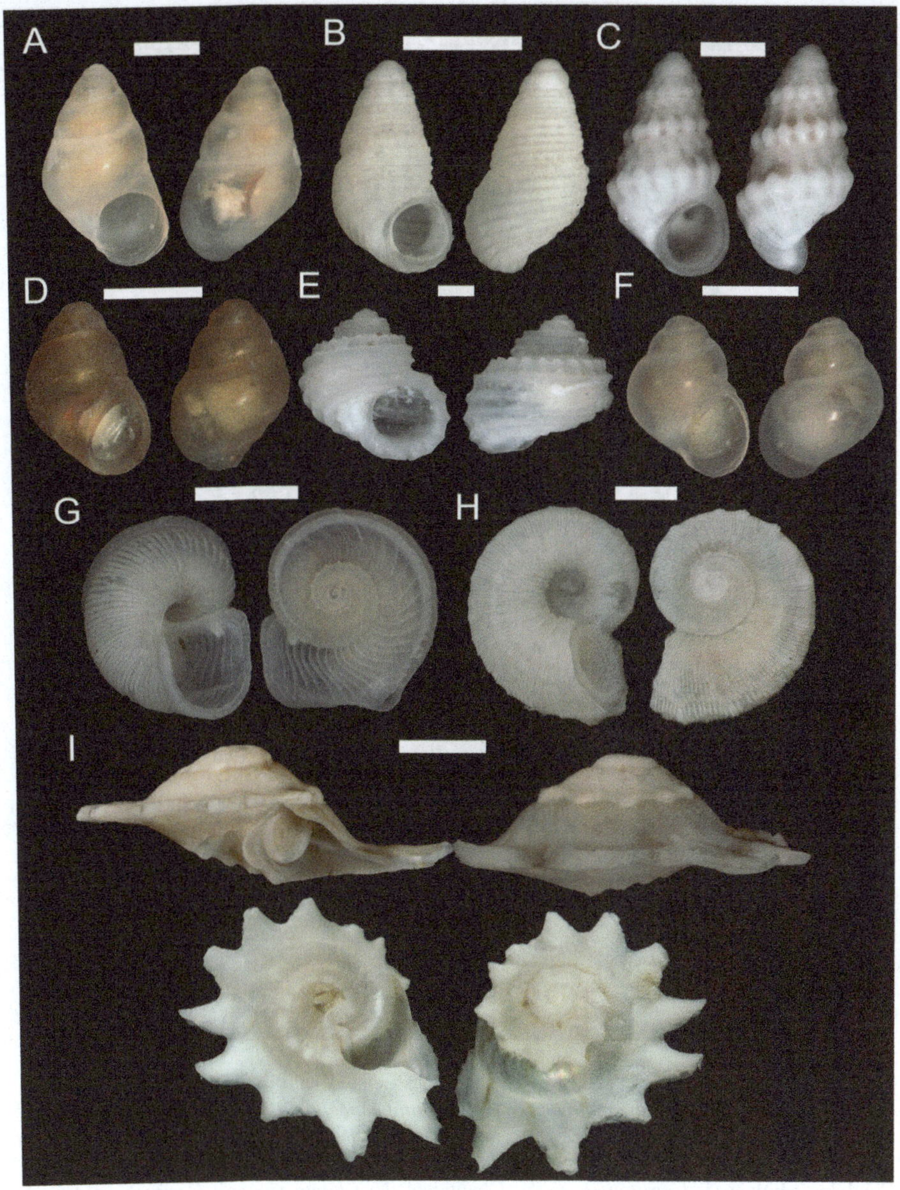

Fig. 5.6 Photomicrographs of the gastropods species associated with rhodoliths of Brazilian coast islands. Barleeiidae: (**a**) *Pseudodiala puncturina*, (**b**) *Caelatura spirocordata*, (**c**) *Caelatura barcellosi*; Anabathridae: (**d**) *Amphithalamus glabrus*; Tornidae: (**e**) *Parviturbo weberi*; Rissoellidae: (**f**) *Rissoella ornata*; Scissurellidae: (**g**) *Scissurella* sp.; Skeneidae: (**h**) *Cyclostrema* sp.; Turbinidae: (**i**) *Astralium latispina*. Scale bars: **a–g** = 500 μm; **h** = 1000 μm; **i** = 2000 μm

Fig. 5.7 Photomicrographs of the bivalves species associated with rhodoliths of Brazilian coast islands. Arcidae: (**a**) *Barbatia domingensis*, (**b**) *Barbatia cancellaria*, (**c**) Arcidae 1, (**d**) Arcidae 2; Spondylidae: (**e**) *Spondylus americanus*; Pteriidae: (**f**) *Pinctada imbricata*; Pectinidae: (**g**) *Leptopecten bavayi*; Limidae: (**h**) *Lima caribaea*; Crassatellidae: (**i**) *Crassinella lunulata*; Condylocardiidae: (**j**) *Carditopsis smithi*; Chamidae: (**k**) *Chama sinuosa*; Gastrochaenidae: (**l**) *Lamychaena hians*. Scale bars: **a–e** = 1000 μm; **f** = 2000 μm; **g–i** = 1000 μm; **j** = 500 μm; **k** and **l** = 1000 μm

The last couple of decades have been providential for the characterization of the macrofaunal groups associated with rhodolith beds. For instance, the few existing studies regarding polychaetes (another important group inhabiting rhodoliths) from SWA rhodolith beds were responsible for important discoveries such as the

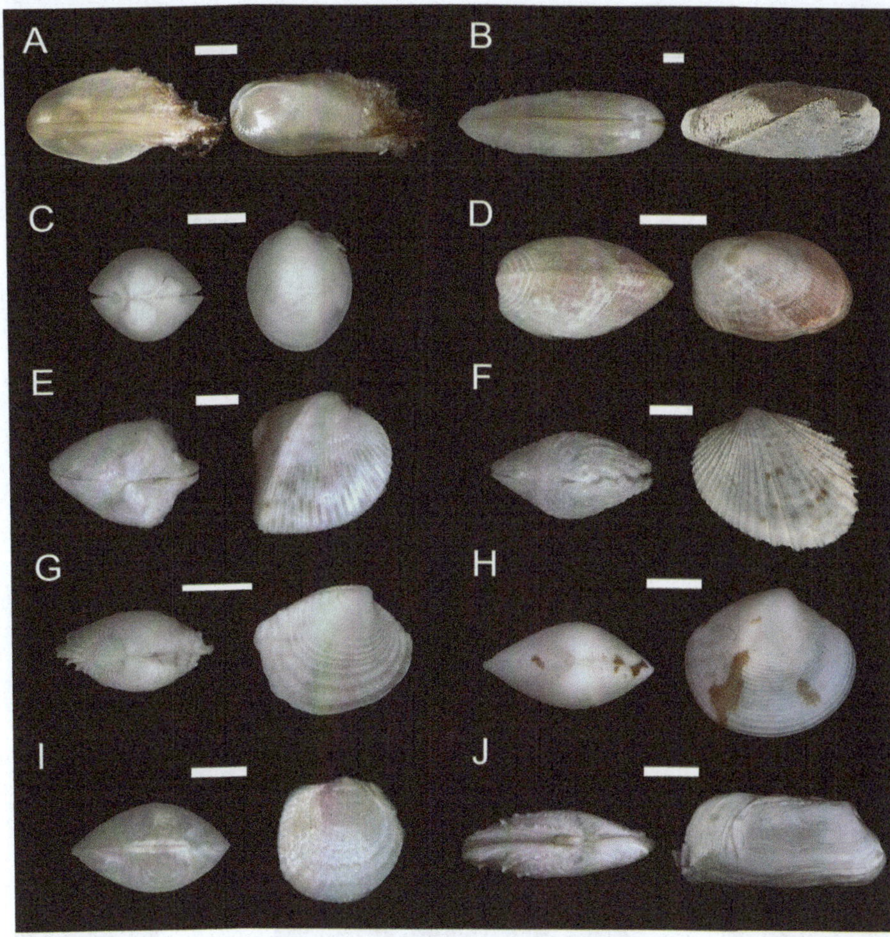

Fig. 5.8 Photomicrographs of the bivalves species associated with rhodoliths of Brazilian coast islands. Mytilidae: (**a**) *Gregariella* sp., (**b**) *Leiosolenus bisulcatus*, (**c**) *Crenella decussata*, (**d**) *Musculus lateralis*; Cardiidae: (**e**) *Americardia media*, (**f**) *Papyridea semisulcata;* Veneridae: (**g**) Chioninae, (**h**) *Gouldia cerina*; (**i**) Philobryidae; (**j**) Lyonsiidae: *Lyonsiidae* 1. Scale bars: **a**–**d** = 1000 μm; **e** and **f** = 500 μm; **g** and **h** = 1000 μm; **i** = 500 μm; **j** = 1000 μm

description of a new endemic species sabellariid, *Sabellaria corallinea* (Santos et al., 2011), the recording of four polychaete families exclusive to rhodolith beds from Espírito Santo state and of nine families exclusive to the Abrolhos Bank (Berlandi et al., 2012) and, more recently, the occurrence of one specimen of *Nuchalosyllis maiteae* in a rhodolith bed in the Fernando de Noronha Archipelago (Santos et al., 2016), when the only other individuals of that species ever sampled were from deep rhodolith offshore areas from Rio de Janeiro and Bahia states (Fukuda & Nogueira, 2013). Further, as predicted, mapping of recently discovered rhodolith beds is on the move. Among the newest findings are rhodolith beds in the

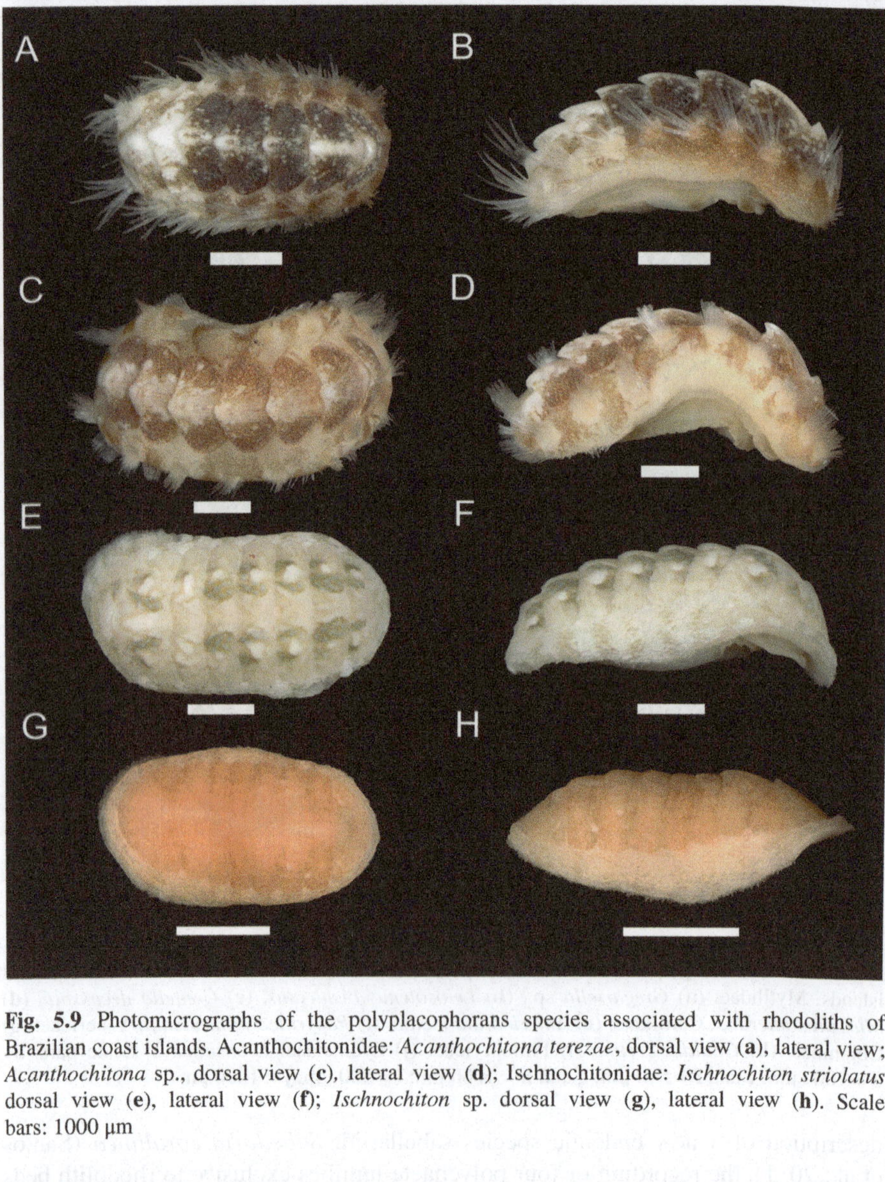

Fig. 5.9 Photomicrographs of the polyplacophorans species associated with rhodoliths of Brazilian coast islands. Acanthochitonidae: *Acanthochitona terezae*, dorsal view (**a**), lateral view; *Acanthochitona* sp., dorsal view (**c**), lateral view (**d**); Ischnochitonidae: *Ischnochiton striolatus* dorsal view (**e**), lateral view (**f**); *Ischnochiton* sp. dorsal view (**g**), lateral view (**h**). Scale bars: 1000 μm

region of the Amazon River mouth occurring from 23 to 120 m deep, inside a heterogeneous mosaic of benthic megahabitats (Moura et al., 2016; Lavagnino et al., 2020). Despite recurrent efforts to characterize the area, little is the information acquired so far on the diversity of organisms inhabiting it. Therefore, future research conducted there is very promising in this sense.

5.4 Threats

Just like coral reefs, rhodolith beds are also subject to global threats such as ocean acidification. Rhodolith beds contribute to world $CaCO_3$ production, and just the beds from the Abrolhos Bank themselves have a production comparable to that of Caribbean coral reefs (Amado-Filho et al., 2012a, b). However, rhodoliths may be even more sensible than corals to the effects of ocean acidification, for the form of $CaCO_3$ calcareous algae secrete (high-magnesium calcite) is the most soluble form of $CaCO_3$ minerals (see Chap. 8). Decreasing in ocean's pH levels may then lead to a great loss of habitats provided by rhodolith beds culminating in loss of biodiversity.

But just as problematic as global threats are the local threats South Atlantic rhodolith beds face. Just as rhodolith beds, oil exploitation also occurs along the Brazilian coast and poses an imminent risk to these habitats. There is a huge mismatch between the speed at which rhodolith bed areas are being exploited and the development of strategies to help protect them (Marques & Araújo, 2019), mainly because of the lack of information on the ecological aspects involved. For instance, the most oil and gas productive area in Brazil, Campos Basin located offshore Rio de Janeiro state, has been being explored since the 1970s, but much more recent is the discovery of mesophotic rhodolith beds (~100 m deep) in this area (Amado-Filho et al., 2017). The diversity associated with these habitats is still being studied, and there is much more to know, but from what has been already discovered, it is possible to say that it refers to a biodiverse area (Tâmega et al., 2013). But the rhodolith beds located in Campos Basin are not the only ones threatened by the ghost of oil and gas exploitation, so are the ones from the Abrolhos complex (Vilardo et al., 2020) and the recent-found ones from the Amazon River mouth, the latter being a major concern for its promising oil prospection against a disproportionately small knowledge on the area's biological and oceanographic characteristics (Vale et al., 2018; Lavagnino et al., 2020). Recently, Brazil experienced a major catastrophe when thousands of tons of crude oil were spilt and washed up on the Northeastern coast. The spill threatens several SWA marine habitats, including rhodolith beds, and its consequences will take years to be even accurately measured (Sissini et al., 2020).

5.5 Final Remarks

Although it feels like we are still far behind in acknowledging the whole ecological aspects of rhodolith beds, little by little gaps on this matter are being filled once new sites are being mapped and studied (e.g., Pereira-Filho et al., 2019, 2023). Thanks to the development of safer diving and less aggressive sampling techniques, the extensive Brazilian mesophotic rhodolith beds have been reached and better studied (e.g., Lavagnino et al., 2020; Holz et al., 2020; Veras et al., 2020). However, the absence of accurate data concerning associated biodiversity still hovers above many environmental licensing, which often seems to underestimate or even ignore how many forms of life lie on rhodolith beds.

Here, we bring the first detailed description of the malacofauna associated with rhodolith beds from the SWA, a singular effort to exemplify how rich and specific these habitats can be in terms of their associated communities. Moreover, we hope to foster and raise awareness on how much still there is to be learned regarding the biodiversity from the South Atlantic rhodolith beds. We reinforce that these habitats should not be overlooked in the decision-making process regarding the management and conservation of Brazilian coastal-marine ecosystems, once they can be considered up to now the neglected treasures of the Brazilian continental shelf.

References

Amado-Filho, G. M., & Pereira-Filho, G. H. (2012). Rhodolith beds in Brazil: A new potential habitat for marine bioprospection. *Revista Brasileira de Farmacognosia, 22*(4), 782–788.

Amado-Filho, G. M., Pereira-Filho, G. H., Bahia, R. G., Abrantes, D. P., Veras, P. C., & Zaira, M. (2012a). Occurrence and distribution of rhodolith beds on the Fernando de Noronha archipelago of Brazil. *Aquatic Botany, 101*, 41–45.

Amado-Filho, G. M., Moura, R. L., Bastos, A. C., Salgado, L. T., Sumida, P. Y. G., Guth, A. Z., Francini-Filho, R. B., Pereira-Filho, G. H., Arantes, D. P., Brasileiro, P. S., Bahia, R. G., Leal, R. N., Kaufman, L., Kleypas, J., Farina, M., & Thompson, F. L. (2012b). Rhodolith beds are major CaCO3 bio-factories in the tropical South West Atlantic. *PLoS One, 7*(4), e3517.

Amado-Filho, G. M., Moura, R. L., Bastos, A. C., Francini-Filho, R. B., Pereira-Filho, G. H., Bahia, R., Moraes, F. C., & Motta, F. S. (2016). Mesophotic ecosystems of the unique South Atlantic atoll are composed by rhodolith beds and scattered consolidated reefs. *Marine Biodiversity: International Journal of Marine Science, XX*, 1–4.

Amado-Filho, G. M., Bahia, R., Pereira Filho, G. H., & Longo, L. L. (2017). South Atlantic Rhodolith beds: Latitudinal distribution, species composition, structure and ecosystem functions, threats and conservation status. In R. Riosmena-Rodríguez, W. Nelson, & J. Aguirre (Eds.), *Rhodolith/Maërl beds: A global perspective* (Vol. 15, pp. 299–317). Springer.

Barroso, C., Lotufo, T., & Matthews-Cascon, H. (2016). Biogeography of Brazilian prosobranch gastropods and their Atlantic relationships. *Journal of Biogeography*. https://doi.org/10.1111/jbi.12821

Berlandi, R., Figueiredo, M., Figueiredo, O., & Paiva, P. (2012). Rhodolith morphology and the diversity of Polychaetes off the southeastern Brazilian coast. *Journal of Coastal Research., 1*, 280–287. https://doi.org/10.2112/11T-00002.1

Brasileiro, P. S., Pereira-Filho, G. H., Bahia, R. G., Abrantes, D. P., Guimarães, S. M. P. B., Moura, R. L., Francini-Filho, R. B., Bastos, A. C., & Amado-Filho, G. M. (2015). Macroalgal composition and community structure of the largest rhodolith beds in the world. *Marine Biodiversity: International Journal of Marine Science, 46*, 407–420.

Brown, K. M., & Lydeard, C. (2010). Mollusca: Gastropoda. In J. H. Thorp & A. P. Covich (Eds.), *Ecology and classification of North American freshwater invertebrates* (3rd ed., pp. 277–307). Academic.

Buckley, R. C. (2015). Grand challenges in conservation research. *Frontiers in Ecology and Evolution, 3*. https://doi.org/10.3389/fevo.2015.00128

Carvalho, V. F., Assis, J., Serrão, E. A., Nunes, J. M., Batista, A. A., Batista, M. B., et al. (2019). Environmental drivers of rhodolith beds and epiphytes community along the South Western

Atlantic coast. *Marine Environmental Research, 154*, 104827. https://doi.org/10.1016/j.marenvres.2019.104827

de Cerqueira Veras, P., Pierozzi-Jr, I., Lino, J. B., Amado-Filho, G. M., de Senna, A. R., Santos, C. S. G., de Moura, R. L., Passos, F. D., Giglio, V. J., & Pereira-Filho, G. H. (2020). Drivers of biodiversity associated with rhodolith beds from euphotic and mesophotic zones: Insights for management and conservation. *Perspectives in Ecology and Conservation, 18*, 37. https://doi.org/10.1016/j.pecon.2019.12.003

Foster, M.; Amado-Filho, G.; Kamenos, N.; Riosmena-Rodriguez, R.; Steller, D. (2013). Rhodoliths and Rhodolith beds.

Fukuda, M., & Nogueira, J. (2013). On a new species of *Nuchalosyllis* (Polychaeta: Syllidae), a rare syllid genus only known from Brazilian waters. *Journal of the Marine Biological Association of the United Kingdom, 93*. https://doi.org/10.1017/S0025315412000069

Gherardi, D. F. M. (2004). Community structure and carbonate production of a temperate rhodolith bank from Arvoredo Island, southern Brazil. *Brazilian Journal of Oceanography, 52*(3/4), 207–224.

Giribet, G. (2008). Bivalvia. In W. F. Ponder & D. R. Lindberg (Eds.), *Phylogeny and evolution of the Mollusca* (pp. 105–141). University of California Press.

Haimovici, M., Santos, R. A., & Fischer, L. G. (2009). Class Cephalopoda. In E. C. de Rios (Ed.), *Compendium of Brazilian sea shells* (pp. 610–649). Evangraf.

Holz, V., Bahia, R., Karez, C., Vieira, F., Moraes, F., Vale, N., Sudatti, D., Salgado, L., Moura, R., Amado-Filho, G., & Bastos, A. (2020). Structure of Rhodolith beds and surrounding habitats at the Doce River shelf (Brazil). *Diversity, 12*, 75. https://doi.org/10.3390/d12020075

Jones, N. S., Ridgwell, A., & Hendy, E. J. (2015). Evaluation of coral reef carbonate production models at a global scale. *Biogeosciences, 12*, 1339–1356. https://doi.org/10.5194/bg-12-1339-2015

Kaas, P., & Van Belle, R. (1985). *Monograph of living chitons (Mollusca: Polyplacophora), volume 1 order Neoloricata: Lepidopleurina*. Brill.

Kamenos, N. A., Moore, P. G., & Hall-Spencer, J. M. (2003). The heterogeneity of dredged versus un-dredged maerl grounds. *Journal of the Marine Biological Association of the United Kingdom, 83*, 411–413.

Kamenos, N. A., Moore, P. G., & Hall-Spencer, J. M. (2004). Nursery-area function of maerl (subtidal red coralline algae) grounds for juvenile queen scallops *Aequipecten opercularis* and other invertebrates. *Marine Ecology Progress Series, 274*, 183–189.

Lavagnino, A. C., Bastos, A. C., Amado Filho, G. M., de Moraes, F. C., Araujo, L. S., & de Moura, R. L. (2020). Geomorphometric seabed classification and potential Megahabitat distribution in the Amazon continental margin. *Frontiers in Marine Science, 7*. https://doi.org/10.3389/fmars.2020.00190

Lemer, S., Bieler, R., & Giribet, G. (2019). Resolving the relationships of clams a cockles: Dense transcriptomic sampling drastically improves the bivalve tree of life. *Proceedings of the Royal Society B, 286*, 20182684.

Marques, S., & de Araújo, T. C. M. (2019). Survey and assessment of seabed resources from the Brazilian continental shelf by the law of the sea: From national to international jurisdictions. *Ocean & Coastal Management, 178*, 104858.

Moura, R. L., Secchin, N. A., Amado-Filho, G. M., Francini-Filho, R. B., Freitas, M. O., Minte-Vera, C. V., Teixeira, J. B., Thompson, F. L., Dutra, G. F., Sumida, P. Y. G., & Guth, A. Z. (2013). Spatial patterns of benthic megahabitats and conservation planning in the Abrolhos Bank. *Continental Shelf Research, 1*, 1.

Moura, R. L., Amado-Filho, G. M., Moraes, F. C., Brasileiro, P. S., Salomon, P. S., Mahiques, M. M., et al. (2016). An extensive reef system at the Amazon River mouth. *Science Advances, 2*(4), e1501252–e1501252. https://doi.org/10.1126/sciadv.1501252

Otero-Ferrer, F., Mannarà, E., Cosme, M., Falace, A., Montiel-Nelson, J. A., Espino, F., Haroun, R., & Tuya, F. (2019). Early-faunal colonization patterns of discrete habitat units: A case study with rhodolith-associated vagile macrofauna. *Estuarine, Coastal and Shelf Science, 218*, 9–22. https://doi.org/10.1016/j.ecss.2018.11.020

Pereira-Filho, G. H., Amado-Filho, G. M., Guimarães, S. M., Moura, R. L., Sumida, P. Y., Abrantes, D. P., Bahia, R. G., Güth, A. Z., Jorge, R. R., & Francini Filho, R. B. (2011). Reef fish and benthic assemblages of the Trindade and Martin Vaz Island group, southwestern Atlantic. *Brazilian Journal of Oceanography, 59*, 201–212.

Pereira-Filho, G. H., Amado-Filho, G. M., de Moura, R. L., Bastos, A. C., Guimaraes, S. M., Salgado, L. T., Francini-Filho, R. B., Bahia, R. G., Abrantes, D. P., Guth, A. Z., & Brasileiro, P. S. (2012). Extensive Rhodolith beds cover the summits of southwestern Atlantic Ocean seamounts. *Journal of Coastal Research, 28*, 261–269.

Pereira-Filho, G. H., Shintate, G. S., Kitahara, M. V., Moura, R. L., Amado-Filho, G. M., Bahia, R. G., Moraes, F. C., Neves, L. M., Francini, C. L. B., Gibran, F. Z., & Motta, F. S. (2019). The southernmost Atlantic coral reef is off the subtropical Island of Queimada Grande (24°S), Brazil. *Bulletin of Marine Science, 95*.

Pereira-Filho, G. H., Candido, L. C., Moreira, P. S., Laranja, D. H., Motta, F. D. S., & Comin, E. (2023). *Nota técnica conjunta n 001/2023 IMar/Unifesp—Fundação Florestal de São Paulo: Descoberta do Banco de Rodolitos na Ilha das Couves*, Ubatuba, SP. Disponível em: https://repositorio.unifesp.br/handle/11600/67268

Reid, R. G. B., & Reid, A. M. (1975). The carnivorous habit of members of the septibranch genus Cuspidaria (Mollusca:Bivalvia). *Sarsia, 56*, 47–56.

Reynolds, P. D. (2002). The scaphopoda. *Advances in Marine Biology, 137–236*. https://doi.org/10.1016/s0065-2881(02)42014-7

Santos, A. S., Riul, P., Brasil, A. C. S., & Christoffersen, M. L. (2011). Encrusting Sabellariidae (Annelida: Polychaeta) in rhodolith beds, with description of a new species of Sabellaria from the Brazilian coast. *Journal of the Marine Biological Association of the United Kingdom, 91*(2), 425–438.

Santos, C. S. G., Lino, J. B., Veras, P. C., Amado-Filho, G. M., Francini-Filho, R. B., Motta, F. S., Moura, R. L., & Pereira-Filho, G. H. (2016). Environmental licensing on rhodolith beds: Insights from a worm. *Natureza & Conservação, 14*(2), 1–5.

Scheltema, A. H. (1996). Phylogenetic position of Sipuncula, Mollusca and the progenetic Aplacophora. In J. Taylor (Ed.), *Origin and evolutionary radiation of the Mollusca* (pp. 53–58). Oxford University Press.

Schubert, N., Salazar, V. W., Rich, W. A., Bercovich, M. V., Saá, A. A., Fadigas, S. D., et al. (2019). Rhodolith primary and carbonate production in a changing ocean: The interplay of warming and nutrients. *Science of the Total Environment, 676*, 455–468.

Sissini, M. N., Berchez, F., Hall-Spencer, J., Guilardi-Lopes, N., Carvalho, V. F., Schubert, N., Koerich, G., Diaz-Pulido, G., Silva, J., Serrão, E., Assis, J., Santos, R., Floeter, S. R., Rorig, L., Barufi, J. B., Bernardino, A. F., Francini-Filho, R., Turra, A., Hoffmann, L. C., Aguirre, J., Le Gall, L., Peña, V., Nash, M. C., Rossi, S., Soares, M., Pereira-Filho, G., Tâmega, F., & Horta, P. A. (2020). Brazil oil spill response: Protect rhodolith beds. *Science, 367*(6474), 156. https://doi.org/10.1126/science.aba2582

Steller, D. L., & Cáceres-Martínez, C. (2009). Coralline algal rhodoliths enhance larval settlement and early growth of the Pacific calico scallop Argopecten ventricosus. *Marine Ecology Progress Series, 396*, 49–60.

Steller, D. L., & Foster, M. S. (1995). Environmental factors influencing distribution and morphology of rhodoliths in Bahía Concepción, B.C.S., Mexico. *Journal. Experimental Marine Biology and Ecology, 194*, 201–212.

Steller, D. L., Riosmena-Rodríguez, R., Foster, M. S., & Roberts, C. (2003). Rhodolith bed diversity in the Gulf of California: The importance of rhodolith structure and consequences of anthropogenic disturbances. *Aquatic Conservation Marine and Freshwater Ecosystems, 13*, 5–20.

Stöger, I., Kocot, K. M., & Poustka, J. A. (2016). Monoplacophoran mitochondrial genomes: Convergente gene arrangements and little phylogenetic signal. *BMC Evolutionary Biology, 16*, 274.

Tâmega, F. T. S., Spotorno-Oliveira, P., & Figueiredo, M. A. O. (2013). *Catalogue of the benthic marine life from Peregrino oil field, Campos Basin, Brazil* (p. 140). Instituto Biodiversidade Marinha.

Vale, N. F., Amado-Filho, G. M., Braga, J. C., Brasileiro, P. S., Karez, C. S., Moraes, C. F., Bahia, R. G., Bastos, A. C., & Moura, R. L. (2018). Structure and composition of rhodoliths from the Amazon River mouth, Brazil. *Journal of South American Earth Sciences, 84*, 149–159.

Vilardo, C., La Rovere, E. L., Evora, J. E. M., & Montaño, M. (2020). Lost at SEA? Environmental assessment and offshore oil and gas planning in Brazil. *Impact Assessment and Project Appraisal, 38*, 261–268.

Villas-Boas, A. B., Riosmena-Rodriguez, R., & de Oliveira Figueiredo, M. A. (2014). Community structure of rhodolith-forming beds on the central Brazilian continental shelf. *Helgoland Marine Research, 2014*(68), 27–35. https://doi.org/10.1007/s10152-013-0366-z

Chapter 6
Fossil and Recent Coralline Algal Bioconstructions and Rhodoliths in the Brazilian Shelf

Frederico T. S. Tâmega, Alex C. Bastos, Davide Bassi, Julio Aguirre, and Juan C. Braga

Abstract In Brazil, studies on fossil crustose coralline algae (CCA) are scarce and limited to a few taxonomic accounts and to palaeoenvironmental interpretation of deposits with abundant CCA. Here, the most relevant reports on fossil and recent CCA bioconstructions and rhodolith beds are summarized from the oldest records to Holocene and living examples. The oldest CCA recorded in Brazil are from middle Albian (Lower Cretaceous) deposits from the Sergipe and Potiguar basins (NE Brazil). CCA are widespread and abundant in Holocene deposits and on the modern Brazilian shelf forming extensive rhodolith beds. Boundstones of CCA are more common on the outer shelf/shelf edge. The high abundance of CCA on the Brazilian shelf is probably related to the widespread mesotrophic conditions and the relatively low input of fine-grained riverine sediment. In recent years, new molecular and morpho-anatomical studies suggest that living CCA have higher diversity and higher levels of endemicity than the ones previously reported based on morpho-anatomical taxonomy. Future studies are needed to better understand the fossil CCA occurrences along the Brazilian coast and their taxonomic and palaeoecological implications.

Keywords Cretaceous · Holocene · Fossil coralline algae · Brazil

---

F. T. S. Tâmega (✉)
Departamento de Biotecnologia Marinha, Instituto de Estudos do Mar Almirante Paulo Moreira, Arraial do Cabo, RJ, Brazil

A. C. Bastos
Departamento de Oceanografia, Universidade Federal do Espírito Santo, Vitória, ES, Brazil

D. Bassi
Dipartimento di Fisica e Scienze della Terra, Università degli Studi di Ferrara, Ferrara, Italy

J. Aguirre · J. C. Braga
Departamento de Estratigrafía y Paleontología, Universidad de Granada, Granada, Spain

P. A. Horta, M. N. Sissini (eds.), *Brazilian Rhodolith Beds*, Brazilian Marine Biodiversity, https://doi.org/10.1007/978-3-031-61449-1_6

Modern calcareous coralline algae (CCA) are known since the mid-eighteenth century (e.g., Linnaeus, 1758; Ellis & Solander, 1786) although they were seen as animals resembling plants (zoophytes) until Philippi (1837) identified them as photosynthetic organisms. Regarding fossil CCA, Michelin (1840–1847), in a series of papers dealing with fossil cnidarians from France, reported five species attributed to the genus *Nullipora* defined by Lamarck (1801) that were actually coralline algal species (reviewed by Woelkerling, 1998). After this first report, studies describing fossil CCA species started to be published by the second half of the nineteenth century (e.g., Gümbel, 1861, 1871; Früh, 1890; Rothpletz, 1891), with a significant increase in the number of publications during the twentieth century (Aguirre & Braga, 2005).

Despite this high number of works on fossil coralline algae, most of them were concentrated in the Western Tethyan/Mediterranean region (Aguirre et al., 2000; Braga, 2017). In Brazil, studies on fossil CCA are actually very rare, limited to a few taxonomic accounts (Maury, 1936; Granier et al., 1991, 2008; Woelkerling et al., 2014; Tâmega et al., 2016) and to palaeoecology of deposits with abundant CCA (Maury, 1929; Tâmega et al., 2014; Spotorno-Oliveira et al., 2016; Bastos et al., 2018; Brasileiro et al., 2018; Vale et al., 2018).

The most relevant reports on fossil and recent CCA bioconstructions and rhodolith beds are summarized in the following sections, from the oldest records to the Holocene and living examples (Fig. 6.1). In this synopsis, the original taxon names applied by the authors of the reports are kept, and therefore, the same taxon might appear here under different names.

6.1 Cretaceous

The opening of the South Atlantic started at about 135 Ma (Lower Cretaceous) and was probably completed by ca. 95 Ma ago (Cenomanian) (Larson & Ladd, 1973; Ghidella et al., 2007; Granot & Dyment, 2015). During the Albian and Cenomanian, South America was entirely separated from Africa, and shallow-water marine carbonates were extensively deposited along the northeastern Brazilian coasts (Arai et al., 2000). Although in some inland basins in NE Brazil, such as Parnaíba, Araripe, Tucano, and Jatobá basins, there are older, upper Aptian marine sediments (Arai et al., 2000; Arai, 2014; Assine et al., 2016), and the oldest CCA recorded in Brazil is from middle Albian (Lower Cretaceous) deposits from Sergipe and Potiguar basins (NE Brazil; Maury, 1929, 1936).

The first references to fossil CCA in Brazil are by Maury (1929, 1936). Maury (1929) made a brief discussion of the Jatobá-Buique series (Cretaceous), where fossil CCA were collected along the road from Jatobá to Espírito Santo. Maury (1936) described the Cretaceous rocks of the State of Sergipe, in which middle Albian CCA, molluscs, including ammonites, and sea urchins were found. Regarding CCA, two new species, *Lithothamnium mauryae* Howe in Maury (1936) and *Lithothamnium regonis* Howe in Maury (1936), were established.

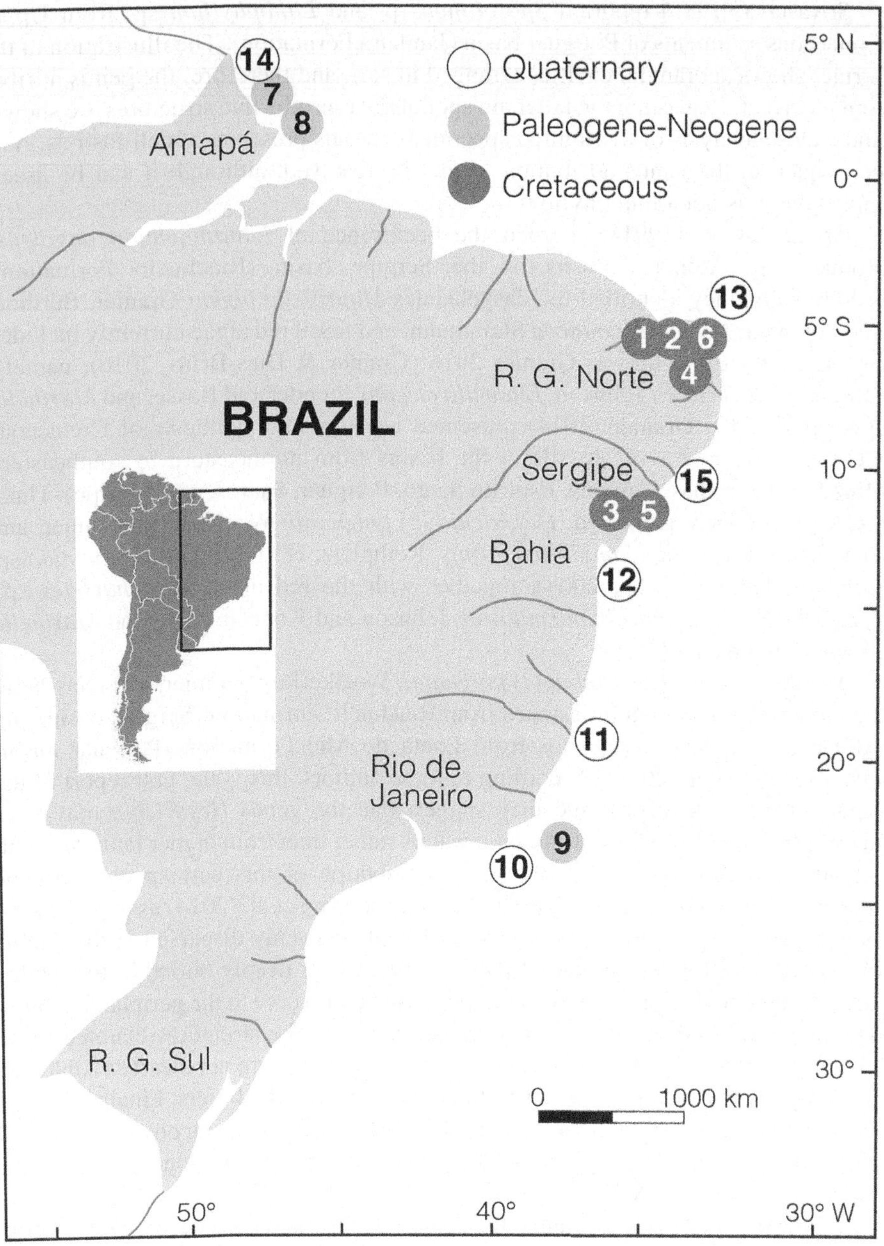

Fig. 6.1 Geographic locations of the most relevant reports on Cretaceous to Recent CCA bioconstructions and rhodolith beds. 1. Maury (1929, 1936); 2. Srivastava (1982); 3. Granier et al. (1991); 4. Granier et al. (2008); 5. Woelkerling et al. (2014); 6. De Castro et al. (1988); 7. Wolff and Carozzi (1984), Figueiredo et al. (2007), Cruz et al. (2019); 8. Sousa et al. (2009); 9. Waisman (2002), Della Giustina (2006), BouDagher-Fadel et al. (2010), Tâmega et al. (2014), Carelli et al. (2019); 10. Spotorno-Oliveira et al. (2016), Tâmega et al. (2016); 11. Amado-Filho et al. (2012a), Moura et al. (2013), Bastos et al. (2016, 2018), Brasileiro et al. (2018), Holz et al. (2020); 12. Kikuchi and Leão (1998); 13. Gherardi and Bosence (2001), Amado-Filho et al. (2012b); 14. Vale et al. (2018); 15. Vale et al. (2022b)

Srivastava (1982) reported *Sporolithon* sp. and *Lithophyllum* sp. from Upper Cretaceous sediments of Potiguar basin (Jandaíra Formation). The illustration of the former shows sporangial cavities arranged in sori, and therefore, the genus attribution is correct. Regarding the latter, no sporangial reproductive structures are shown, and a close analysis of the figured specimens reveals presence of cell fusions. As a consequence, the genus attribution cannot be resolved, although it can be ascertained that it is not a lithophylloid.

Granier et al. (1991) recorded the occurrence of *Lithothamnion angolense* Romanes in Albian deposits of the Sergipe basin (Riachuelo Formation). Additionally, they identified the dasycladales *Maurinella olgeni* Granier, Berthou, and Poignant; *Neomeris cretacea* Steinmann; and fossil red algae currently included in the family Elianellaceae Granier 2016 (Granier & Dias-Brito, 2016), namely, *Parachaetetes keitheii* Johnson, *Elianella elegans* Pfender and Basse, and *Marinella lugeoni* Pfender. Granier (2014) presented a taxonomic synthesis of Cretaceous CCA and *Incertae sedis* fossils in the basins from northeastern to southeastern Brazil (Barreirinhas, Campos, Espírito Santo, Potiguar, Santos, and Sergipe). Three CCA species were identified: *Heydrichia* (?) *poignantii* Woelkerling, Granier, and Dias-Brito; *Sporolithon amphiroaeformis* (Rothpletz, 1891); and *Lithothamnion* sp. (see also Granier et al., 2008), together with the red algae *Parachaetetes* sp., *Elianella* sp., *Pycnoporidium sinuosum* Johnson and Konishi 1960, and *Marinella lugeoni* Pfender 1939.

The new species *Heydrichia* (?) *poignantii* Woelkerling, Granier, and Dias-Brito was described in Albian limestones from Riachuelo Formation (Sergipe basin) and Albian-Cenomanian carbonates from Ponta do Mel Formation (Potiguar basin) (Woelkerling et al., 2014). According to these authors, this is the first report of the genus in the rock record, and they suggest that the genus *Heydrichia* may have arrived in Brazil from Central Atlantic waters rather than from higher latitude South Atlantic waters. Nonetheless, the genus attribution of this new species remains unsolved. The void cavities interpreted by Woelkerling et al. (2014) as putative sporangial reproductive structures are isolated voids unevenly dispersed in the thallus but not arranged in sori. Some of these cavities occur deeply buried in the thallus such as at the transition from the basal plumose ventral core to the peripheral region. This is an atypical position for sporangial cavities. The structures claimed to be putative involucres (Figs. 7–8 in Woelkerling et al. 2014) are not clearly visible and probably are fine-grained remnants of cell-wall grinding by borers. Finally, most of the cavities cut the cell filaments. According to these features, the observed cavities can be due to bioerosion by etching sponges (i.e., producing the ichnogenus *Entobia* Bronn; e.g., Bromley & D'Alessandro, 1989).

De Castro et al. (1988) detailed the sedimentology and palaeoecology of marine Cretaceous (Albian–Maastrichtian) deposits of the Potiguar basin, Northeastern Brazilian continental margin. Within the six distinguished marine depositional sequences, the Sequence II, late Albian in age, includes fluvio-estuarine clastic deposits of the Açu Formation and marine carbonates of the Ponta do Mel Formation. The carbonate platform formed a narrow and elongate belt developed parallel to the modern shelf. The bioclastic to detrital limestones are formed mainly by CCA, molluscs, green algae, oncolite, and benthic foraminifers.

6.2 Paleogene–Neogene

Detailed reports of CCA jump from the Lower Cretaceous to the Quaternary, with no precise descriptions of the CCA found in Upper Cretaceous, Paleogene, and Neogene deposits. However, CCA occur as facies-dominant components together with larger benthic foraminifers in the upper Paleocene to middle Miocene deposits, off the coast of Northern and Northeast Brazil (Schaller et al., 1971). CCA are also common rock-forming components of the extensive Paleocene to lower Pliocene deposits in the Amapá carbonate platform (Amazon continental margin, Brazil; Wolff & Carozzi, 1984; Figueiredo et al., 2007; Cruz et al., 2019). In the Amazonas basin fragments of non-geniculate and geniculate CCA occur in uppermost Paleocene and lower Eocene larger foraminiferal and red algal grainstone to packstone facies localized in inner ramp settings (Sousa et al., 2009).

In the Campos basin, southeastern Brazil, the Oligocene-lower Miocene Siri Member of the Emborê Formation yielded detrital calcirudites and bioclastic calcarenites composed of CCA fragments and rhodoliths deposited in a reefal carbonate platform (Waisman, 2002; BouDagher-Fadel et al., 2010). In the same basin carbonate deposits of the Grussaí member (Emborê Formation), ranging from the Eocene to the Pleistocene (Carelli et al., 2019), have CCA (*Lithothamnion* sp.) as major components, occurring as rhodoliths and encrusting forms (Della Giustina, 2006).

6.3 Quaternary to Recent

Rhodolith facies from the beach deposits of Abaí and Praia do Forte, northern Bahia area, recorded the latemost Pleistocene and Early Holocene sea-level rise that drowned preexisting coral reefs in the area (Kikuchi & Leão, 1998). Nonetheless, Pleistocene records are still rare for the Brazilian margin (Bastos et al., 2016).

In the Abrolhos continental shelf, the presence of sinkholes along the mid and outer shelf revealed horizontal beds of limestones with conspicuous borings and dissolution cavities, showing distinct phases of boring and filling (Bastos et al., 2013, 2016). The limestones from two studied sinkholes consist of alternations of wackestone, grainstone, packstone, and rudstone. Throughout the stratigraphic section, CCA are one of the main components, together with coral fragments, bryozoans, serpulids, benthic foraminifers, vermetids, ascidian and sponge spicules, echinoderms, and bivalves. *Porolithon onkodes*, *Titanoderma* sp., *Mesophyllum* sp., *Sporolithon* sp., and *Paulsilvella* sp. were the main CCA identified. The limestones forming the sinkhole walls comprise two stratigraphic units separated by an erosion surface developed during Marine Isotopic Stages MIS4 to MIS2. Unit 1 formed during the Late Pleistocene, possibly in the MIS 5. The presence of *P. onkodes* encrusting coral fragments and the CCA assemblage indicate a shallow-water depositional setting, near to coral reefs. This unit was exposed after MIS 5 and affected by one phase of dissolution and meteoric cement filling (cement was radiocarbon dated

29,000 years BP). Marine benthic growth or marine sedimentation during MIS3 cannot be excluded, considering that the deep sinkhole is at 60 m water depth. Unit 2 represents the post-last glacial maximum carbonate deposition (i.e., the transgressive deposits formed during deglaciation). The base of the limestone accumulated in a low-energy marine setting characterized by a wackestone with rare CCA record. The top of Unit 2 in the deep sinkhole is a boundstone composed of encrusting CCA, bryozoans, foraminifers, corals, serpulids, and cemented gastropods, which resemble the assemblages in the modern rhodolith bed of the outer Abrolhos shelf. The bulk rock revealed a radiocarbon age of 8500–8100 cal years BP.

Bastos et al. (2018) investigated Late Holocene reef growth in the unique mushroom-shaped pinnacles in the Abrolhos Shelf. The pinnacles consist of a complex framework dominated by bryozoans (mainly *Celleporaria atlantica* (Busk 1884)), which account for up to 44% of the reef framework, while CCA and corals are subordinate components (28% and 23%, respectively) in the analyzed interval (last 2000 years BP). *Porolithon onkodes* (Foslie) is the main CCA at the tops and illuminated sides of the pinnacles, whereas *Sporolithon* sp. predominates in the shaded lower parts. Pinnacle accretion rates are similar to typical coralgal reefs (2.7 to 0.9 mm years^{-1}).

Calcareous coralline algae dominate in intertidal settings in Holocene shallow-water carbonate and mixed siliciclastic-carbonate deposits in Arraial do Cabo, Rio de Janeiro State, southeastern Brazil (Tâmega et al., 2016). The CCA assemblage includes four non-geniculate species, *Lithophyllum pustulatum* (J.V. Lamouroux) Foslie, *Spongites fruticulosus* Kützing, *Spongites yendoi* (Foslie) Y.M. Chamberlain, and *Mesophyllum engelharti* (Foslie) W.H. Adey and undetermined geniculate corallines. The identified species occur in present-day coralline assemblages of Brazil and represent the first fossil records of these species along the southwestern Atlantic Ocean coasts.

In the outcrops in Arraial do Cabo, Spotorno-Oliveira et al. (2016) provided a microfacies analysis of Holocene supratidal to the subtidal carbonate and mixed siliciclastic-carbonate deposits. The age/depth model is based on ^{14}C dating and revealed a basal age of ~7000 cal. Years BP. In the six distinguished microfacies, the biogenic components are dominated by vermetids and coralline algae. A number of subordinate components are also present, including thin-shelled bivalves. Fossil vermetids are represented by monospecific clusters of overgrowing *Petaloconchus varians* (d'Orbigny 1841). *L. pustulatum*, *S. fruticulosus*, *S. yendoi*, *M. engelharti*, and unidentified geniculate are the main corallines. In the upper subtidal zone of the bay, the substrate of the benthic community changed according to the relative sea level. The loose and coarse substrate formed during the rapid transgression changed into a fine and soft bottom during the slow decline to present sea level after the Holocene sea-level maximum. At ~7000 years ago, the coarse substrate was characterized by fruticose corallines and quartz grains. After the postglacial sea level, maximum vermetids and corallines, together with thin-shelled bivalves, thrived in low-sedimentation and high-nutrient level environments (Spotorno-Oliveira et al., 2016).

Reef frameworks and rhodolith internal structure are the focus of main studies on CCA in Holocene and modern carbonate deposits along the Brazilian margin. These studies are geographically distributed from Amapá to São Paulo (e.g., Leão & Ginsburg, 1997; Gherardi & Bosence, 2001; Tâmega et al., 2014; Bastos et al., 2018; Brasileiro et al., 2018; Vale et al., 2018).

The largest modern rhodolith beds are found in Australia (Harvey et al., 2016) and Brazil (Kempf, 1970; Foster, 2001; Amado-Filho et al., 2012a). In Brazil, rhodolith beds (Fig. 6.2a, b) extend over almost the entire continental shelf from Maranhão (0°50′S) to Santa Catarina state (27°17′S; Horta et al. 2016), covering a potential area of 230,000 km^2 (Carvalho et al., 2020). These beds are mostly found in relation with seagrass and close to coral reefs (Fig. 6.2c) in some areas (Leão, 2002), from nearshore down to 100 m depths (Henriques et al., 2012; Tâmega et al., 2013, 2014; Vieira et al., 2019). They are thriving from clean waters far from riverine influence and to inshore turbid waters (Dutra et al., 2006). However, these beds are not uniform since they differ in the rhodolith-forming CCA species, associated macroalgal assemblages, and density of nodules, which vary in size from millimeters to tens of centimeters (Kempf, 1970; Testa & Bosence, 1999; Amado-Filho et al., 2010, 2012a, b, 2016; Pereira-Filho et al., 2012; Pascelli et al., 2013; Brasileiro et al. 2015, 2018).

On the northeast Brazilian continental shelf, there are two types of CCA deposits: maërl (Fig. 6.2d) and rhodoliths, with dominance of maërl beds. Besides CCA, large deposits formed exclusively by the green alga *Halimeda* occur in quartz sand sediment, between 12 and 25 m water depths on the Paraíba and Ceará shelf, between Fortaleza and Icapuí (Kempf, 1970). In the inner shelf of NE Brazil, rhodoliths and branching coralline maërls grow in troughs of siliciclastic, large-scale submarine dunes and also form belts perpendicular to long (5–10 km) sand ribbons and to small-scale bedforms (Testa & Bosence, 1998, 1999).

The Amazon outer-shelf carbonate province, in the Brazilian equatorial margin, is a mesophotic ecosystem in the outer continental shelf from Amazonas to Maranhão states (Moura et al., 2016; Francini-Filho et al., 2018). In this region, the rhodolith beds and reefs are one of the main habitats due to their wide extent and provision of ecosystem services (Moura et al., 2016). This reef ecosystem, studied by Collette and Rutzler (1977) and described as "reef," is made up of sessile organisms (CCA, sponges, gorgonians, polychaete worms, vermetid molluscs, and bryozoans). Moura et al. (2016) confirmed the results achieved by Collette and Rutzler (1977) and mapped an extensive area of ~9500 km^2 of carbonate reef system from French Guiana to Maranhão. The relief of "reefs," however, was due to erosion of older Pleistocene sandstones during low sea levels, and the actual accretion potential of the living communities is very reduced (Vale et al., 2022a).

The structure and composition of rhodoliths and macroids with CCA in the Amazon River mouth vary along depth gradients (23–120 m) and their location relative to the river plume (Vale et al., 2018). In the Northern Zone of the Amazon mouth, the deepest rhodoliths, at 120 m water depth, consist of a thin algal/bryozoan/encrusting foraminiferal cover around relatively large nuclei formed by fragments of invertebrate boundstone or rudstone with ooids. Macroids predominantly

Fig. 6.2 (**a**) Rhodolith bed in the subtidal zone at a depth of 8–10 m off Armação dos Búzios, Rio de Janeiro State, southeastern Brazil. (**b**) Rhodolith from the bed illustrated in A. (**c**) Calcareous coralline algae at 4 m water depth in a fringing reef in the Abrolhos Marine National Park, northeastern Brazil. (**d**) Maërl bed at 10–12 m water depth in the Abrolhos Marine National Park. (**e** and **f**) Dropcam images of rhodolith beds, 70 m water depth, Espírito Santo continental shelf (Marine Protected Area Costa das Algas); bottom frame size is 60 × 60 cm

built up of bryozoans with subordinate CCA around small bioclastic nuclei extend on the seafloor at 95 m water depth. In the Central Zone, at 100 m water depth, the rhodoliths consist of thin CCA covers on nuclei of sandstone clasts. Boundstone and sandstone nuclei are fragments of sedimentary rocks that accumulated on the outer shelf during significantly lower sea level. In the Central Zone, at 50–55 m water depths, the benthic habitats are dominated by sub-spheroidal rhodoliths mainly made up of CCA. In the Southern Zone, far from the influence of the river plume, rhodoliths are common in shallower waters, at 20–30 m, and show a higher diversity of CCA as well as other components. Some rhodoliths are relatively recent in age (hundreds of years), whereas others grew during several thousand years in two or more phases with long algal growth lapses in between. This is probably due to high burial rates caused by sediment from the river plume, which changed in time. The temporal decrease of sediment influx made possible the exhumation after burial of rhodoliths that grew in several phases (Vale et al., 2018).

Gherardi and Bosence (2001) showed that the reef framework in the Atol das Rocas, the only atoll in the South Atlantic, is dominated by CCA (*Porolithon pachydermum* (Foslie) Foslie, *Lithophyllum* sp., *Sporolithon* sp., and *Lithoporella* sp.) and vermetids. The atoll framework was established around 3490 ± 45 years BP and vertical accretion rates range from 3.2 to 18.4 mm/years. The framework composition, although dominated by *P. pachydermum,* varies between leeward and windward sides of the atoll. This variation in reef framework assemblage was interpreted as an indication that hydrodynamic energy is possibly main controlling factor on the reef composition.

Rhodoliths and crusts of CCA thrive in large mesophotic areas, where seabed disturbance can be very low, episodic, or absent, leading, for instance, to nodule lateral growth, acquiring discoid shapes, or lateral and vertical coalescence of rhodoliths or any other rudstones, forming reef bioconstructions. Stabilized rhodoliths can grow and fuse together by encrusting CCA (Pereira-Filho et al., 2015; Amado-Filho et al., 2016). This process in rhodolith-dominated mesophotic areas is well documented along the Brazilian outer shelf. Rhodolith beds mainly occur from 25 to 55 m water depths on the Sergipe-Alagoas outer shelf in areas sheltered from sediment influx (Vale et al., 2022b). The optimal growth of CCA off Rio de Janeiro occurs at water depths between 60 and 90 m (Della Giustina, 2006). The highest rhodolith abundance on the Abrolhos shelf is observed at similar depths (Amado-Filho et al., 2012a), whereas on the Espírito Santo shelf, rhodoliths are dominant from 40 m water depth (Vieira et al., 2019; Holz et al., 2020). Considering this extensive dominance of rhodoliths along the Brazilian outer shelf (water depths >40–50 m), the formation of CCA-dominated carbonate crusts or ledges along the mid-outer shelf can be considered as an important boundstone-forming process.

On the southeastern Brazilian continental shelf, at 60–80 m water depth, between Abrolhos bank and Cabo de São Tomé, surveys using biological trawlers (REVIZEE Project) evidenced the existence of CCA bioconstructions on the sea floor. Dias and Jeveaux (1996) identified three different CCA formations: (1) thin laminar crusts, interspersed with siliciclastic mud (São Tomé Cape), (2) thick laminar crusts, and (3) nodules made up by *Titanoderma* sp., found at the edge of the platform, off Itapemirim, Espírito Santo State. The thick laminar crusts are similar to deep-water algal buildups found in the NE Australian platform between 80 and 120 m water depth, which range in thickness from centimeters to several meters, and formed entirely during the Holocene transgression (Davies et al., 2004). These buildups consist of a CCA (mainly *Mesophyllum* sp.) boundstone and subordinate bryozoans, encrusting foraminifers and serpulids, partially filled by mudstone to packstone sediments.

Extensive rhodolith beds spread along the Abrolhos continental shelf (~26,000 km^2) (Amado-Filho et al., 2012a; Moura et al., 2013) and the Espírito Santo shelf (Vieira et al., 2019). A total of 13 species of rhodolith-forming CCA were described for the Abrolhos and Espírito Santo shelf (Brasileiro et al., 2018). Rhodolith coverage can reach more than 200 nodules per m^2 in Abrolhos (Amado-Filho et al., 2012a; Leal et al., 2012) and range from 40 to 120 nodules per m^2 on the Espírito Santo shelf (Fig. 6.2e, f). Rhodoliths in South Espírito Santo State, extending from 10 to 60 m water depth, are built mainly up of hapalidialean corallines, with subordinate bryozoans and encrusting foraminifers. They are highly porous and contain a relatively high amount of siliciclastics (up to 8%) in constructional voids and borings. Radiocarbon ages indicate that rhodoliths are relatively young (less than 700 years BP; Brasileiro et al., 2018). In a different environmental context, on a 200 km wide continental shelf, far from the shoreline and riverine influence, rhodoliths from the Abrolhos Continental Shelf, (20–75 m water depth), are mainly made up of a structureless carbonate mass produced by multiphase boring and boring infilling in the original CCA boundstone. The massive carbonate is encrusted by a veneer of living CCA. The siliciclastic proportions are low. Radiocarbon dating of the innermost parts of deeper rhodoliths yielded ages of ~7000 years BP (75 m) and ~ 2000 years BP (65 m water depth) (Brasileiro et al., 2018). The siliciclastic sediment influx favors burial of rhodoliths, which have relatively young ages in South Espírito Santo State. By contrast, the low sediment input in the rhodolith beds in the Abrolhos Continental Shelf led to long residence times of rhodoliths on the seafloor. The algal nodules underwent a pervasive destruction of the original CCA/invertebrate framework by bioerosion. The same processes were observed by Franco (2018) in rhodoliths from the mesophotic habitat (60–70 m deep) of the Jaseur Seamount, along the Vitória Trindade Seamount Chain. Rhodoliths are characterized by different phases of borings, filling, cementation, and new perforations replacing the original inner structure of the nodules. Six CCA genera were described in the rhodoliths in the Jaseur Seamount: *Sporolithon*, *Titanoderma*, *Hydrolithon*, *Harveylithon*, *Mesophyllum*, and *Lithothamnion,* in addition to the aragonitic alga *Peyssonnelia* (Franco, 2018).

In a rhodolith bed at 13–15 m water depth in the Espírito Santo continental shelf, Basso et al. (2012) described the occurrence of a *coralligène de plateau*, produced by concretions of irregular dome-shaped CCA, such as *Hydrolithon* sp., *Lithophyllum* cf. *corallinae* (P. Crouan and H. Crouan) Heydrich, *Lithothamnion muelleri* Lenormand ex Rosanoff, *Mesophylum* sp., and *Titanoderma pustulatum* (J.V. Lamouroux) Nägeli. The radiocarbon age of 2504 ± 44 years BP of the concretions indicates a CCA vertical growth of about 20 μm/year, alternating with episodical burial and partial erosion. Carbonate concretions or hardground ledges with a main framework of CCA (*Lithophyllum* sp., *Peyssonnelia* sp.) and bryozoans (*Stylopoma hastata*) occur in the central-northern outer Espírito Santo shelf (Marine Protected Area Costa das Algas) (Holz et al., 2020). In this shelf, Bastos et al. (2015) also indicate the presence of CCA carbonate concretions along the margin of shelf-incised valleys or palaeochannels. High-relief reefs formed by CCA, bryozoans, ascidians, and octocorals also occur in deeper water (80–120 m) (Holz et al., 2020).

Della Giustina (2006) studied the sedimentology and morphology of the middle and outer continental shelf of the Campos basin, southeastern Brazil, at 65–300 m water depth. In this area, deposits mainly consist of carbonate with subordinate mixed siliciclastic–carbonate. Six sedimentary facies were identified: (1) siliciclastic sand with bioclasts, (2) grainstones with rhodoliths and siliciclastic sands, (3) boundstone ridges, (4) boundstone mounds, (5) boundstone banks with CCA and grainstones, and (6) mixed facies. The highest abundance of CCA occurs from 45 to 150 m water depth. The occurrence of boundstones from 170 to 230 m indicates that during lower sea level (about −120 m) in the last glacial maximum, the carbonate factory was active in a paleodepth range of 50–110 m.

Tâmega et al. (2014) reported a large occurrence of living and fossil rhodolith beds on the central Brazilian continental shelf within the Peregrino oil field, Campos basin. A deep rhodolith bed covers about 15 km^2 of a 40 km^2 area sampled at 100–106 m water depth. Coralline algae are the dominant components in the living rhodolith surfaces associated with subordinate bryozoans, cnidarians, brachiopods, and porifers. In the inner parts of some CCA nodules, encrusting acervulinid foraminifera are the main components. There is a difference of ca. 4700 years in radiocarbon ages between the living outer rhodolith and 3 mm inward. This suggests that a proportion of rhodoliths had been recolonized after long periods of burial and/or erosion.

6.4 Final Remarks

Fossil CCA are recorded in Brazilian onshore basins since the middle Albian (Maury, 1936). The opening of the South Atlantic favored the expansion of shallow marine areas and CCA colonization from the western Tethys (Arai, 2014). However, no detailed descriptions of algae have been published. CCA are rock-forming components in carbonate formations from the Upper Cretaceous to the Neogene in offshore basins along the Brazilian shelf (Figueiredo et al., 2007), indicating that they thrived in areas sheltered from high terrigenous influence.

Crustose coralline algae are widespread and abundant in post-glacial deposits and modern seafloor along the Brazilian shelf. Boundstones of CCA crusts mainly occur on the outer shelf/shelf edge, frequently growing on rocky substrates subaerially exposed during the last glacial maximum (Dias & Jeveaux, 1996; Della Giustina, 2006; Bastos et al., 2016; Moura et al., 2016). CCA are most common forming transgressive and highstand rhodolith bed deposits along the Brazilian inner to outer shelf (Kempf, 1970; Amado-Filho et al., 2012a, b; Tâmega et al., 2014; Vieira et al., 2019). These beds show a high variability in density of nodules, siliciclastic input influence, depth, and distance to the shoreline (Brasileiro et al., 2018; Vale et al., 2018). Accordingly, rhodoliths display a high variability in size, shape, internal structure, and components, as well as in growth rates and continuity (Brasileiro et al., 2018). In this regard, dated examples range from relatively continuous growth for tens to several hundred years to multiphase rhodoliths, which started to grow in the mid Holocene and continue having a living CCA cover today, after one of several periods of arrested growth, probably due to temporal burial (Tâmega et al., 2014; Brasileiro et al., 2018; Vale et al., 2018).

The main environmental drivers for rhodolith beds and CCA are, basically, water temperature, irradiance, nutrients, and current velocity (e.g., Carvalho et al., 2020). However, the high abundance of CCA in the Brazilian shelf is also probably related to the widespread mesotrophic conditions, enhanced by upwelling during spring/summer, mainly along the southeast coast (Ghisolfi et al., 2015; Palóczy et al., 2016). The CCA abundance is also favored by the relatively low levels of fine-grained riverine input in large areas of the mid/outer shelf, mainly along the east-northeast coast (Vieira et al., 2019). In specific cases, such as the Abrolhos shelf, shelf width and distance to the coast are also important drivers.

In the last years, new molecular and morpho-anatomical studies suggest that living CCA have higher diversity and higher levels of endemicity than the ones previously reported based on morpho-anatomical taxonomy (Bahia, 2014; Sissini et al., 2014; Costa et al., 2014, 2019; Jesionek et al., 2016; Tâmega et al., 2016). Future studies are needed to better understand the fossil CCA occurrences along the Brazilian coast and their taxonomic and palaeoecological implications.

References

Aguirre, J., & Braga, J. C. (2005). The citation of nongeniculate fossil coralline red algal species in the twentieth century literature: An analysis with implications. *Revista Española de Micropaleontología, 37*, 57–62.

Aguirre, J., Riding, R., & Braga, J. C. (2000). Diversity of coralline red algae: Origination and extinction patterns from the early cretaceous to the Pleistocene. *Paleobiology, 26*, 651–667.

Amado-Filho, G. M., Maneveldt, G. W., Pereira-Filho, G. H., Manso, R. C. C., Bahia, R. G., Barros-Barreto, M. B., & Guimarães, S. M. P. B. (2010). Seaweed diversity associated with a Brazilian tropical rhodolith bed. *Ciencias Marinas, 36*, 371–391.

Amado-Filho, G. M., Moura, R. L., Bastos, A. C., Salgado, L. T., Sumida, P. Y., Guth, A. Z., Francini-Filho, R. B., Pereira-Filho, G. H., Abrantes, D. P., Brasileiro, P. S., Bahia, R. G., Leal, R. N., Kaufman, L., Kleypas, J. A., Farina, M., & Thompson, F. L. (2012a). Rhodolith beds are major CaCO3 bio-factories in the tropical south west Atlantic. *PLoS One, 7*(4), e35171. https://doi.org/10.1371/journal.pone.0035171

Amado-Filho, G. M., Pereira-Filho, G. H., Bahia, R. G., Abrantes, D. P., Veras, C. P., & Matheus, Z. (2012b). Occurrence and distribution of rhodolith beds on the Fernando de Noronha archipelago of Brazil. *Aquatic Botany, 101*, 41–45.

Amado-Filho, G. M., Moura, R. L., Bastos, A. C., Francini-Filho, R. B., Pereira-Filho, G. H., Bahia, R. G., Moraes, F. C., & Motta, F. S. (2016). Mesophotic ecosystems of the unique South Atlantic atoll are composed by rhodolith beds and scattered consolidated reefs. *Marine Biodiversity, 46*, 933–936. https://doi.org/10.1007/s12526-015-0441-6

Arai, M. (2014). Aptian/Albian (early cretaceous) paleogeography of the South Atlantic: A paleontological perspective. *Brazilian Journal of Geology, 44*, 339–350.

Arai, M., Botelho Neto, J., Cunha Lana, C., & Pedrão, E. (2000). Cretaceous dinoflagellate provincialism in Brazilian marginal basins. *Cretaceous Research, 21*, 351–366.

Assin, M. L., Quaglio, F., Warren, L. V., & Simões, M. G. (2016). Comments on paper by M. Arai "Aptian/Albian (early cretaceous) paleogeography of the South Atlantic: A paleontological perspective". *Brazilian Journal of Geology, 46*(1), 3. https://doi.org/10.1590/2317-4889201620150046A

Bahia, R. G. (2014). *Algas coralináceas formadoras de rodolitos da plataforma continental tropical e ilhas oceânicas do Brasil: levantamento florístico e taxonomia.* Ph.D. thesis. Escola Nacional de Botânica Tropical, Rio de Janeiro.

Basso, D., Tavares de Macedo, D. G., & Henriques, M. C. (2012). The occurrence of coralligène de plateau in Brazilian coastal waters. In *Abstracts of the 4th international Rhodolith workshop*. Universidad de Granada.

Bastos, A. C., Moura, R. M., Amado-Filho, G. M., Dagostini, D. P., Secchin, N. A., Francini-Filho, R. B., Güth, A. Z., Sumida, P. Y. G., Mahiques, M., & Thompson, F. L. (2013). Buracas: Novel and unusual sinkhole-like features in the Abrolhos Bank. *Continental Shelf Research, 70*, 118–125. https://doi.org/10.1016/j.csr.2013.04.035i

Bastos, A. C., Quaresma, V. S., Marangoni, M. B., D'Agostini, D. P., Bourguignon, S. N., Cetto, P. H., Silva, A. E., Filho, G. M. A., Moura, R. L., & Collins, M. (2015). Shelf morphology as an indicator of sedimentary regimes: A synthesis from a mixed siliciclasticecarbonate shelf on the eastern Brazilian margin. *Journal of South American Earth Sciences, 63*, 125–136.

Bastos, A. C., Amado-Filho, G. M., Moura, R. M., Sampaio, F. M., Bassi, D., & Braga, J. C. (2016). Origin and sedimentary evolution of sinkholes (buracas) in the Abrolhos continental shelf, Brazil. *Palaeogeography Palaeoclimatology Palaeoecology, 462*, 101–111.

Bastos, A. C., Moura, R. L., Moraes, F. C., Vieira, L. S., Braga, J. C., Ramalho, L. V., Amado-Filho, G. M., Magdalena, U. R., & Webster, J. M. (2018). Bryozoans are major modern builders of South Atlantic oddly shaped reefs. *Scientific Reports, 8*, 1–11. https://doi.org/10.1038/s41598-018-27961-6

BouDagher-Fadel, M. K., Price, G. D., & Koutsoukos, E. A. M. (2010). Oligocene–Miocene carbonate succession in Campos basin, southeastern Brazil. *Stratigraphy, 7*, 283–299.

Braga, J. C. (2017). Neogene rhodoliths in the Mediterranean basins. In R. Riosmena-Rodríguez, W. Nelson, & J. Aguirre (Eds.), *Rhodolith/maërl beds: A global perspective* (pp. 169–193). Springer, Series, Coastal Research Library. https://doi.org/10.1007/978-3-319-29315-8_7

Brasileiro, P. S., Pereira-Filho, G. H., Bahia, R. G., Abrantes, D. P., Guimarães, S. M. P. B., Moura, R. L., Francini-Filho, R. B., Bastos, A. C., & Amado-Filho, G. M. (2015). Macroalgal composition and community structure of the largest rhodolith beds in the world. *Marine Biodiversity, 406*, 407–420.

Brasileiro, P. S., Braga, J. C., Amado-Filho, G. M., Leal, R. N., Bassi, D., Franco, T., Bastos, A. C., & Moura, R. L. (2018). Burial rate determines Holocene rhodolith development on the Brazilian shelf. *PALAIOS, 33*, 464–477.

Bromley, R. G., & D'Alessandro, A. (1989). Ichnological study of shallow marine endolithic sponges from the Italian coast. *Rivista Italiana di Paleontologia e Stratigrafia, 95*, 279–314.

Carelli, T. G., Plantz, J. B., Mello, C. L., & Borghi, L. (2019). Fácies e paleoambientes do membro São Tomé (Formação Emborê) na porção emersa da bacia de Campos. *Anu do Inst de Geociênc, 42*, 454–470.

Carvalho, V. F., Assis, J., Serrão, E. A., Nunes, J. M., Batista, A. A., Batista, M. B., Barufi, J. B., Silva, J., Pereira, S. M. B., & Horta, P. A. (2020). Environmental drivers of rhodolith beds and epiphytes community along the South Western Atlantic coast. *Marine Environmental Research, 154*, 104827. https://doi.org/10.1016/j.marenvres.2019.104827

Collette, B. B., & Rutzler, K. (1977). Reef fishes over sponge bottoms of the mouth of Amazon River. In *Abstracts of the 3rd international coral reef symposium*.

Costa, I. O., Horta, P. A., Bergstrom, E. R., & Nunes, J. M. C. (2014). Taxonomic study of crustose coralline algae off the northeastern Brazilian coast. *Phytotaxa, 190*, 130–161.

Costa, I. A., de Jesus, P. B., de Jesus, T. S., Souza, P. S., Horta, P. A., & Nunes, J. M. C. (2019). Reef-building coralline algae from the Southwest Atlantic: Filling gaps with the recognition of *Harveylithon* (Corallinaceae, Rhodophyta) on the Brazilian coast. *Journal of Phycology, 55*, 1370–1385.

Cruz, A. M., Reis, A. T., Suc, J. P., Silva, C. G., Praeg, D., Granjeon, D., Rabineau, M., Popescu, S. M., & Gorini, C. (2019). Neogene evolution and demise of the Amapá carbonate platform, Amazon continental margin, Brazil. *Marine and Petroleum Geology, 105*, 185–203. https://doi.org/10.1016/j.marpetgeo.2019.04.009

Davies, P. J., Braga, J. C., Lund, M., & Webster, J. M. (2004). Holocene deep water algal buildups on the eastern Australian shelf. *PALAIOS, 19*, 598–609.

De Castro, J. C., Viviers, M. C., & Regali, M. S. P. (1988). Stratigraphic analysis of the marine cretaceous in the eastern margin of Potiguar basin. *Revista Brasileira de Geociencias, 18*, 231–236.

Della Giustina, I. D. (2006). *Sedimentação carbonática de algas vermelhas coralináceas da plataforma continental da bacia de Campos: um modelo carbonático análogo para o Terciário*. Dissertation. Universidade Federal do Rio Grande do Sul.

Dias, G. T. M., & Jeveaux, J. L. (1996). Tipos de fundo da plataforma continental sudeste revelados em levantamentos por sonar de varredura lateral e dragagens durante a expedição REVIZEE central I 2ª etapa. In: Abstracts of the 39° Congresso Brasileiro de Geologia, Salvador, 1–6 September 1996.

Dutra, L. X. C., Kikuchi, R. K. P., & Leão, Z. M. A. N. (2006). Effects of sediment accumulation on reef corals from Abrolhos, Bahia, Brazil. *Journal of Coastal Research, 39*, 633–638.

Ellis, J., & Solander, D. (1786). *The natural history of many curious and uncommon zoophytes, collected from various parts of the globe*. Benjamin White and Son.

Figueiredo, J. J. P., Zalán, P. V., & Soares, E. F. (2007). Bacia da Foz do Amazonas. *Bol geociênc Petrobras, 15*, 299–309.

Foster, M. S. (2001). Rhodoliths: Between rocks and soft places. *Journal of Phycology, 37*, 659–667.

Francini-Filho, R. B., Asp, N. E., Siegle, E., Hocevar, J., Lowyck, K., D'Avila, N., Vasconcelos, A. A., Baitelo, R., Rezende, C. E., Omachi, C. Y., Thompson, C. C., & Thompson, F. L. (2018). Perspectives on the great Amazon reef: Extension, biodiversity, and threats. *Frontiers in Marine Science, 5*, 142. https://doi.org/10.3389/fmars.2018.00142

Franco, T. (2018). *Análise da composição e estrutura interna de rodolitos da Cadeia Vitória-Trindade*. Ph.D. thesis. Universidade Federal do Espirito Santo.

Früh, J. J. (1890). Zur Kenntniss der gesteinbildenden Algen der Schweizer-Alpen: mit besonderer Berücksichtigung des Säntisgebietes. *Abh Schweiz Paläontol Ges, 17*, 1–32.

Gherardi, D. F. M., & Bosence, D. W. J. (2001). Composition and community structure of the coralline algal reefs from Atol das Rocas, South Atlantic, Brazil. *Coral Reefs, 19*, 205–219. https://doi.org/10.1007/s003380000100

Ghidella ME, Lawver LA, Gahagan LM (2007) Break-up of Gondwana and opening of the South Atlantic: Review if existing plate tectonic models. *US Geological Survey and the National Academies: USGS OF-2007-1047, Short Research Paper*, 55 https://doi.org/10.3133/of2007-1047.srp055.

Ghisolf, R. D., Pereira da Silva, M., Thomaz dos Santos, F., Servino, R. N., Cirano, M., & Thompson, F. L. (2015). Physical forcing mechanisms controlling the variability of chlorophyll-a over the Royal-Charlotte and Abrolhos Banks–Eastern Brazilian Shelf. *PLoS One, 10*(2), e0117082. https://doi.org/10.1371/journal.pone.0117082

Granier, B. (2014). Algas calcárias marinhas bentônicas no Cretáceo do Brasil. In: D. Dias-Briton & P. Tibana (Eds.), *Atlas dos calcários do Cretáceo do Brasil: um Atlas* (pp. 519–567), Obra 1 Publisher: UNESP-IGCE-UNESPetro, Rio Claro.

Granier, B., & Dias-Brito, D. (2016). On the fossil alga *Elianella elegans* Pfender and Basse, 1948, and its so-called lookalikes, with description of *Elianella brasiliana* n.sp. revision of the Juliette Pfender collection. Part 1. *Carnets de Géologie, 16*, 213–229.

Granier, B., Berthou, P. Y., & Poignant, A. F. (1991). Constructions bio-sédimentaires laminées, *Lithothamnium* et Parachaetetes de la Formation Riachuelo (Albien) du bassin de Sergipe (Nord–Est du Brésil). *Geociências, 10*, 169–181.

Granier, B., Dias-Brito, D., & Bucur, I. (2008). Calcareous algae from upper Albian-Cenomanian strata of the Potiguar basin (NE Brazil). *Geologia Croatica, 61*, 311–320.

Granot, R., & Dyment, J. (2015). The cretaceous opening of the South Atlantic Ocean. *Earth and Planetary Science Letters, 414*, 156–163.

Gümbel, C. W. (1861). *Geognostische Beschreibung des bayerischen Alpengebirges und seines Vorlandes*. Perthes, Munich.

Gümbel, C. W. (1871). Die sogenennten Nulliporen (*Lithothamnium* und *Dactylopora*) und ihre Betheiligung an der Zusammensetzung der Kalkgesteine. Erster Theil: Die Nulliporen des Pflanzenreichs (*Lithothamnium*). *ABAW, 11*, 1–42.

Harvey, A. S., Harvey, R. M., & Merton, E. (2016). The distribution, significance and vulnerability of Australian rhodolith beds: A review. *Marine and Freshwater Research, 68*, 411–428. https://doi.org/10.1071/MF15434

Henriques, M. C. M. O., Villas Boas, A., & Riosmena Rodrigues, R. (2012). New records of rhodolith–forming species (Corallinales, Rhodophyta) from deep water in Espírito Santo state, Brazil. *Helgoland Marine Research, 66*, 219–231.

Holz, V. L., Bahia, R. G., Karez, C. S., Vieira, F. V., Moraes, F. C., Vale, N. F., Sudatti, D. B., Salgado, L. T., Moura, R. L., Amado-Filho, G. M., & Bastos, A. C. (2020). Structure of rhodolith beds and surrounding habitats at the Doce river shelf (Brazil). *Diversity, 12*, 75. https://doi.org/10.3390/d12020075

Horta, P. A., Riul, P., Amado-Filho, G. M., Gurgel, C. F. D., Berchez, F., Nunes, J. M. C., Scherner, F., Pereira, S., Lotufo, T., Peres, L., Sissini, M., Bastos, E. O., Rosa, J., Munoz, P., Matins, C., Gouvêa, L., Carvalho, V., Bergstrom, E., Schubert, N., Bahia, R. G., Rodrigues, A. C., Rörig, L., Barufi, J. B., & Figueiredo, M. (2016). Rhodoliths in Brazil: Current knowledge and potential impacts of climate change. *Brazilian Journal of Oceanography, 64*, 117–136. https://doi.org/10.1590/S1679-875920160870064sp2

Jesionek, M. B., Bahia, R. G., Hernández-Kantún, J. J., Adey, W. H., Yoneshigue-Valentin, Y., Longo, L. L., & Amado-Filho, G. M. (2016). A taxonomic account of non-geniculate coralline algae (Corallinophycidae, Rhodophyta) from shallow reefs of the Abrolhos Bank, Brazil. *Algae, 31*, 317–340.

Kempf, M. (1970). Notes on the bentic bionomy of the N–NE Brazilian shelf. *Marine Biology, 5*, 213–224.

Kikuchi, R. K. P., & Leão, Z. M. A. N. (1998). The effects of Holocene sea level fluctuation on reef development and coral community structure, northern Bahia, Brazil. *Anais da Academia Brasileira de Ciências, 70*, 159–171.

Lamarck, J. B. (1801). *Systême des animaux sans vertebres, ou tableau général des classes, des ordres et des genres de ces animaux*. Deterville.

Larson, R. L., & Ladd, J. W. (1973). Evidence for the opening of the South Atlantic in the early cretaceous. *Nature, 246*, 209–212.

Leal, R. N., Bassi, D., Posenato, R., & Amado-Filho, G. M. (2012). Tomographic analysis for bioerosion signatures in shallow-water rhodoliths from the Abrolhos Bank, Brazil. *Journal of Coastal Research, 28*, 306–309.

Leão, Z. M. A. N. (2002). Abrolhos, BA. O complexo recifal mais extenso do Atlântico Sul. In: C. Schobbenhaus, D. A. Campos, E. T. Queiroz, M. Winge, & M. L. C. Berbert–Born (Eds.), *Sítios geológicos e paleontológicos do Brasil* (pp. 345–359). 1° edição: DNPM/CPRM—Comissão Brasileira de Sítios Geológicos e Paleobiológicos (SIGEP), Brasflia.

Leão, Z. M. A. N., & Ginsburg, R. N. (1997). Living reefs surrounded by siliciclastics sediments: The Abrolhos coastal reefs, Bahia, Brazil. In H. A. Lessios, & I. G. Mcintre (Eds.), *Proc 8th Int Coral Reef Sym*, vol. 2 (pp. 1767–1772).

Linnaeus, C. (1758). *Systema Naturae*. Laurentii Salvi, Holmiae.

Maury, C. J. (1929). Novas coleções paleontológicas do serviço geológico do Brasil. *SGMB, 33*, 6–11.

Maury, C. J. (1936). O Cretáceo de Sergipe. *SGMB, 11*, 1–283.

Michelin, H. (1840–1847). Iconographie zoologique, description par localités et terrains des polypiers fossiles de France et pays environnants. Bull Soc geol Fr, Paris, p. 346, 79 pl.

Moura, R. L., Secchin, N. A., Amado-Filho, G. M., Francini Filho, R. B., Freitas, M. O., Minte-Vera, C. V., Teixeira, J. B., Thompson, F. L., Dutra, G. F., Sumida, P. Y. G., Guth, A. Z., Lopes, R. M., & Bastos, A. C. (2013). Spatial patterns of benthic megahabitats and conservation planning in the Abrolhos Bank. *Continental Shelf Research, 70*, 109–117.

Moura, R. M., Amado-Filho, G. M., Moraes, F. C., Brasileiro, P. S., Salomon, P. S., Mahiques, M. M., Bastos, A. C., Almeida, M. G., Silva-Jr, J. M., Araujo, B. F., Brito, F. P., Rangel, T. P., Oliveira, B. C. V., Bahia, R. G., Paranhos, R. P., Dias, R. J. S., Siegle, E., Figueiredo-Jr, A. G., Pereira, R. C., Leal, C. V., Hajdu, E., Asp, N. E., Gregoracci, G. B., Neumann-Leitão, S., Yager, P. L., Francini-Filho, R. B., Fróes, A., Campeão, M., Silva, B. S., Moreira, A. P. B., Oliveira, L., Soares, A. C., Araujo, L., Oliveira, N. L., Teixeira, J. B., Valle, R. A. B., Thompson, C. C., Rezende, C. E., & Thompson, F. L. (2016). An extensive reef system at the Amazon River mouth. *Science Advances, 2*, 1–11.

Palóczy, A., Brink, K. H., da Silveira, I. C. A., Arruda, W. Z., & Martins, R. P. (2016). Pathways and mechanisms of offshore water intrusions on the Espírito Santo basin shelf (18°S–22°S, Brazil). *Journal of Geophysical Research, Oceans, 121*, 5134–5163.

Pascelli, C., Riul, P., Riosmena-Rodriguez, R., Scherner, F., Nunes, M., Hall-Spencer, J. M., Oliveira, E. C., & Horta, P. (2013). Seasonal and depth-driven changes in rhodolith bed structure and associated macroalgae off Arvoredo Island (southeastern Brazil). *Aquatic Botany, 111*, 62–65.

Pereira-Filho, G. H., Amado-Filho, G. M., Moura, R. L., Bastos, A. C., Guimarães, S. M. P. B., Salgado, L. T., Francini-Filho, R. B., Bahia, R. G., Abrantes, D. P., Guth, A. Z., & Brasileiro, P. S. (2012). Extensive rhodolith beds cover the summits of southwestern Altantic Ocean. *Journal of Coastal Research, 28*, 261–269.

Pereira-Filho, G. H., Francini-Filho, R. B., Pierozzi-Jr, I., Pinheiro, H. T., Bastos, A. C., Moura, R. L., Moraes, F. C., Matheus, Z., Bahia, R. G., & Amado-Filho, G. M. (2015). Sponges and fish facilitate succession from rhodolith beds to reefs. *Bulletin of Marine Science, 91*, 45–46.

Philippi, R. A. (1837). Beweis, dafs die Nulliporen Pflanzen sind. *Arch Naturgesch, 3*, 387–393.

Rothpletz, H. (1891). Fossile kalkagen aus den familien der Codiaceen und der Corallineen. *Zeitschrift der Deutschen Geologischen Gesellschaft, 43*, 295–322.

Schaller, H., Vasconcelos, D. N., & Castro, J. C. (1971). Estratigrafia preliminar da Bacia Sedimentar da foz do rio Amazonas. In: 25 Congresso Brasileiro de Geologia, Sociedade Brasileira Geologia, São Paulo, 1–4 Setembro 1971.

Sissini, M. N., Oliveira, M. C., Gabrielson, P. W., Robinson, N. R., Okolodkov, Y. B., Riosmena-Rodríguez, R., & Horta, P. A. (2014). *Mesophyllum erubescens* (Corallinales, Rhodophyta)—So many species in one epithet. *Phytotaxa, 190*, 299–319.

Sousa, S. H. M., Rossetti, T. R., Fairchild, T. R., Burone, L., Mahiques, M. M., & Tibana, P. (2009). Microfacies and sequence stratigraphy of the Amapá formation, late Paleocene to early Eocene, Foz do Amazonas basin, Brazil. *Palaeogeography Palaeoclimatology Palaeoecology, 280*, 440–455.

Spotorno-Oliveira, P., Tâmega, F. T. S., Oliveira, C. A., Castro, J. W. A., Coutinho, R., Iryu, Y., & Bassi, D. (2016). Effects of Holocene sea level changes on subtidal palaeoecosystems, southeastern Brazil. *Marine Geology, 381*, 17–28. https://doi.org/10.1016/j.margeo.2016.08.007

Srivastava, N. K. (1982). Calcareous algae from Jandaira formation (Potiguar basin), Brazil. *Anais da Academia Brasileira de Ciências, 54*, 219–231.

Tâmega, F. T. S., Spotorno-Oliveira, P., & Figueiredo, M. A. O. (2013). *Catalogue of the benthic marine life from Peregrino oil field, Campos basin, Brazil*. Instituto Biodiversidade Marinha.

Tâmega, F. T. S., Bassi, D., Figueiredo, M. A. O., & Cherkinsky, A. (2014). Deep-water rhodolith bed from central Brazilian continental shelf, Campos basin: Coralline algal and faunal taxonomic composition. *Galaxea, 16*, 21–31.

Tâmega, F. T. S., Spotorno-Oliveira, P., Coutinho, R., & Bassi, D. (2016). Taxonomic assessment of fossil Holocene coralline red algae (Rhodophyta, Corallinales, Hapalidiales) from southwestern Atlantic. *Phytotaxa, 254*, 237–250. https://doi.org/10.11646/phytotaxa.245.4.1

Testa, V., & Bosence, D. W. J. (1998). Carbonate-siliciclastic sedimentation on a high-energy, ocean-facing, tropical ramp, NE Brazil. In V. P. Wright, & Z. P. Burchette (Eds.), *Carbonate Ramps*, vol. 149 (pp. 55–71). Geological Society, London, Special Publications.

Testa, V., & Bosence, D. W. J. (1999). Physical and biological controls on the formation of carbonate and siliciclastic bed forms on the northeast Brazilian shelf. *Sedimentology, 46*, 279–301.

Vale, N. F., Amado-Filho, G. M., Braga, J. C., Brasileiro, P. S., Karez, C. S., Moraes, F. C., Bahia, R. G., Bastos, A. C., & Moura, R. L. (2018). Structure and composition of rhodoliths from the Amazon river mouth, Brazil. *Journal of South American Earth Sciences, 84*, 149–159. https://doi.org/10.1016/j.jsames.2018.03.01484

Vale, N. F., Braga, J. C., Moura, R. L., Salgado, L. T., Moraes, F. C., Karez, C., Carvalho, R. T., Salomon, P. S., Menandro, P. S., Amado-Filho, G. M., & Bastos, A. C. (2022a). Distribution, morphology and composition of mesophotic 'reefs' on the Amazon continental margin. *Marine Geology, 447*, 106779. https://doi.org/10.1016/j.margeo.2022.106779

Vale, N. F., Braga, J. C., Bastos, A. C., Moraes, F. C., Karez, C. S., Bahia, R. G., Leão, L. A., Pereira, R. C., Amado-Filho, G. M., & Salgado, L. T. (2022b). Structure and composition of rhodolith beds from the Sergipe-Alagoas Basin (NE Brazil, southwestern Atlantic). *Diversity, 14*, 282. https://doi.org/10.3390/d14040282

Vieira, F., Bastos, A. C., Quaresma, V. S., Leite, M. D., Costa, A., Jr., Oliveira, K. S. S., Dalvi, C., Bahia, R. G., Holz, V. L., Moura, R. L., & Amado-Filho, G. M. (2019). Along-shelf changes in mixed carbonate-siliciclastic sedimentation patterns. *Continental Shelf Research, 187*, 1039642. https://doi.org/10.1016/j.csr.2019.103964

Waisman, G. (2002). *Análise estratigráfica de seqüências terciárias com ênfase nos carbonatos do Membro Siri. Região centro-sul da Bacia de Campos*. Dissertation. Universidade Federal do Rio Grande do Sul.

Woelkerling, W. J. (1998). Type collections of non-geniculate corallines housed at the Laboratoire de Cryptogamie (PC). In W. J. Woelkerling & D. Lamy (Eds.), *Non-geniculate coralline red algae and the Paris muséum: Systematics and scientific history* (pp. 279–404). MNHN.

Woelkerling, W. J., Granier, B., & Dias-Brito, D. (2014). *Heydrichia* (?) *poignantii*, sp. nov. (Sporolithaceae, Sporolithales, Rhodophyta), a 100 million year old fossil coralline red alga from North-Eastern Brazil, and a new Hauterivian record of *Sporolithon* from Switzerland. *Carnets de Geologie, 14*, 139–158. https://doi.org/10.4267/2042/53736

Wolff, B., & Carozzi, A.V. (1984). *Microfacies, depositional environments, and diagenesis of the Amapá carbonates (Paleocene–Middle Miocene), Foz do Amazonas basin, offshore NE Brazil*. Petrobrás, Rio de Janeiro.

Chapter 7
Rhodolith Beds Mapping Along Brazilian Continental Shelf

Vanessa Freire de Carvalho, Lidiane P. Gouvêa, Eduardo Bastos, Marina Nasri Sissini, and Paulo A. Horta

Abstract Rhodolith beds play a crucial role in enhancing local biodiversity by providing habitat for a wide variety of marine species. Despite ecosystem services provided, these species have been lost due to disturbances associated with ongoing environmental changes. Their calcification and slow growth of these organisms make them more vulnerable to increasing ocean acidification and warming. The predictive modelling is one technique to find the most suitable places to search these habitats by selecting environmental variables that are known to influence the rhodoliths physiology, associated with occurrence and abundance data. The settlement, distribution and abundance of rhodolith beds throughout the world have been associated with environmental factors such as temperature, light, nutrients and currents.

V. F. de Carvalho
Bióloga, Departamento de Pesquisa e Conservação de Fauna (MAPCF), Prefeitura Municipal de Curitiba, Florianópolis, SC, Brazil

Programa de Pós-Graduação em Ecologia, Universidade Federal de Santa Catarina, Florianópolis, SC, Brazil
e-mail: vancarvalho@curitiba.pr.gov.br

L. P. Gouvea
Centre of Marine Sciences, University of Algarve, Faro, Portugal

E. Bastos
Laboratório de Ficologia, Departamento de Botânica, Universidade Federal de Santa Catarina, Florianópolis, SC, Brazil

Programa de Pós-Graduação em Fungos, Algas e Plantas, Universidade Federal de Santa Catarina, Florianópolis, SC, Brazil

M. N. Sissini
Programa de Pós-Graduação em Ecologia, Universidade Federal de Santa Catarina, Florianópolis, SC, Brazil

P. A. Horta (✉)
Laboratório de Ficologia, Departamento de Botânica, Universidade Federal de Santa Catarina, Florianópolis, SC, Brasil

P. A. Horta, M. N. Sissini (eds.), *Brazilian Rhodolith Beds*, Brazilian Marine Biodiversity, https://doi.org/10.1007/978-3-031-61449-1_7

To find the most suitable areas on the Brazilian coast, a recent study used these environmental factors together with results of biomass of rhodolith bed and occurrence data. The models available to Brazil indicate 167,379–229,718 km^2 of niche suitability from south to north of Brazilian coast. Despite the large suitable areas of occurrence along the coast, only a few beds are in environmental protection areas, such as Fernando de Noronha, Abrolhos and Santa Catarina and the Vitória-Trindade chain, and most of them are threatened by coastal pollution and oil exploitation. Conservation policies are needed to protect these organisms and respective environments that are responsible for ecological cascades that find in Brazilian coast their planetary largest refuge.

Keywords Rhodolith beds · Marine species · Environmental changes · Ocean acidification · Oil exploitation · Ecosystem services · Marine biodiversity · Conservation policies · Ecological cascades

7.1 Introduction

One branch of macroecology has dedicated attention on the understanding of how environmental resources that vary along latitude can influence the distribution and abundance of ecosystem structuring species, such as forests (Huang et al., 2010; Reich et al., 2014), mangroves (Rovai et al., 2016), coral reefs (Lajeunesse et al., 2004; Ulstrup et al., 2006), kelps (Graham et al., 2007; Bernardes Batista et al., 2018) and rhodolith beds (Hernandez-Kantun et al., 2017a, b).

Among these habitats, rhodolith beds are one of the most abundant in the world (Foster et al., 2013; Riosmena-Rodríguez et al., 2017). Despite being one of the less studied, they structure environments that are described as biogenic reefs, built by free-living coralline red alga (Woelkerling, 1998; Bosence & Wilson, 2003). These beds are distributed from tropical regions (Cavalcanti et al., 2014) to polar (Jørgensbye & Halfar, 2017) and from shallow areas to more than 200 metres deep (Amado-Filho et al., 2017). These algae provide habitat for various marine organisms and are responsible for increasing the diversity and abundance of flora (Gherardi, 2004; Ordines et al., 2015) and fauna of the environments where they occur (Harvey & Bird, 2008; Konar et al., 2006). Some species of associated fauna appear to be rhodolith-specific (Steller et al., 2003), being influenced by the morphology of rhodoliths and their forming species (Hinojosa-Arango et al., 2009; Sañé et al., 2016). Rhodoliths have recently been found to harbour previously unknown endolytic life stages of ecologically important dinoflagellate microalgae (Krayesky-Self et al., 2017).

Despite their wide distribution, the occurrence and abundance of these species are conditioned to abiotic and biotic factors (Pascelli et al., 2013; McConnico et al., 2017), which will be further detailed below. In this context, rhodolith beds worldwide have experienced declines due to bottom fishing disturbances, and this trend is expected to continue as environmental changes intensify (Cornwall et al., 2019). In light of the aforementioned factors, it is urgent to establish strategic conservation areas for rhodolith beds, given the crucial ecosystem services they provide, as well as their key role in maintaining marine biodiversity (Fragkpoulou et al., 2021). However, mapping rhodolith beds can be costly and time-consuming, so it is necessary to adopt criteria to find the most suitable places to search. One of the techniques is predictive modelling (Martin et al., 2014). By selecting environmental variables that are known to influence the physiology of rhodoliths, associated with occurrence and abundance data, it is possible to predict on a large scale the most promising areas for bed development (Carvalho et al., 2020).

7.2 Ecological Importance and Environmental Aspects on Rhodolith Physiology and Morphology

The settlement, distribution and abundance of rhodolith beds throughout the world have been associated with environmental factors such as temperature, light, salinity and currents (Wilson et al., 2004a, b; Teichert et al., 2012). To a certain physiological limit, irradiance and temperature are positively related to photosynthesis and calcification (Kamenos & Law, 2010; Burdett et al., 2014), which are closely linked processes (Johansen, 1981; Hill et al., 2015). Although not requiring high irradiance values to reach the maximum photosynthetic rate, the higher frequency of light exposure may favour photosynthesis and growth of rhodoliths (Teichert & Freiwald, 2014). In addition, the increase in temperature (up to 30 °C), can favour a high metabolic rate (Vásquez-Elizondo & Enríquez, 2016; Hofmann et al., 2016). As such, the tropics' seas, having warmer temperatures and light availability all year round, have more favourable environmental conditions for the development of rhodolith beds.

While the largest rhodolith beds are distributed in the tropical regions, with the largest continuous formation of the South Atlantic located in Abrolhos, on the Brazilian coast (Amado-Filho et al., 2012a), there is a decrease in registrations towards the arctic region (Fig. 7.1). In addition, the Abrolhos bed has a high rate of calcium carbonate production (1000 ± 700 g m^{-2} $year^{-1}$: Amado-Filho et al., 2012a, b) compared to other rhodolith beds in the world, such as of temperate regions (Brazil—Deserta Island—SC, 55–136.3 g $CaCO_3$ m^{-2} $year^{-1}$, Gherardi, 2004; France—490 g $CaCO_3$ m^{-2} $year^{-1}$, Martin et al., 2007) and polar (Norway 200.3 g $CaCO_3$ m^{-2} $year^{-1}$, Teichert & Freiwald, 2014). However, despite the lower diversity and abundance of rhodolith beds in polar regions, the forming species are adapted to low temperature and light restriction conditions for much of the year, and they are fundamental in the maintenance of health of the local environment (Teichert & Freiwald, 2014).

In the Arctic regions, the occurrence of registered rhodolith beds is limited, with only a few identified locations in Norway (Freiwald & Henrich, 1994; Teichert et al., 2012, Teichert & Freiwald, 2014), Greenland (Jørgensbye & Halfar, 2017),

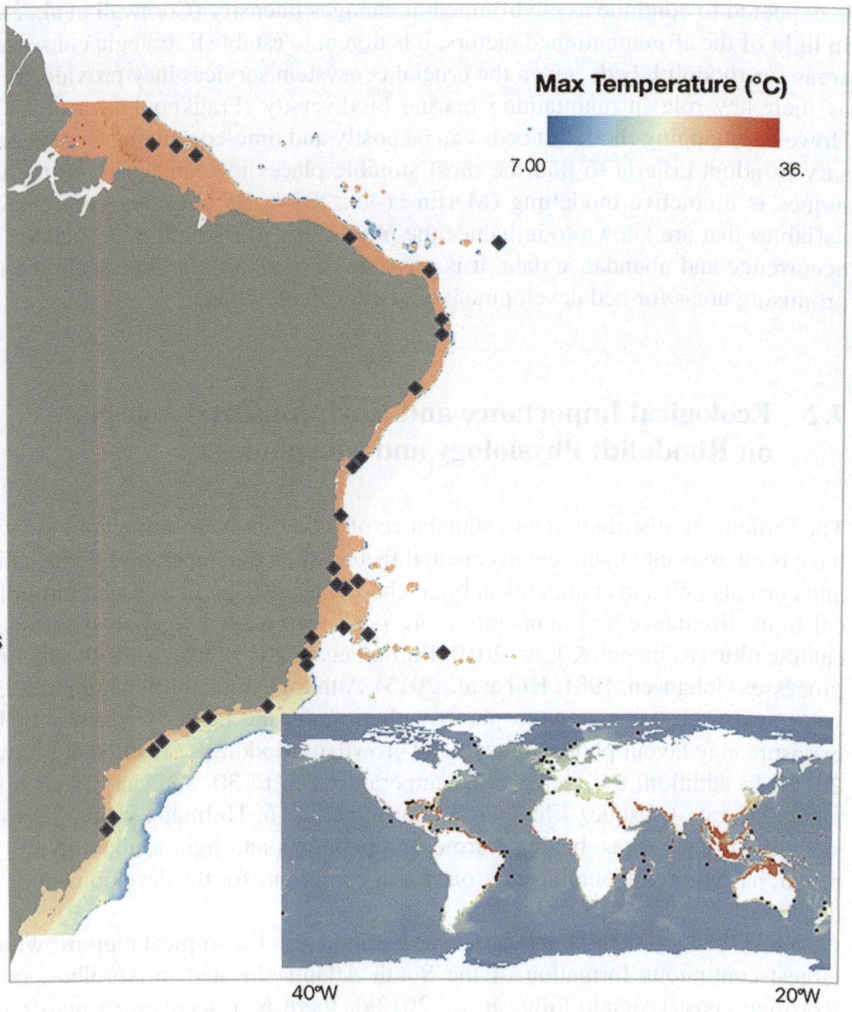

Fig. 7.1 Distribution of registered rhodolith beds throughout the world. The scale refers to the average background of bottom temperature (°C, extracted from Bio-ORACLE). Occurrence data for rhodolith-forming species were extracted from the websites of the organisations Ocean Biogeographic Information System (www.iobis.org) and Global Biodiversity Information Facility (www.gbif.org) and confirmed by scientific publications (black dots)

Canada (Gagnon et al., 2012) and Alaska (Konar et al., 2006). In Norway, *Lithothamnion glaciale* Kjellman and *Phymatolithon tenue* (Rosenvinge) Düwel & Wegeberg calcareous algae occur in the form of rhodoliths at depths of 27–50 metres, in a form of incrustation 70–80 metres, and provide habitat for a benthic fauna composed of 59 species (Teichert et al., 2012, Teichert & Freiwald, 2014). This distribution is mainly due to the availability of light, as at depths of >80 metres, it is insufficient for photosynthesis of fouling algae, and at <30 metres, it is high enough for the development of fast growing red filamentous algae that could compete with the rhodoliths (Teichert et al., 2012). At the other three points of occurrence in the Arctic, the composition of rhodoliths is even more restricted, with only *Clathromorphum compactum* (Kjellman) Foslie making up almost all rhodoliths in Greenland (and few areas covered by *L. glaciale*) and only *Phymatolithon calcareum* (Pallas) Adey & McKibbin ex Woelkering & Irvine for Alaska (Konar et al., 2006; Jørgensbye & Halfar, 2017) and *L. glaciale* for Canada (Gagnon et al., 2012).

In the tropical regions, the Caribbean coastlines (Littler et al., 1991), Africa (Bosence, 1983) and, mainly, Brazil (Foster et al., 2013; Amado-Filho et al., 2017) stand out for their large number of rhodolith beds records. On the Brazilian coast, these beds occur from latitude 5°N, in the region under the influence of the mouth of the Amazon River (Moura et al., 2016) to 27°S, on the coast of the state of Santa Catarina (Gherardi, 2004; Pascelli et al., 2013), a region considered by some authors as subtropical (Vieira-Pinto et al., 2014; Anderson et al., 2014) and by others as warm temperate (Horta et al., 2001; Pascelli et al., 2013). Santa Catarina's rhodolith beds are referred to as the southern limit of distribution of this type of environment to the Atlantic (Pascelli et al., 2013).

In the Abrolhos rhodolith bed, one of the largest in the South Atlantic, 14 rhodolith-forming species were identified, with *Hydrolithon rupestre* (Foslie) Penrose and *L. crispatum* as the most frequents (Brasileiro et al., 2016). At low latitudes, rhodoliths harbour a large diversity of macroalgae species adapted to high temperature and irradiance values (Riul et al., 2009; Bahia et al., 2010; Amado-Filho et al., 2012a, 2017; Foster et al., 2013; Carneiro et al., 2021). The first biological survey of the Royal Charlotte Bank (Negrão et al., 2021) revealed another massive region of the Brazilian continental shelf with the benthic community constructed by moderately dense to very dense rhodolith beds, colonised by seaweed forests, with frequent domain of *Sargassum* and Dictyotales (Brown algae).

At medium latitudes, at depths of 45–120 metres, the rhodolith beds provide substrate for the endemic kelp-forming species *Laminaria abyssalis* Joly & Oliveira (Amado-Filho et al., 2007; Rothman et al., 2017). In the southernmost latitudes, these species become rare or are replaced by species more tolerant to colder temperatures (Pascelli et al., 2013). Although tropical species dominate these environments, *L. glaciale* has already been found at depths of 60 metres, where temperatures below 15 °C allow its development (Henriques et al., 2012).

Despite the wide occurrence of rhodolith beds at various latitudes, abiotic and biotic local-scaled factors also seem to be determinant for the formation and abundance of these habitats (Hernandez-Kantun et al., 2017a, b). Current velocity, waves and depth are among the local factors that most influence the formation of rhodoliths (Hinojosa-Arango et al., 2009; Sañé et al., 2016). Under very weak currents, the calcareous algae would not come off and would not roll. Under very strong currents, the thallus will be fragmented (Foster, 2001; Hinojosa-Arango et al., 2009). The morphology of the rhodoliths is also influenced by the degree of water movement: in more agitated regions, they have a spherical and smooth shape, while in calmer regions, they will be flatter and more discoid (Sañé et al., 2016; Cabanellas-Reboredo et al., 2018). The intensity of the currents and waves influence not only the rhodoliths but also the associated community. Regions with greater movement tend to have less abundance of mobile organisms and an increase of organisms with stronger fixation structures (Maughan & Barnes, 2000; Hinojosa-Arango et al., 2009). Other authors have reported a seasonal difference in epiphyte coverage, with its decrease in winter due to storms occurring in this season (Amado-Filho et al., 2007; Pascelli et al., 2013).

The influence of bathymetry on the abundance of rhodoliths is variable, as different responses have been reported in the literature. While some authors note a negative correlation between rhodolith size and depth (Riul et al., 2009; Bahia et al., 2010; Pascelli et al., 2013; Vale et al., 2018), others report a positive relationship (Queiroz et al., 2016). The degree of water movement along a depth gradient may have a greater influence on the bathymetric distribution of the rhodolith beds than light penetration, as these organisms have a high tolerance to different light intensities (Sañé et al., 2016; Cabanellas-Reboredo et al., 2018). The predominance of rhodoliths at a given depth may be related to the degree of sedimentation of the area (Brasileiro et al., 2016; Vale et al., 2018), which is further explored below. The composition of the associated community, mainly macroalgae, is strongly influenced by bathymetry. Authors have reported a decrease in epiphyte algae coverage with increasing depth due to decreased light (Riul et al., 2009; Brasileiro et al., 2016).

In addition to water movement, proximity to nutrient inputs also influences the occurrence of rhodoliths and associated communities (Björk et al., 1995; Figueroa et al., 2014). Regions near the discharge of rivers and sewage generate greater sedimentation, reducing light availability and photosynthetic efficiency of rhodoliths (Martin et al., 2014). According to Vale et al. (2018), in areas very close to the Amazon River plume, the biogenic nodules are mostly bryozoans. The authors relate this composition to the absence of light due to sedimentation and excess nutrients caused by discharges. Similarly, Brasileiro et al. (2016) pointed to the proximity to the Doce River (south of Abrolhos) discharge as a probable cause of the absence of rhodoliths at depths less than 40 metres. The reduction in the coverage of calcareous algae in regions with large sediment supply may be related to the reduction in photosynthesis. One experiment showed that even a 30% reduction in light caused by a thin layer of sediment can decrease the photosynthesis of rhodolites by 70% (Riul et al., 2008). Other shows that 70% coverage of the sediment rhodolith surface can cause a 30% reduction in the photosynthetic efficiency of *Mesophyllum engelhartii* (Foslie) Adey and *Lithothamnion* sp. (Figueiredo et al., 2015). Increasing nutrient input from estuaries or pollutants may also alter the composition of epiphytes, causing the proliferation of rapidly growing algae on the surface of rhodoliths (Fong et al., 1994; Hardison et al., 2011; Neil & Nelson, 2016). On the other hand, small rivers that provide only seasonal freshwater inputs, or local conditions that reduce the risk of burial (such as discontinuities in the morphology of the seabed), may allow the accumulation and even favour the growth of rhodoliths (Carneiro et al., 2021).

7.3 Oceanographic Conditions of Brazilian Coast and Rhodolith Bed Distribution

As seen before, the Brazilian coast has environmental conditions that favour rhodolith bed developments. However, some regions have particular conditions that favour the development of rhodolith-forming species in higher quantities. Horta et al. (2001) and Spalding et al. (2007) sort this coast in tropical and warm temperate regions (Fig. 7.2). The first authors based their classification on phycological composition and refer to the zone between the two regions as a transition zone, which assembles traits of both. The second authors consider additional marine organisms and forcing agents (e.g. upwelling, nutrients, temperature, currents and bathymetry).

The tropical region has the major part of the mapped rhodolith beds of the coast, namely, the beds of Amazon River (Vale et al., 2018), Fernando de Noronha (Amado-Filho et al., 2012a), Rio Grande do Norte (Testa & Bosence, 1999), Paraíba (Riul et al., 2009), Salvador (Bahia et al., 2010) and at Abrolhos, which is considered the world's largest rhodolith bed (Brasileiro et al., 2016). The tropical region receives an important system of western boundary currents, which originates from the bifurcation of the westward-flowing South Equatorial Current (SEC) around 15°S. After this bifurcation, the sea surface temperature (SST) varies around 27 °C, reaching up to 29 °C (Duarte et al., 2020). The Brazilian coastal shelf in this region is relatively narrow, varying from 8 km offshore of Salvador but reaching up to 200 km in Abrolhos (Knoppers et al., 1999). Most of the SEC water is carried north, across the entire equatorial coast of Brazil and into the northern hemisphere, by the North Brazil Current (NBC) (Dossa et al., 2021). A smaller branch of the SEC flows southward, forming the Brazil Current (BC), which borders the South American continent to the region of the Subtropical Convergence, located around 38 °S ± 2° (Olson et al., 1988). Both BC and NBC flow off the Brazilian continental slope, and near the surface, they carry the warm (>20 °C), saline (>36) and oligotrophic tropical water (TW) (Emilson, 1961; Silveira et al., 2020). On the continental shelf, the TW usually encounters colder and less-saline coastal and river water masses, forming different shelf circulation systems across the country (Matano et al., 2010; Amorim et al., 2011; Dias et al., 2013). And along the BC, there are also intrusions of the colder and nutrient-rich South Atlantic Central Water (SACW) (Cirano et al., 2006).

At the transition between the tropical and warm temperate regions, under the influence of CW and by BC (Castro & Miranda, 1998), rhodolith beds were mapped in Espírito Santo state from 4 m to 58 m depth (Amado-Filho et al., 2007; Henriques et al., 2012) and at 90–100 m depths offshore from the Cabo Frio region (Villas-Bôas et al., 2014). Despite the predominance of oligotrophic and warm waters, with SSTs varying from 21 to 27 °C, the SACW reaches the southern margin of this region during summer (Schmid et al., 1995; Matano et al., 2010).

The rhodolith beds in the warm temperate zone are observed between temperatures extreme, with maximum SST values up to 33 °C, observed in the Rio de

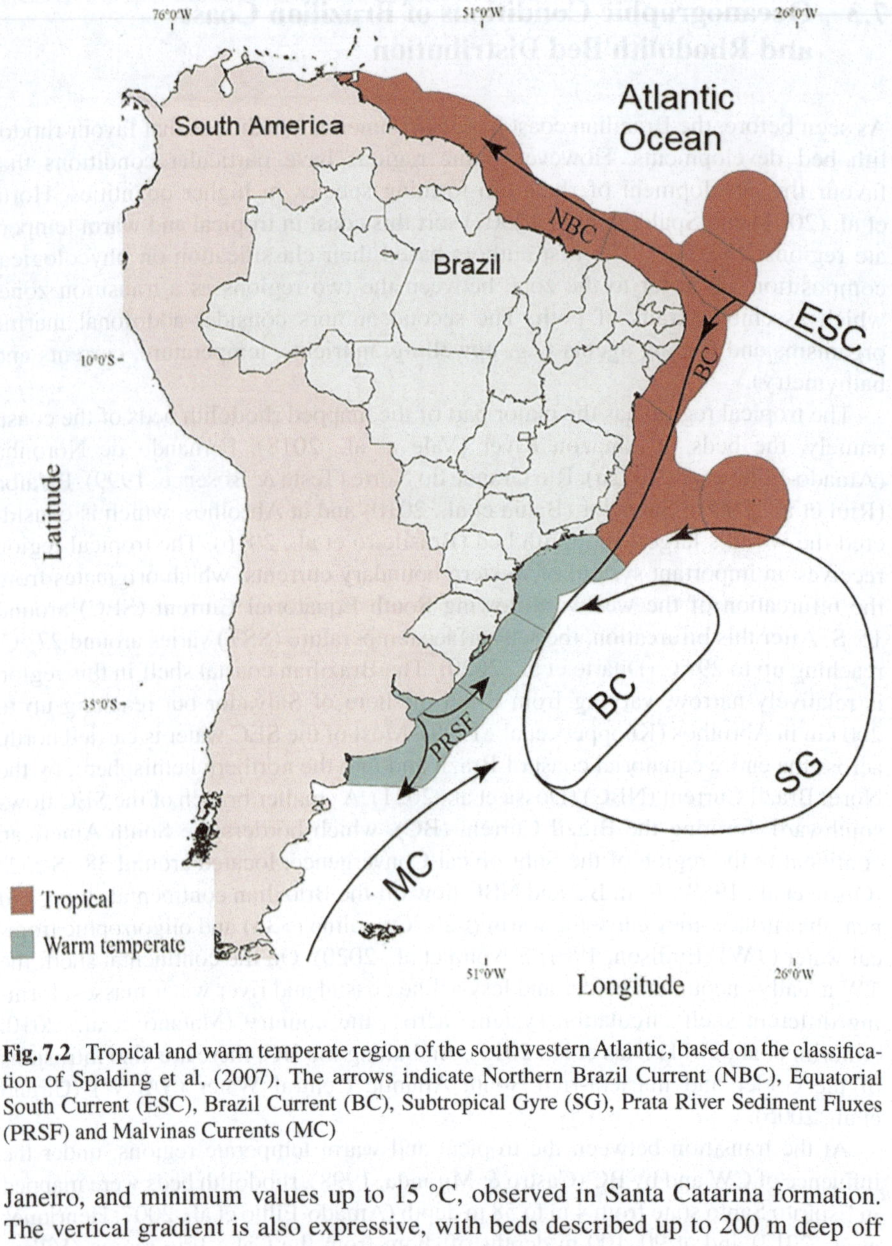

Fig. 7.2 Tropical and warm temperate region of the southwestern Atlantic, based on the classification of Spalding et al. (2007). The arrows indicate Northern Brazil Current (NBC), Equatorial South Current (ESC), Brazil Current (BC), Subtropical Gyre (SG), Prata River Sediment Fluxes (PRSF) and Malvinas Currents (MC)

Janeiro, and minimum values up to 15 °C, observed in Santa Catarina formation. The vertical gradient is also expressive, with beds described up to 200 m deep off São Paulo (dos Santos Filho et al., 2022), but also in shallow areas as in the Arvoredo Biological Marine Reserve (Rebio Arvoredo), where they thrive between 4 m and 18 m depth (Pascelli et al., 2013). These rhodolith beds represent the southernmost limit of distribution of rhodoliths in the Western Atlantic (Gherardi, 2004; Pascelli et al., 2013), where there is complex oceanographic dynamics due to strong winds, subtropical convergence and frequent presence of SACW, also influenced by the La Plata plume, with low salinities and enrichment of nutrient (Möller et al., 2008; Strub et al., 2015; Freire et al., 2017) and currents with temperature of 8–12 °C (Bernardes Batista et al., 2018).

7.4 Modelling Tools Used in Prediction Niche Suitability in Brazil

Mapping using the acoustic surveys or direct sampling are timing consuming and expensive approaches. Habitat suitability modelling is an important alternative method to estimate their potential distribution and abundance. Recent studies of niche modelling of rhodolith beds on Brazilian coast features suggested potential distribution and main environmental drivers. Carvalho et al. (2020) estimated around 230,000 km^2 with high niche suitability, while dos Santos et al. (2023) predicted the occurrence of rhodolith beds over ~167,379 km^2 of the Brazilian continental shelf (Fig. 7.3). Both works highlighted temperature and irradiance as main environmental drivers. Carvalho et al. (2020) revealed areas with high suitability and high rhodolith potential abundance, where later, as in the Royal Charlotte bank (Negrão et al., 2021), rhodolith beds were identified and were corroborated as dominant feature of the bottom physiognomy. However, evaluating both niche suitability in relation to ground truthing provided by dos Santos Filho et al. (2022) or Pereira-Filho et al. (2023), we can speculate that rhodolith beds can have higher distribution than the value proposed by Santos et al. (2023), once areas absent in their proposal presented rhodolith beds.

Besides providing information about potential occurrence, niche modelling can also help forecasting potential abundance. The projection available to the Brazilian coast reveals higher values in tropical sites decreasing towards warm temperate province (Carvalho et al., 2020). These authors described high rhodolith dry mass, volume and more dense rhodolith beds. This prediction is corroborated by higher calcification rates in lower latitudes observed throughout the Atlantic (Schubert et al., 2022). This physiological feature can be associated with stronger cell walls and consequently reduction in bioerosion, reducing internal cavities and increasing density (Nalin et al., 2008).

7.5 Environmental Drivers of Bed Distribution

Different techniques of methodology, as well as different data sets, will produce different.

niche suitability and main ecological drivers as demonstrated by Santos et al. (2023). Maximum entropy (MAXENT), random forest (RF) and boosted regression

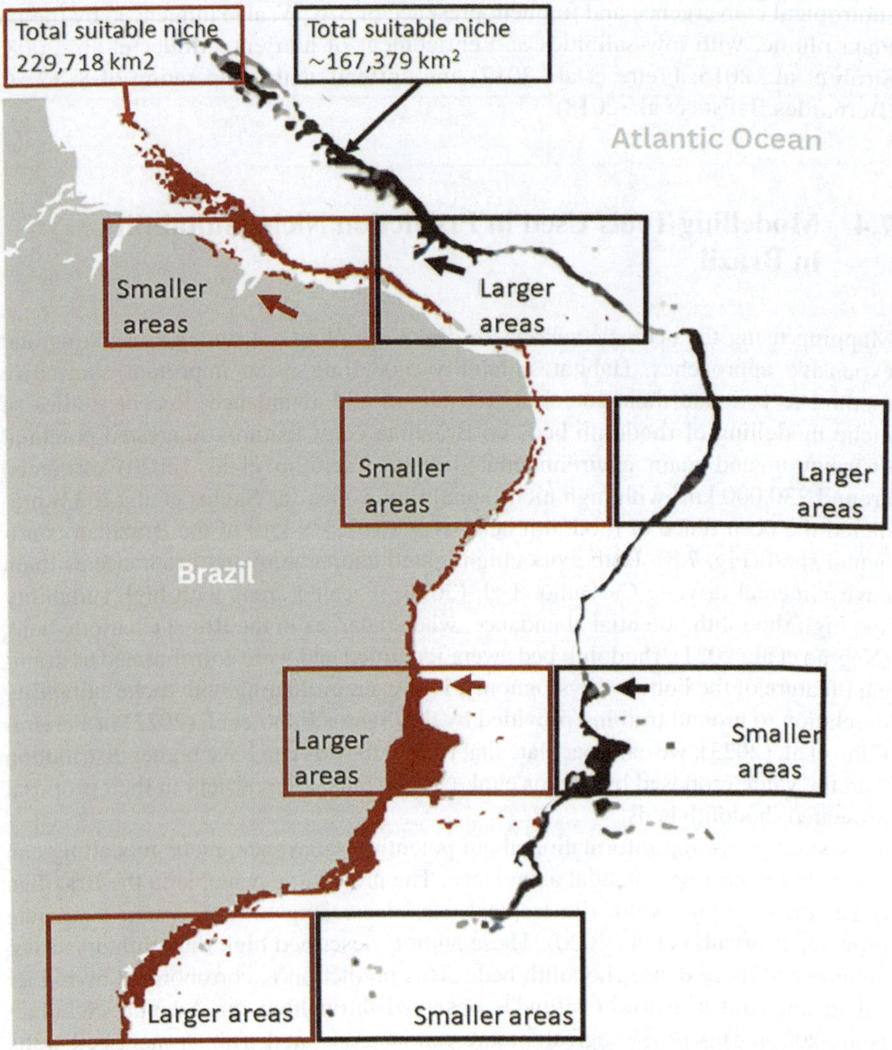

Fig. 7.3 Comparison between the predictions of Carvalho et al. (2020) in red and dos Santos et al. (2023) in black (displaced to the right), in grey add areas where there were discrepancies, to illustrate differences of niche suitability in the southern Brazilian coast regarding different approaches

trees (BRT) represent algorithms that have been used to evaluate the ecological niche modelling of rhodolith beds in Brazil. While maximum light was the most important factor in the BRT and RF evaluation, minimum light was the most important factor to MAXENT. Saturation state of calcite, temperature and salinity appear among the most important diverse of the rhodolith bed occurrence. It is important to remember that the effect of temperature is not yet well known at global scales on calcification, photosynthesis and respiration of rhodolith species and associated communities.

Despite some advances in the last years, most of the available information comes from different regions (Comeau et al., 2016; Schubert et al., 2022) and macroecological conditions (King & Schramm, 1982; Martin et al., 2013). However, the thresholds observed elsewhere, with temperatures around 31.5 °C (Comeau et al., 2016), or the increases of photosynthesis during the summer (Martin et al., 2013), corroborate the importance of the abovementioned environmental drivers to the rhodolith niche suitability. Schubert et al. (2022) reinforced that the mechanisms associated with photosynthesis and calcification present high latitudinal influence and display a large variability among species, suggesting potential different community responses to environmental drivers.

Regarding the available modelling for rhodolith abundance, it is highlighted the importance of nutrients as one of the main drivers, what agrees with Ichiki et al. (2000) and Dutertre et al. (2015) that reported positive relation between nitrate concentration and rhodolith growth and abundance, respectively. However, extreme values of nutrients can also play a negative effect on these organisms, reducing niche suitability. Vale et al. (2018) reported a reduction of rhodoliths near Amazon mouth, areas with high organic matter loads. This may also explain the absence (Santos et al., 2023) or the predicted lower dry mass at southern Brazilian coast (Carvalho et al., 2020), which receive La Plata plume, South Atlantic Central Water, both with high nutrient concentrations (Möller et al., 2008).

Currents were also important to explain the rhodolith abundance. Currents provide the movement to avoid burial (Martin et al., 2014), as well as the renovation of the boundary layer around the rhodolith, facilitating gas and nutrient exchanges with the water column (Schubert et al., 2021). Epibionts overgrowth can also be influenced by disturbances imposed by water movements (Steller & Foster, 1995; Sañé et al., 2016), resulting in a scenario where moderate hydrodynamics is the optimum condition to favour occurrence of rhodolith (Cabanellas-Reboredo et al., 2018; Melbourne et al., 2018).

Besides the oceanographic conditions that influence the development of rhodolith beds, biotic interactions also have an important role in maintaining a healthy ecosystem. Some species of epiphyte algae may compete with rhodoliths for light causing shadow and absorbing much of the available light and nutrients on the surface (Dodds, 1991; Wahl, 2008). Conversely, encrusting calcareous algae have strategies that prevent overgrowth of epiphytic algae on their surface, such as peeling epithelial cells (Hofmann et al., 2016) and production of allelopathic substances (Gross, 2003; Kim et al., 2004). In addition, some associated herbivorous animals such as fish, crabs, polychaetes and molluscs may be beneficial as they would feed on epiphytic algae spores and control their growth (Scherner et al., 2010; Legrand et al., 2017). Herbivory is one of the factors that most influence the composition of the macroalgal community (Poore et al., 2012; Bonaldo & Hay, 2014; Longo et al., 2015). Therefore, since facilitation or competition may have an impact on the preferences of an organism (Boulangeat et al., 2012), including biotic interactions on future modelling techniques could bring a more assertive approach to mapping rhodolith beds.

7.6 Final Remarks

The understanding of patterns and processes involved in structuring ecosystems, the study of key species responses to current environmental variations, allows us to predict their responses to future scenarios, supporting conservation and management (Smale & Wernberg, 2013). The United Nations recognized for decades that the increasing emission of CO_2 into the atmosphere has caused an increase in ocean temperature and a decrease in pH (IPCC, 2014; Pörtner et al., 2022). Conservador scenarios for 2100 predict temperatures 1.8 to 6.5 warmer than the preindustrial values and CO_2 concentrations reaching up to 1000 ppm (Jiang et al., 2023). This increasing temperature has led to the tropicalization of temperate environments and the weakening of habitat-forming coralline organisms in tropical regions (Vergés et al., 2014). As a tropical country with a large diversity of ecosystems, Brazil is one of the countries that will suffer the most from climate change (Horta et al., 2016). Some ecosystems located on the Brazilian continental shelf are more subject to ocean acidification because they are in areas that act as CO_2 sinks at specific seasons (Kerr et al., 2016). These areas include regions near the Amazon River plume and southeast continental margin (~ 20°S to 28°S) (Kerr et al., 2016).

Due to increasing climate change, some studies predict the tropicalization of temperate and subtropical climate environments and the disappearance of various species at tropical communities in low latitudes (Vergés et al., 2014; Wernberg et al., 2016; Araújo et al., 2018). Increasing acidification and ocean temperature due to the emissions of CO_2 will cause a decrease in calcium carbonate organisms in its composition, with consequent decrease in other dependent species (Hooidonk et al., 2014; Legrand et al., 2017). Anthropogenic alterations of local impacts such as nutrient discharge and sedimentation will result in habitat destruction and species disappearance (Steller et al., 2003; Wilson et al., 2004a, b).

Considered "ecosystem engineers" (Basso, 2012), Brazilian rhodolith beds are diverse and play an important role in the provision of ecosystem services and in maintenance of marine biodiversity and socioeconomic activities (Pereira-Filho et al., 2012). However, only a few are in marine protected areas, such as in Fernando de Noronha (Amado-Filho et al., 2012a), Abrolhos (Brasileiro et al., 2016), Santa Catarina (Pascelli et al., 2013) and the Vitória-Trindade Chain (Pereira-Filho et al., 2012), the latter being created only recently under Decree 9312 of 03/19/2018.

To provide subsidies for a possible expansion of conservation areas, it is essential to identify areas where the beds are more likely to occur as well as their socio-environmental functions are strategic for sustainable coastal management (Magris et al., 2021).

References

Amado-Filho, G., Maneveldt, G., Manso, R. C. C., Marins-Rosa, B. V., Pacheco, M. R., & Guimarães, S. M. P. B. (2007). Structure of rhodolith beds from 4 to 55 meters deep along the southern coast of Espírito Santo state, Brazil. *Ciencias Marinas, 33*, 399–410.

Amado-Filho, G. M., et al. (2012a). Occurrence and distribution of rhodolith beds on the Fernando de Noronha archipelago of Brazil. *Aquatic Botany, 101*, 41–45.

Amado-Filho, G. M., Moura, R. L., Bastos, A. C., Salgado, L. T., Sumida, P. Y., Guth, A. Z., Francini-Filho, R. B., Pereira-Filho, G. H., Abrantes, D. P., Brasileiro, P. S., Bahia, R. G., Leal, R. N., Kaufman, L., Kleypas, J. A., Farina, M., & Thompson, F. L. (2012b). Rhodolith beds are major $CaCO_3$ bio-factories in the tropical South West Atlantic. *PLoS One, 7*, e35171.

Amado-Filho, G. M., et al. (2017). South Atlantic Rhodolith beds: Latitudinal distribution, species composition, structure and ecosystem functions, threats and conservation status. In *Rhodolith/Maërl beds: A global perspective* (pp. 299–317). Springer.

Amorim, F. N. D., Cirano, M., Soares, I. D., & Lentini, C. A. D. (2011). Coastal and shelf circulation in the vicinity of Camamu Bay (14°S), eastern Brazilian shelf. *Continental Shelf Research., 31*, 108–119.

Anderson, A. B., et al. (2014). Recovery of grouper assemblages indicates effectiveness of a marine protected area in Southern Brazil. *Marine Ecology Progress Series, 514*, 207–215.

Araújo, F. G., et al. (2018). Shifts in the abundance and distribution of shallow water fish fauna on the southeastern Brazilian coast: A response to climate change. *Hydrobiologia, 814*, 205–218.

Bahia, R. G., et al. (2010). Rhodolith bed structure along a depth gradient on the northern coast of Bahia state, Brazil. *Brazilian Journal of Oceanography., 58*, 323–337.

Basso, D. (2012). Carbonate production by calcareous red algae and global change. *Geodiversitas, 34*, 13–33.

Bernardes Batista, M., Batista Anderson, A., Franzan Sanches, P., Simionatto Polito, P., Lima Silveira, T. C., Velez-Rubio, G. M., Scarabino, F., Camacho, O., Schmitz, C., Martinez, A., Ortega, L., Fabiano, G., Rothman, M. D., Liu, G., Ojeda, J., Mansilla, A., Barreto, L. M., Assis, J., Serrão, E. A., Santos, R., & Horta, P. A. (2018). Kelps' long-distance dispersal: Role of ecological/oceanographic processes and implications to marine forest conservation. *Diversity, 10*, 11.

Björk, M., Mohammed, S. M., Bjorklund, M., & Semesi, A. (1995). Coralline algae, important coral-reef builders threatened by pollution. *Oceanographic Literature Review, 7*, 734.

Bonaldo, R. M., & Hay, M. E. (2014). Seaweed-coral interactions: Variance in seaweed allelopathy, coral susceptibility, and potential effects on coral resilience. *PLoS One, 9*, e85786.

Bosence, D., & Wilson, J. (2003). Maerl growth, carbonate production rates and accumulation rates in the northeast Atlantic. *Aquatic Conservation, 13*, S21–S32.

Bosence, D. W. J. (1983). The occurrence and ecology of recent rhodoliths—A review. In *Coated grains* (pp. 225–242). Springer.

Boulangeat, I., Dominique, G., & Thuiller, W. (2012). Accounting for dispersal and biotic interactions to disentangle the drivers of specie distributions and their abundances. Ecology Letters, 15, 584–593.

Brasileiro, P. S., et al. (2016). Macroalgal composition and community structure of the largest rhodolith beds in the world. *Marine Biodiversity, 46*, 407–420.

Burdett, H. L., et al. (2014). Dynamic photoinhibition exhibited by red coralline algae in the red sea. *BMC Plant Biology, 14*, 139.

Cabanellas-Reboredo, M., et al. (2018). Morpho-demographic traits of two maërl-forming algae in beds with different depths and fishing histories. *Aquatic Conservation: Marine and Freshwater Ecosystems, 28*, 133–145.

Carneiro, P. B. M., de Lima, J. P., Bandeira, Ê. V. P., Neto, A. R. X., Barreira, C. D. A. R., de Souza Tâmega, F. T., et al. (2021). Structure, growth and $CaCO_3$ production in a shallow rhodolith bed from a highly energetic siliciclastic-carbonate coast in the equatorial SW Atlantic Ocean. *Marine Environmental Research., 166*, 105280.

Carvalho, V. F., et al. (2020). Environmental drivers of rhodolith beds and epiphytes community along the South Western Atlantic coast. *Marine Environmental Research., 154*, 104827.

Castro, B. M., & Miranda, L. B. (1998). Physical oceanography of Western Atlantic continental shelf located between 48N and 348S. In A. R. Robinson & K. H. Brink (Eds.), *The sea* (pp. 209–251). Wiley.

Cavalcanti, G. S., et al. (2014). Physiologic and metagenomic attributes of the rhodoliths forming the largest CaCO 3 bed in the South Atlantic Ocean. *The ISME Journal, 8*, 52.

Cirano, M., Mata, M. M., Campos, E. J. D., & Deiró, N. F. R. (2006). A circulação oceânica de larga-escala na região oeste do Atlântico Sul com base no modelo de circulação Global OCCAM. *Revista Brasileira de Geofisica, 24*, 209–230.

Comeau, S., Carpenter, R. C., Lantz, C. A., & Edmunds, P. J. (2016). Parameterization of the response of calcification to temperature and pCO2 in the coral *Acropora pulchra* and the alga *Lithophyllum kotschyanum. Coral Reefs, 35*, 929–939.
Cornwall, C. E., Diaz-Pulido, G., & Comeau, S. (2019). Impacts of ocean warming on coralline algal calcification: Meta-analysis, knowledge gaps, and key recommendations for future research. *Frontiers in Marine Science, 6*, 1–10.
Dias, F. J. S., Castro, B. M., & Lacerda, L. (2013). Continental shelf water masses off the Jaguaribe River (4S), northeastern Brazil. *Continental Shelf Research., 66*, 123–135.
Dodds, W. K. (1991). Community interactions between the filamentous alga *Cladophora glomerata* (L.) Kuetzing, its epiphytes, and epiphyte grazers. *Oecologia, 85*, 572–580.
dos Santos Filho, J. R., Figueiredo Jr, A. G., Carneiro, J. C., de Macedo Dias, G. T., & Ramalho, A. S. (2022). Mesophotic bioclastics and bioconstructions at the continental shelf of Santos Basin, Brazil. Sedimentary Geology, 442, 106294.
dos Santos, V. S., de Moura, R. L., Magdalena, U. R., Hovey, R., Kendrick, G., Bahia, R. G., ... & de Siqueira, M. F. (2023). Spatial modeling reveals a growing threat to the world's largest rhodolith beds. Ocean & Coastal Management, 232, 106441.
Dossa, A. N., Silva, A. C., Chaigneau, A., Eldin, G., Araujo, M., & Bertrand, A. (2021). Near-surface western boundary circulation off Northeast Brazil. *Progress in Oceanography., 190*, 102475.
Duarte, G. A., Villela, H. D., Deocleciano, M., Silva, D., Barno, A., Cardoso, P. M., et al. (2020). Heat waves are a major threat to turbid coral reefs in Brazil. *Frontiers in Marine Science, 7*, 179.
Dutertre, M., et al. (2015). Environmental factors affecting maerl bed structure in Brittany (France). *European Journal of Phycology, 50*, 371–383.
Emflsson, I. (1961). The shelf and coastal waters off southern Brazil. *Boletim do Instituto Oceanográfico., 11*, 101–112.
Figueiredo, M. A. O., et al. (2015). The effect of sediment mimicking drill cuttings on deep water rhodoliths in a flow-through system: Experimental work and modeling. *Marine Pollution Bulletin., 95*, 81–88.
Figueroa, F. L., Barufi, J. B., Malta, E. J., Conde-Álvarez, R., Nitschke, U., Arenas, F., Mata, M., Connan, S., Abreu, M. H., Marquardt, R., Vaz-Pinto, F., Konotchick, T., et al. (2014). Short-term effects of increasing CO_2, nitrate and temperature on three Mediterranean macroalgae: Biochemical composition. *Aquatic Biology, 22*, 177–193.
Fong, P., Donohoe, R. M., & Zedler, J. B. (1994). Nutrient concentration in tissue of the macroalga *Enteromorpha* as a function of nutrient history: An experimental evaluation using field microcosms. *Marine Ecology Progress Series., 106*, 273–281.
Foster, M. S. (2001). Mini-review: Rhodoliths, between rocks and soft places. *Journal of Phycology, 37*, 659–657.
Foster, M. S., Amado Filho, G. M., Kamenos, N. A., Riosmena-Rodríguez, R., & Steller, D. L. (2013). Rhodoliths and rhodolith beds. *Research and discoveries: The revolution of science through SCUBA*.
Fragkopoulou, E., Serrão, E. A., Horta, P. A., Koerich, G., & Assis, J. (2021). Bottom trawling threatens future climate refugia of rhodoliths globally. *Frontiers in Marine Science, 7*, 594537.
Freire, A. S., Varela, A. R. D., Fonseca, A. L. O., Menezes, B. S., Fest, C. B., Obata, S. C., Gorri, C., & Pereira, T. L. (2017). O ambiente oceanográfico. In B. Segal, A. S. Freire, A. Lindner, J. P. Krajewski, & M. Soldateli (Eds.), *Maare: Monitoramento Ambiental da Reserva Marinha do Arvoredo e Entorno* (p. 268). UFSC/Maare.
Freiwald, A., & Henrich, R. (1994). Reefal coralline algal build-ups within the Arctic circle: Morphology and sedimentary dynamics under extreme environmental seasonality. *Sedimentology, 41*, 963–984.
Gagnon, P., Matheson, K., & Stapleton, M. (2012). Variation in rhodolith morphology and biogenic potential of newly discovered rhodolith beds in Newfoundland and Labrador (Canada). *Botanica Marina., 55*, 85–99.
Gherardi, D. F. M. (2004). Community structure and carbonate production of a temperate rhodolith bank from Arvoredo Island, southern Brazil. *Brazilian Journal of Oceanography., 52*, 207–224.

Graham, M. H., Kinlan, B. P., Druehl, L. D., Garske, L. E., & Banks, S. (2007). Deep-water kelp refugia as potential hotspots of tropical marine diversity and productivity. *Proceedings of the National Academy of Sciences, 104*, 16576–16580.

Gross, E. M. (2003). Allelopathy of aquatic autotrophs. *Critical Reviews in Plant Sciences, 22*, 313–339.

Hardison, A. K., et al. (2011). Carbon and nitrogen dynamics in shallow photic systems: Interactions between macroalgae, microalgae, and bacteria. *Limnology and Oceanography., 56*, 1489–1503.

Harvey, A. S., & Bird, F. L. (2008). Community structure of a rhodolith bed from cold-temperate waters (southern Australia). *Australian Journal of Botany., 56*, 437–450.

Henriques, M. C., et al. (2012). New records of rhodolith-forming species (Corallinales, Rhodophyta) from deep water in Espírito Santo state, Brazil. *Helgoland Marine Research., 66*, 219.

Hernandez-Kantun, J. J., et al. (2017a). North Atlantic Rhodolith beds. In *Rhodolith/Maërl beds: A global perspective* (pp. 265–279). Springer.

Hernandez-Kantun, J. J., Hall-Spencer, J. M., Grall, J., Adey, W., Rindi, F., Maggs, C. A., Bárbara, I., & Peña, V. (2017b). North Atlantic Rhodolith beds. In R. Riosmena-Rodríguez, W. Nelson, & J. Aguirre (Eds.), *Rhodolith/Maërl beds: A global perspective* (pp. 265–279). Springer.

Hill, R., et al. (2015). Can macroalgae contribute to blue carbon? *An Australian Perspective. Limnology and Oceanography, 60*, 1689–1706.

Hinojosa-Arango, G., Maggs, C. A., & Johnson, M. P. (2009). Like a rolling stone: The mobility of maerl (Corallinaceae) and the neutrality of the associated assemblages. *Ecology, 90*, 517–528.

Hofmann, L. C., Koch, M., & De Beer, D. (2016). Biotic control of surface pH and evidence of light-induced H+ pumping and Ca2+-H+ exchange in a tropical crustose coralline alga. *PLoS One, 11*, 1–24.

Hooidonk, R., et al. (2014). Opposite latitudinal gradients in projected ocean acidification and bleaching impacts on coral reefs. *Global Change Biology., 20*, 103–112.

Horta, P. A., et al. (2001). Considerações sobre a distribuição e origem da flora de macroalgas marinhas brasileiras. *Hoehnea., 28*, 243–265.

Horta, P. A., Riul, P., Amado Filho, G. M., Gurgel, C. F. D., Berchez, F., Nunes, J. M. D. C., & Sissini, M. (2016). Rhodoliths in Brazil: Current knowledge and potential impacts of climate change. *Brazilian Journal of Oceanography, 64*, 117–136.

Huang, J., et al. (2010). Radial growth response of four dominant boreal tree species to climate along a latitudinal gradient in the eastern Canadian boreal forest. *Global Change Biology, 16*(2), 711–731.

Ichiki, S., Mizuta, H., & Yamamoto, H. (2000). Effects of irradiance, water temperature and nutrients on the growth of sporelings of the crustose coralline alga *Lithophyllum yessoense* Foslie (Corallinales, Rhodophyceae). *Phycological Research, 48*, 115–120.

IPCC. (2014). Climate Change 2014: Synthesis Report. In *Contribution of working groups I, II and II to the fifth assessment report of the intergovernmental panel on climate change* (p. 151).

Jiang, L. Q., Dunne, J., Carter, B. R., Tjiputra, J. F., Terhaar, J., Sharp, J. D., et al. (2023). Global surface ocean acidification indicators from 1750 to 2100. *Journal of Advances in Modeling Earth Systems, 15*(3), e2022MS003563.

Johansen, H. W. (1981). *Coralline algae: A first synthesis*. CRC Press.

Jørgensbye, H. I. Ø., & Halfar, J. (2017). Overview of coralline red algal crusts and rhodolith beds (Corallinales, Rhodophyta) and their possible ecological importance in Greenland. *Polar Biology., 40*, 517–531.

Kamenos, N. A., & Law, A. (2010). Temperature controls on coralline algal skeletal growth. *Journal of Phycology, 46*, 331–335.

Kerr, R., et al. (2016). The Western South Atlantic Ocean in a high-CO_2 world: Current measurement capabilities and perspectives. *Environmental Management., 57*, 740–752.

Kim, J., Choi, J. S., Kang, S. E., Cho, J. Y., Jin, H. J., Chun, B. S., & Hong, Y. K. (2004). Multiple allelopathic activity of the crustose coralline alga *Lithophyllum yessoense* against settlement and germination of seaweed spores. *Journal of Applied Phycology, 16*, 175–179.

King, R. J., & Schramm, W. (1982). Calcification in the maerl coralline alga *Phymatolithon calcareum*: Effects of salinity and temperature. *Marine Biology, 70*, 197–204.
Knoppers, B., Ekau, W., & Figueiredo, A. G. (1999). The coast and shelf of East and Northeast Brazil and material transport. *Geo-Marine Letters, 19*, 171–178.
Konar, B., Riosmena-Rodriguez, R., & Iken, K. (2006). Rhodolith bed: A newly discovered habitat in the North Pacific Ocean. *Botanica Marina., 49*, 355–359.
Krayesky-Self, S., et al. (2017). Eukaryotic life inhabits Rhodolith-forming coralline algae (Hapalidiales, Rhodophyta), Remarkable Marine Benthic Microhabitats. *Scientific Reports, 7*, 45850.
Lajeunesse, T. C., et al. (2004). Closely related *Symbiodinium* spp. differ in relative dominance in coral reef host communities across environmental, latitudinal and biogeographic gradients. *Marine Ecology Progress Series., 284*, 147–161.
Legrand, E., Riera, P., Lutier, M., Coudret, J., Grall, J., & Martin, S. (2017). Species interactions can shift the response of a maerl bed community to ocean acidification and warming. *Biogeosciences Discussions, 14*, 5359.
Littler, M. M., Littler, D. S., & Hanisak, M. D. (1991). Deep-water rhodolith distribution, productivity, and growth history at sites of formation and subsequent degradation. *Journal of Experimental Marine Biology and Ecology., 150*, 163–182.
Longo, G. O., et al. (2015). Between-habitat variation of benthic cover, reef fish assemblage and feeding pressure on the benthos at the only atoll in South Atlantic: Rocas Atoll, NE Brazil. *PLoS One, 10*, e0127176.
Magris, R. A., Costa, M. D., Ferreira, C. E., Vilar, C. C., Joyeux, J. C., Creed, J. C., ... & Floeter, S. R. (2021). A blueprint for securing Brazil's marine biodiversity and supporting the achievement of global conservation goals. *Diversity and Distributions, 27*(2), 198–215.
Martin, S., Clavier, J., Chauvaud, L., & Thouzeau, G. (2007). Community metabolism in temperate maerl beds. I. Carbon and carbonate fluxes. *Marine Ecology Progress Series, 335*, 19–29.
Martin, S., Cohu, S., Vignot, C., Zimmerman, G., & Gattuso, J. P. (2013). One-year experiment on the physiological response of the Mediterranean crustose coralline alga, *Lithophyllum cabiochae*, to elevated p CO_2 and temperature. *Ecology and Evolution, 3*, 676–693.
Martin, C. S., et al. (2014). Coralligenous and maërl habitats: Predictive modelling to identify their spatial distributions across the Mediterranean Sea. *Scientific Reports, 4*, 5073.
Matano, R. P., Palma, E. D., & Piola, A. R. (2010). The influence of the Brazil and Malvinas currents on the southwestern Atlantic shelf circulation. *Ocean Science Journal, 7*, 1–35.
Maughan, B. C., & Barnes, D. K. A. (2000). Epilithic boulder communities of lough Hyne, Ireland: The influences of water movement and sediment. *Journal of the Marine Biological Association of the United Kingdom., 80*, 767–776.
McConnico, L. A., et al. (2017). Temporal variation in seaweed and invertebrate assemblages in shallow rhodolith beds of Baja California Sur, México. *Aquatic Botany., 139*, 37–47.
Melbourne, L. A., Denny, M. W., Harniman, R. L., Rayfield, E. J., & Schmidt, D. N. (2018). The importance of wave exposure on the structural integrity of rhodoliths. *Journal of Experimental Marine Biology and Ecology, 503*, 109–119.
Möller, O. O., Jr., Piola, A. R., Freitas, A. C., & Campos, E. J. D. (2008). The effects of river discharge and seasonal winds on the shelf off southeastern South America. *Continental Shelf Research, 28*, 1607–1624.
Moura, R. L., et al. (2016). An extensive reef system at the Amazon River mouth. *Science Advances., 2*, e1501252.
Nalin, R., Nelson, C. S., Basso, D., & Massari, F. (2008). Rhodolith-bearing limestones as transgressive marker beds: Fossil and modern examples from North Island, New Zealand. *Sedimentology, 55*(2), 249–274.
Negrão, F., Lacerda, C. H. F., Melo, T. H., Bianchini, A., Calderon, E. N., Castro, C. B., et al. (2021). The first biological survey of the Royal Charlotte Bank (SW Atlantic) reveals a large and diverse ecosystem complex. *Estuarine, Coastal and Shelf Science, 255*, 107363.
Neill, K. F., & Nelson, W. A. (2016). Soft sediment macroalgae in two New Zealand harbours: Biomass, diversity and community composition. *Aquatic Botany, 129*, 9–18.

Olson, D. B., Podestá, G. P., Evans, R. H., & Brown, O. B. (1988). Temporal variations in the separation of Brazil and Malvinas currents. *Deep-Sea Research., 35*, 1971–1990.

Ordines, F., et al. (2015). Red algal beds increase the condition of nekto-benthic fish. *Journal of Sea Research., 95*, 115–123.

Pascelli, C., et al. (2013). Seasonal and depth-driven changes in rhodolith bed structure and associated macroalgae off Arvoredo Island (southeastern Brazil). *Aquatic Botany., 111*, 62–65.

Pereira-Filho, G. H., et al. (2012). Extensive rhodolith beds cover the summ its of southwestern Atlantic Ocean seamounts. *Journal of Coastal Research., 28*, 261–269.

Pereira-Filho, G. H., Candido, L. C., Moreira, P. S., Laranja, D. H., Motta, F. D. S., & Comin, E. (2023). Nota Técnica conjunta n° 001/2023 IMar/Unifesp–Fundação Florestal de São Paulo: descoberta do banco de rodolitos na Ilha das Couves, Ubatuba, SP.

Poore, A. G. B., et al. (2012). Global patterns in the impact of marine herbivores on benthic primary producers. *Ecology Letters, 15*, 912–922.

Pörtner, H. O., Roberts, D. C., Poloczanska, E. S., Mintenbeck, K., Tignor, M., Alegría, A., … & Okem, A. (2022). IPCC, 2022: Summary for policymakers.

Queiroz, E. V., et al. (2016). Morphological characteristics of rhodolith and correlations with associated sediment in a sandstone reef: Northeast Brazil. *Regional Studies in Marine Science., 8*, 133–140.

Reich, P. B., et al. (2014). Temperature drives global patterns in forest biomass distribution in leaves, stems, and roots. *Proceedings of the National Academy of Sciences, 111*, 13721–13726.

Riosmena-Rodríguez, R., Nelson, W., & Aguirre, J. (Eds.). (2017). *Rhodolith/Maërl beds: A global perspective*. Springer International Publishing.

Riul, P., Targino, C. H., Farias, J. D. N., Visscher, P. T., & Horta, P. A. (2008). Decrease in *Lithothamnion* sp. (Rhodophyta) primary production due to the deposition of a thin sediment layer. *Journal of the Marine Biological Association of the United Kingdom, 88*, 17–19.

Riul, P., et al. (2009). Rhodolith beds at the easternmost extreme of South America: Community structure of an endangered environment. *Aquatic Botany., 90*, 315–320.

Rothman, M. D., et al. (2017). A phylogeographic investigation of the kelp genus laminaria (Laminariales, Phaeophyceae), with emphasis on the South Atlantic Ocean. *Journal of Phycology, 53*, 778–789.

Rovai, A. S., Riul, P., Twilley, R. R., Castañeda-Moya, E., Rivera-Monroy, V. H., Williams, A. A., et al. (2016). Scaling mangrove aboveground biomass from site-level to continental-scale. *Global Ecology and Biogeography, 25*, 286–298.

Sañé, E., et al. (2016). Environmental factors controlling the distribution of rhodoliths: An integrated study based on seafloor sampling, ROV and side scan sonar data, offshore the W-pontine archipelago. *Continental Shelf Research., 129*, 10–22.

Scherner, F., Riul, P., Bastos, E., Bouzon, Z. L., & Horta, P. A. (2010). Herbivory in a rhodolith bed: A structuring factor. *Pan-American Journal of Aquatic Sciences, 5*, 358–366.

Schmid, C., Sharfer, H., Podestá, G., & Zenk, W. (1995). The Vitória eddy and its relation to the Brazil current. *Journal of Physical Oceanography, 25*, 2532–2546.

Schubert, N., Hofmann, L. C., Almeida Saá, A. C., Moreira, A. C., Arenhart, R. G., Fernandes, C. P., et al. (2021). Calcification in free-living coralline algae is strongly influenced by morphology: Implications for susceptibility to ocean acidification. *Scientific Reports, 11*(1), 1–14.

Schubert, N., Peña, V., Salazar, V. W., Horta, P. A., Neves, P., Ribeiro, C., et al. (2022). Rhodolith physiology across the Atlantic: Towards a better mechanistic understanding of intra-and interspecific differences. *Frontiers in Marine Science, 16648714*, 112035.

Silveira, I. C. A., Napolitano, D. C., & Farias, I. U. (2020). Water masses and oceanic circulation of the Brazilian continental margin and adjacent abyssal plain. In *Brazilian deep-sea biodiversity* (pp. 7–36). Springer.

Smale, D. A., & Wernberg, T. (2013). Extreme climatic event drives range contraction of a habitat-forming species. *Proceedings of the Royal Society B, 280*, 20122829.

Spalding, M. D., Fox, H. E., Allen, G. R., Davidson, N., Ferdaña, Z. A., Finlayson, M., Halpern, B. S., Jorge, M. A., Lombana, A., Lourie, S. A., Martin, K. D., McManus, E., Molnar, J., Recchia, C. A., & Robertson, J. (2007). Marine ecoregions of the world: A bioregionalization of coast and shelf areas. *Bio Science., 57*, 573–583.

Steller, D. S., & Foster, M. S. (1995). Environmental factors influencing distribution and morphology of rhodoliths in Bahia Concepcion, B. C. S. Mexico. *Journal of Experimental Marine Biology and Ecology, 194*, 201–212.

Steller, D. L., et al. (2003). Rhodolith bed diversity in the Gulf of California: The importance of rhodolith structure and consequences of disturbance. *Aquatic Conservation: Marine and Freshwater Ecosystems., 13*, S1.

Strub, P. T., James, C., Combes, V., Matano, R., Piola, A., Palma, E., Saraceno, M., Guerrero, R., Fenco, H., & Etcheverry, L. R. (2015). Altimeter derived seasonal circulation on the Southwest Atlantic shelf: 27°–43°S. *Journal of Geophysical Research, Oceans, 120*, 3391–3418.

Teichert, S., & Freiwald, A. (2014). Polar coralline algal CaCO 3-production rates correspond to intensity and duration of the solar radiation. *Biogeosciences, 11*, 833–842.

Teichert, S., et al. (2012). Rhodolith beds (Corallinales, Rhodophyta) and their physical and biological environment at 80 31′ N in Nordkappbukta (Nordaustlandet, Svalbard archipelago, Norway). *Phycologia, 51*, 371–390.

Testa, V., & Bosence, D. W. J. (1999). Physical and biological controls on the formation of carbonate and siliciclastic bedforms on the northeast Brazilian shelf. *Sedimentology, 46*, 297–301.

Ulstrup, K. E., et al. (2006). Variation in bleaching sensitivity of two coral species with contrasting bleaching thresholds across a latitudinal gradient on the GBR. *Marine Ecology Progress Series, 314*, 135–148.

Vale, N. F., et al. (2018). Structure and composition of rhodoliths from the Amazon River mouth, Brazil. *Journal of South American Earth Sciences., 84*, 149–159.

Vásquez-Elizondo, R. M., & Enríquez, S. (2016). Coralline algal physiology is more adversely affected by elevated temperature than reduced pH. *Scientific Reports, 6*, 1–14.

Vergés, A., Steinberg, P. D., Hay, M. E., Poore, A. G., Campbell, A. H., Ballesteros, E., et al. (2014). The tropicalization of temperate marine ecosystems: Climate-mediated changes in herbivory and community phase shifts. *Proceedings of the Royal Society B, 281*, 20140846.

Vieira-Pinto, T., et al. (2014). *Lithophyllum* species from Brazilian coast: Range extension of *Lithophyllum margaritae* and description of *Lithophyllum atlanticum* sp. nov. (Corallinales, Corallinophycidae, Rhodophyta). *Phytotaxa, 190*, 355–369.

Villas-Bôas, A. B., Tâmega, F. T. D. S., Andrade, M., Coutinho, R., & Figueiredo, M. A. D. O. (2014). Experimental effects of sediment burial and light attenuation on two coralline algae of a deep water Rhodolith bed in Rio de Janeiro, Brazil. *Cryptogamie Algologie, 35*, 67–76.

Wahl, M. (2008). Ecological lever and interface ecology: Epibiosis modulates the interactions between host and environment. *Biofouling, 24*, 427–438.

Wernberg, T., et al. (2016). Climate-driven regime shift of a temperate marine ecosystem. *Science, 353*, 169–172.

Wilson, S., et al. (2004a). Environmental tolerances of free-living coralline algae (maerl): Implications for European marine conservation. *Biological Conservation, 120*, 279–289.

Wilson, S., Blake, C., Berges, J. A., & Maggs, C. A. (2004b). Environmental tolerances of free-living coralline algae (maerl): Implications for European marine conservation. *Biological Conservation, 120*, 279–289.

Woelkerling, W. J. (1998). *The coralline red algae: An analysis of the genera and subfamilies of non-geniculate Corallinaceae*. Oxford University Press.

Chapter 8
Major Threats to Rhodolith Beds: Ocean Acidification, Global Warming, and Local Stressors

Gabrielle Koerich, Marina Nasri Sissini, and Paulo Antunes Horta

Abstract Since the industrial revolution, the human population has accelerated its magnitude of impact on the world's oceans. The observed consequences of our rising population and globalization have expanded substantially and are expected to affect even the deepest ecosystems. The extensive rhodolith beds along the Brazilian coastline that are present from the shallows down to 133 m are predicted and already observed to suffer from the consequences of human interference. Rhodolith beds are predicted to experience a daunting diversity of anthropogenic threats, which act at different scales. Global stressors such as ocean acidification and global climate change are shown to affect fundamental metabolic processes, which over time are expected to jeopardize the integrity of these ecosystems. Local stressors such as nutrient runoff, pollution, oil/gas exploitation, predatory fishing as bottom trawling, and direct coralline mining are expected to interact with global stressors and, in multiple cases, exacerbate already negative prognosis.

Keywords Anthropogenic impacts · Physiological responses · Eutrophication · Resilience · Ecosystem services and functions

G. Koerich (✉)
Phycology Laboratory, Botanical Department, Federal University of Santa Catarina, Florianópolis, Brazil

Federal University of Santa Catarina, Florianópolis, Brazil
e-mail: gabrielle.koerich@pg.canterbury.ac.nz

M. N. Sissini
Programa de Pós-Graduação em Ecologia, Universidade Federal de Santa Catarina, Florianópolis, SC, Brazil

P. A. Horta
Laboratório de Ficologia, Departamento de Botânica, Universidade Federal de Santa Catarina, Florianópolis, SC, Brasil

P. A. Horta, M. N. Sissini (eds.), *Brazilian Rhodolith Beds*, Brazilian Marine Biodiversity, https://doi.org/10.1007/978-3-031-61449-1_8

8.1 Changes in Atmospheric CO_2 and Its Consequences

Earth's climate is greatly defined by a relatively delicate equilibrium in the composition of the atmosphere. As radiation enters the atmosphere, it interacts with gases capable of absorbing and emitting the radiant energy. This fraction of gases represents less than 1% of the total atmosphere mass and is composed, among others, by water vapor, ozone, methane, nitrous oxide, and the now widely known carbon dioxide (CO_2) (Hartmann, 2016). CO_2, among the other gases, is a greenhouse gas, trapping heat radiating from Earth back to space. This phenomenon is responsible for keeping Earth warm enough to sustain life. As such, the concentration of this gas in the atmosphere is directly related to how much heat is retained in the planet. The natural CO_2 concentrations between glacial and interglacial periods ranges between 180 and 300 ppm (Ruddiman, 2008). As of March of 2019, CO_2 concentration in the atmosphere reached 400 ppm (NOAA 2019), a value higher than any recorded on Earth in the past 800,000 years (Lüthi et al., 2008). This anthropogenic alteration of the atmosphere has reverberating effects on Earth's climate and biogeochemical cycles and in important ecosystems, such as rhodolith beds.

8.1.1 *Ocean Warming*

The aforementioned increase in CO_2 concentration in the atmosphere has already resulted in an increase of 1 °C in the global average temperature since 1880 (NASA, 2018). By the end of the century, the projected global mean surface temperature can reach up to 4.3 °C if emissions stay business-as-usual (IPCC, 2019). Increasing temperatures have cascading effects not only in the climate system but also in ecosystems and the services provided by them. With a changing climate, species have either to adapt, tolerate, move to a new location, or endure extinction, and what will determine the outcome is their physiological, ecological, and evolutionary response (Berg et al., 2010). As such, investigating the impacts of ocean warming in coralline algae is of extreme importance to determine the future of rhodolith beds.

The distribution and phenology of marine algae has a strong relationship with temperature (Hoek, 1982). Many metabolic processes in algae are affected by temperature, and rhodoliths are no exception. Enzymatic-regulated processes, such as growth (englobing carbon incorporation, calcification), photosynthesis, and respiration, are temperature-sensitive in rhodoliths (Steller et al., 2007). In fact, rhodoliths are often used as climate proxies, as their calcification rates vary according to climatic conditions, and their growth bands can be used to obtain paleotemperature information, as less extensively calcified cells represent fastest growth in warmer temperatures and more-extensively calcified cells are found in colder temperatures (Kamenos & Law, 2010; Martin et al., 2013). The effects of elevated temperatures in the calcification of coralline algae were recently reviewed by Cornwall et al. (2019). The authors found a general trend across multiple studies that coralline algal calcification is affected when temperatures are ~5 °C above the ambient

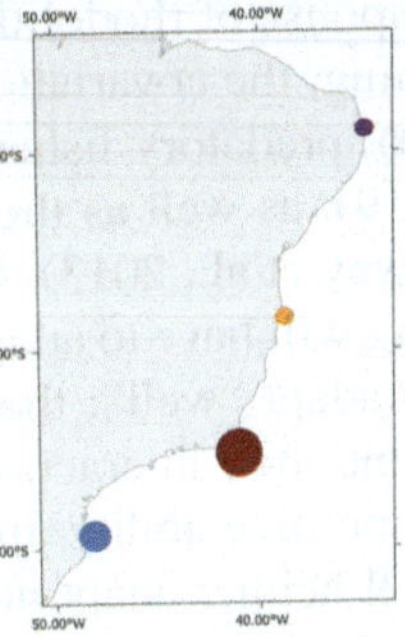

Study	Stressor(s)	Overall trend	Descriptor(s)	Species	Life form
Schubert et al. (2019)	Heat-wave + nutrient enrichment	↓	Photosynthesis + calcification + pigments + others	*Lithothamnion crispatum* + *Melyvonnea erubescens*	Rhodolith
Muñoz et al. (2018)	Ocean acidification + heat-wave	↓	Photosynthetic performance + calcification + others	*Lithothamnion crispatum*	Rhodolith
Scherner et al. (2016)	Ocean acidification + ocean warming	—	Photosynthetic performance	*L. stictaeforme* + *P. conicum* + *P. pachydermum*	CCA
Riul et al. (2008)	Sediment deposition	↓	Photosynthesis	*Lithothamnion* sp.	Rhodolith
Figueiredo et al. (2015)	Sediment deposition	↓	Heat-wave + nutrient enrichment	*Mesophyllum engelhartii* + *Lithothamnion* sp.	Rhodolith
Villas-Bôas (2014)	Sediment deposition + light attenuation	↓	Photosynthetic performance + color changes	*Mesophyllum engelhartii* + *Lithothamnion* sp.	Rhodolith
Reynier et al. (2015)	Sediment deposition	↓	Photosynthesis + Photosynthetic performance	*Mesophyllum engelhartii* + *Lithothamnion* sp.	Rhodolith

Fig. 8.1 Main studies with species of coralline red algae from Brazil. Colors of circles represent the locations of algae samplings; size of circles are equivalent to the number of studies in the region. The overall trend of response from the algae is represented by a downward arrow when negative effects of stressors were observed and by a line when no effects were observed. Photosynthesis refers to measurements of oxygen production (either with respiration measurements or not), and photosynthetic performance refers to measurements of chlorophyll a fluorescence. *CCA* crustose coralline algae

temperatures of both cool and warm temperate regions and tropical reefs. However, each species has a specific response to temperature (and other environmental conditions—see Schubert et al. (2019)), and other physiological and reproductive features may be impaired by ocean warming, even when calcification is not (Cornwall et al., 2019). Thus, to understand the general trends in response of rhodolith-forming species, we need to look at various indicators of the effects of elevated temperatures in species from different environments (Fig. 8.1).

The response of coralline algae to elevated temperatures depends on their thermal range of occurrence. Temperatures within the range that the algae species is historically and evolutionarily adapted to present optimal physiological conditions, while temperatures higher or lower to those will represent suboptimal physiological conditions; temperatures above a certain threshold will be detrimental to the algae (Fortes & Lüning, 1980; Lüning, 1990). This relationship with temperature depends on the distribution of the species (temperate species are adapted to a wider range of temperatures than tropical species) and the depth of their occurrence (e.g., intertidal species are exposed to a wider range of temperatures than benthic species) (Martin & Hall-Spencer, 2017). A small increase in temperature has positive (Martin et al., 2013) or no effects (Noisette et al., 2013a) in growth, calcification, and photosynthesis of rhodoliths. However, temperatures above a certain threshold are damaging for rhodoliths, either from temperate or tropical regions, causing bleaching and declines in photosynthesis and calcification (Anthony et al., 2008). For example, seasonal fluctuations in seawater temperature cause changes in photosynthetic and respiration rates of *Lithophyllum margaritae* (Hariot) Heydrich, but when temperatures surpassed 30 °C, lower photosynthetic and growth rates were observed (Steller et al., 2007). At 32 °C, the rhodolith-forming species *Neogoniolithon* sp. Setchell & L.R.Mason showed declines in photosynthesis and presented photodamage (Vásquez-Elizondo & Enríquez, 2016).

What we can expect by looking at these studies is that the response of rhodoliths to ocean warming will depend on the species, the rate of warming, the covariation of other stressors as ocean acidification (Scherner et al., 2016), predatory fishery (Fragkopoulou et al., 2021), and pollution (Schubert et al., 2019), as well as their communities (Rich et al., 2018), life histories, and habitat (Harvey et al., 2013). In addition, populations that are not extinguished by ocean warming will have to adapt to these new conditions, but this does not mean that they will adapt "well": their fitness will likely be affected, resulting in reduced abundance, and their interaction with other species will be altered, reflecting in a reduced competitive ability and defense against predation (Harley et al., 2012). Species that will neither adapt nor become extinct in the future will have to migrate or rely on stochastic dispersal to new locations. This is a trend already being observed and that is expected to become more frequent in the future, where species experience a range shift to locations with environmental conditions more similar to what they are adapted to (Poloczanska et al., 2013). The bigger expression of this trend is the tropicalization of temperate marine ecosystems, which will also represent changed ecological interactions, as grazers and other organisms will also be under the same shift (Vergés et al., 2014).

8.1.2 Marine Heat Waves: Punctual Events but Lingering Threats

As the ocean absorbs heat from the atmosphere, marine heat waves (periods of extreme warming of the sea surface, lasting from days to months) have increased in frequency, intensity, and extension over the last decades, and this trend is set to persist in the remainder of the century (Frölicher et al., 2018). Extreme events are now being regarded as important drivers of species distributions and in biodiversity patterns, and thus, the increase in their intensity and frequency will lead to changes in communities composition and cause range shifts of species (Wethey et al., 2011; Wernberg et al., 2013; Rilov et al., 2019). In a long-term monitoring of kelp forests and coral reefs in Western Australia, researchers were able to document how an extreme heat wave caused the loss of approximately 2300 km^2 of kelp forests, leading to a community shift to species of subtropical and tropical waters (Wernberg et al., 2016). The loss of benthic cover by kelp and encrusting coralline algae and their replacement by turf-forming algae was also observed after a heat wave in 2011 in Australia (Wernberg et al., 2013).

For coralline algae, there are very few studies investigating the impacts of marine heat waves. Short et al. (2015) measured growth, calcification, and mortality rates of a crustose coralline algae (CCA) subfamily *Hydrolithoideae* A. Kato & M. Baba in response to a heat wave in Western Australia. During summer, higher calcification rates are expected, but in a summer with registered temperature anomalies, increased mortality was observed, and no changes in calcification rates were detectable between seasons (Short et al., 2015). Sordo et al. (2019), in an experiment testing the effects of marine heatwaves in *Phymatolithon lusitanicum* V. Peña from Portugal, described positive effects of increasing temperatures on photosynthesis, calcification, and respiration rates. However, more studies encompassing the responses of rhodolith-forming species to events of anomalous warming are necessary, as this information is virtually nonexistent, and species adapted to different environmental conditions will respond distinctively to heat waves (Cornwall et al., 2019). Schubert et al. (2019) investigated the effects of marine heat waves in rhodoliths from a marine reserve in Southern Brazil (27°S, 48°W). *Melyvonnea erubescens* (Foslie) Athanasiadis and Ballantine (2014) and *Lithothamnion crispatum* Hauck (1878) were exposed to a sudden increase of 5 °C (from 23 °C of the field to 28 °C). *M. erubescens* presented a decrease in net photosynthesis and calcification rates in response to the increase in temperature, while *L. crispatum* upregulated its net photosynthesis but, at the same time, experienced decreased calcification rates. These results indicate that the impacts of heat waves will affect different species composing rhodolith beds distinctively, but the overall losses in primary production and carbonate storage will likely affect the dynamic of these ecosystems. Moreover, the species studied by Schubert et al. (2019) have a wide occurrence in the Brazilian coast, and assessing the response of distinct populations is also important to understand if populations locally adapted to different conditions will be affected differently (Sgrò et al., 2011). Further studies of the impacts of extreme events in the primary and carbonate production by different Brazilian rhodolith populations are necessary to allow the full comprehension of ecosystem's goods and services losses in the future, as well as the response of different populations of the same species.

8.1.3 Ocean Acidification

Not only the ocean absorbs heat, but it is also the main agent in the absorption of atmospheric CO_2, storing approximately 30% of the anthropogenic carbon emitted to the atmosphere (IPCC, 2019). Surely, this exchange comes at a price. The higher uptake of CO_2 by oceans directly impacts water's pH, making it more acidic and lowering calcium carbonate ($CaCO_3$) saturation in surface waters. The decrease in pH and alteration in fundamental chemical balances in oceans are called ocean acidification (Doney et al., 2009). According to the Intergovernmental Panel on Climate Change (IPCC), there is high confidence that if CO_2 emissions continue, ocean acidification will persist and highly affect marine ecosystems.

Before the industrial revolution, the average ocean surface water pH was approximately 8.17 but has now decreased to 8.11, and it can decrease even further depending on the rate of emissions (IPCC, 2014). The ocean carbonate system is governed by a series of chemical reactions:

$$CO_{2(atm)} \leftrightarrow CO_{2(aq)} + H_2O \leftrightarrow H_2CO_3 \leftrightarrow H^+ + HCO_3^- \leftrightarrow 2H^+ + CO_3^{2-}$$

When CO_2 is dissolved in seawater, it generates carbonic acid (H_2CO_3), which then forms bicarbonate (HCO_3^-) and carbonate (CO_3^{2-}) by the loss of hydrogen ions

(H^+). By adding CO_2 to seawater, the concentrations of bicarbonate and hydrogen ions also increase. More hydrogen ions result in lower pH because:

$$pH = -\left[H^+\right]$$

which, at the same time, decreases carbonate ion concentration.

The ability of global oceans to absorb and store CO_2 depends on the extent of calcium carbonate ($CaCO_3$) dissolution in seawater, which in turn depends on the concentration of CO_3^{2-} and Ca. $CaCO_3$ has a biogenic origin, found in shells and skeletons of marine organisms, and, in the focus of this book, coralline red algae. The saturation state (Ω) of $CaCO_3$ is therefore:

$$\Omega = \frac{\left[Ca^{2+}\right]\left[CO_3^{2-}\right]}{K'_{sp}}$$

where K'_{sp} is the apparent solubility product, which depends on seawater's salinity, temperature, pressure, and the particular mineral phase. $CaCO_3$ has six polymorphs, two of which are important components in the formation of shells and skeletons of marine organisms: aragonite and calcite. Aragonite is approximately 50% more soluble than calcite, which makes its saturation state's value lower. Formation of $CaCO_3$ happens where $\Omega > 1.0$ and dissolution occurs where $\Omega < 1.0$. Therefore, the increasing input of CO_2 is shallowing the saturation horizons of both aragonite and calcite, as saturation states are directly related to pressure. Coralline red algae deposit $CaCO_3$ in the polymorph of high magnesium calcite (Mg-calcite, or dolomite), more soluble than calcite and aragonite (50 and 20%, respectively), which makes them especially vulnerable to changes in the saturation state (Ragazzola et al., 2013).

Rhodolith beds are, therefore, important in the global carbon cycle, by the storage of carbon in the form of $CaCO_3$. Only in the Abrolhos Shelf in Brazil, the estimation is that $CaCO_3$ production is of 0.025 Gt year^{-1}, comparable to the production of coral reefs in the Caribbean (Amado-Filho et al., 2012). Considering that rhodolith beds occur from Amazon's mouth to southern Brazil, with 230,000 km^2 of habitat suitability only on the Brazilian coast (Carvalho et al., 2019), their global contribution to carbon storage is immense. However, the solubility of their Mg-calcite-rich skeletons makes them highly susceptible to dissolution under future scenarios of ocean acidification, and their role in the global carbon cycle will be impaired.

The vulnerability of rhodoliths to ocean acidification has been assessed in many experimental studies. While some studies have found negative effects of decreasing pH in coralline species, such as decreasing photosynthetic (Anthony et al., 2008) and calcification rates (Büdenbender et al., 2011; Kato et al., 2014), and loss of structural integrity (Ragazzola et al., 2012), others have found either no effects or positive effects on such physiological rates (Noisette et al., 2013a, b; Sordo et al., 2016). Nonetheless, it is necessary to consider that these responses are species-specific and also depend on the proportion of dolomite found in their skeletons (Mccoy & Kamenos, 2015; Nash et al., 2019). In an investigation of the short-term effects of decreased pH (7.4) in the physiology of *Lithothamnion crispatum* from Southern Brazil, decalcification and reduced photosynthetic yield were observed (Muñoz et al., 2018). CCA from northeastern Brazil also demonstrated a decline in photosynthetic yield after exposure for 23 days to a decrease of 0.9 in pH (Scherner et al., 2016).

Another important factor to be considered is the time of exposition. Sordo et al. (2018) observed that after 11 months of exposure to high CO_2 concentrations, the calcification and photosynthesis rates of the mäerl *Phymatolithon lusitanicum* increased. However, after 20 months of exposure, this pattern was reversed, indicating that although coralline algae can maintain high production rates for some time at low pH, they do not present resistance to it in the long term (Sordo et al. 2018). This lower resistance will facilitate bleaching, diseases, and herbivory and reduce the ability of coralline algae to compete with fleshy algae (Short et al., 2014; Sordo et al., 2018). In this regard, Burdett et al. (2018) described the sensitivity of a community formed by coralline algae to CO_2 enrichment, where a short-term exposure to higher concentrations of CO_2 led a shift to net community dissolution of a local coralline algal bed, composed primarily by *Lithothamnion glaciale* Kjellman. This supports the understanding that in a higher-CO_2 world, rhodolith beds will experience a transition in community composition, with species more tolerant to these conditions becoming the foundation of the community (Celis-Plá et al., 2015; Burdett et al., 2018).

8.2 Bad Things Come in More Than Threes

As we alter global climate and biogeochemical systems, we also are changing environments at a local scale. Everything we add in our water ends up draining to the ocean, from chloride to nutrients derived from agricultural lands, even reaching ecosystems far from the coast. The major contributions of agricultural activities to water pollution are nutrients, pesticides, metals, organic matter, and sediments, among others (FAO, 2017). Urbanization and economic activities such as predatory fishery (using bottom trawling), mining, and petroleum extraction also pose threats to marine ecosystems, considering the history of accidents related to these activities (Sissini et al., 2020; Fragkopoulou et al., 2021) and the impact caused by offshore exploration. But rhodolith beds have their own economic value due to their CO_3-rich skeletons. They have been proposed for agronomic applications as fertilizers and for the correction of acidic soils (Foster, 2001; Horta et al., 2016). This demonstrates that rhodoliths beds are in a conflict zone—between activities with economic value in these ecosystems and human use of the functions and services provided by them—their conservation seems like a complicated puzzle. Herein, we discuss the major local threats to Brazilian rhodolith beds.

8.2.1 Increased Nutrient Runoff

The explosive increase in eutrophication of coastal waters globally is related to the human-controlled creation of reactive nitrogen (Howarth, 2008) and phosphorus, especially for the production of fertilizers (Ngatia et al., 2019). As human populations continue to grow, thus increasing the demand for food and the discharge of wastewater—another important source of nutrients to marine waters—eutrophication in coastal zones related to human activities will continue to increase. Nutrients are important to the growth of macroalgae, but calcareous algae may be sensitive to coastal pollution (Bjork et al., 1995). Brazil has a historical problem with coastal pollution. Half of the population (~100 million people) still does not have access to adequate treatment and disposal of sewage (De Sousa & Barrocas, 2017), and where wastewater treatment is available, it does not remove emerging contaminants, such as hormones, pesticides, and illicit drugs (Campestrini & Jardim, 2017). Moreover, the use of fertilizers in Brazil reached a historical record in 2018, when 35,506,299 tons were delivered in the country, according to the National Association for the use of Fertilizers (ANDA, 2019). It is difficult to estimate the quantity of nutrients from point and nonpoint sources transported by rainfall, runoff, and leaching that reach the coastal zone, but the effects are already being observed. The increased presence of sewage-derived nutrients in coastal areas of Brazil led to decreasing species richness and resilience of benthic communities (Martins et al., 2012). Scherner et al. (2013) investigated the changes in coastal assemblages in the Brazilian coast caused by increased sewage discharge and demonstrated declines in calcareous algae in regions with high urban pressure.

Primary production is limited by nutrient availability, and it is no different in the marine environment. Moderate nutrient enrichment can be beneficial to algae, increasing growth rates and maybe even helping them cope with climate change (Celis-Plá et al., 2015). However, in high concentrations, nutrients can be detrimental and affect important metabolic processes and compromise ecosystem services provided by rhodolith beds (Schubert et al., 2019). For example, phosphorus has been regarded as a "poison" for calcification processes, as it inhibits calcite crystal growth (Simkiss, 1964; Bjork et al., 1995). Moreover, fleshy epiphytes will obtain advantage in a scenario with higher availability of nutrients, as they are fast-growing and can thus rapidly cover calcareous algae, which have much lower growth rates and will likely be affected by eutrophic conditions (Russell et al., 2009). Furthermore, the runoff from coastal zones is accompanied by sediments, increasing the turbidity of the water and reducing light availability for photosynthesis. The congregation of these conditions may lead to reduced fitness of coralline algae and, consequently, a shift to a different equilibrium state of the local community.

Some experimental studies have explored the effects of nutrient enrichment in coralline algae. In an investigation of the impacts of increased nitrate (40 μmol/L) and phosphate (1 μmol/L) concentrations on three species of temperate mäerl, Qui-Minet et al. (2019) found no negative effect in calcification in any of the species. Similarly, positive effects of nitrogen enrichment on the calcification rates and photosynthetic pigments of the reef-building CCA, *Porolithon onkodes* (Heydrich) Foslie (Johnson & Carpenter, 2018). Conversely, nutrient enrichment led to lower CCA (*Lithophyllum* Philippi sp.) dry mass, although no impacts in the photosynthetic yield were observed (Russell et al., 2009). In a study of the impacts of nutrient increase in two species of rhodoliths from Arvoredo's Island (Santa Catarina, Brazil), *Melyvonnea erubescens* presented a reduction in net photosynthesis under increased nutrient conditions, while *Lithothamnion crispatum* did not respond to such conditions (Schubert et al., 2019). The results from these studies show that responses to nutrient enrichment are species-specific, and as rhodolith communities are composed of many algae species, community composition will be changed by local nutrient conditions. Therefore, the productivity and carbon storage of rhodolith beds in coastal areas impacted by nutrient runoff will differ according to the species composing the community (Schubert et al., 2019). However, many more studies of the increasing nutrient overload in coastal areas are necessary for the assessment of the precise impacts and possible changes in rhodolith communities' structure, composition, and functioning.

8.2.2 Experimental Evaluation of the Impacts of Increased Nutrient Concentrations

Coastal and oceanic populations are exposed to different environmental conditions, and as such, their response to the same stressor may be different. Oceanic islands are considered as "model systems" for ecological studies, being less impacted by human activities than the coastal zone (Vitousek, 2002; Williams et al., 2015). Island populations are adapted to less variation in nutrient conditions, differently than coastal populations, which are exposed to high variation in nutrient concentrations due to continental runoff (Scherner et al., 2012). Therefore, island populations may be less resistant to increased nutrient and reduced light conditions and understanding their response to these conditions gives insights about the sensibility of rhodoliths to them. Hence, in order to assess the impact of nutrient increase and reduction of light penetration in the water column in populations of oceanic islands, two field experiments were made, one in Fernando de Noronha (7 days) and one in Trindade Island (15 days). For each experiment, three species were chosen based on their abundance in the island, and each species represented one of the major groups of algae—red, brown, and green. As such, in Trindade Island, *Halimeda* sp., *Dictyota* sp., and *Lithothamnion* sp. were used in the experiment. As for Fernando de Noronha, the species chosen were *Sargassum* sp., *Dictyopteris* sp., and a rhodolith-forming species. The depth of sampling of rhodoliths was 6 m in Trindade and 1–2 m in Fernando de Noronha.

Before the beginning of each experiment, algae were collected, cleaned from epiphytes, and weighted to 5 grams of fresh biomass. They were then placed in cylinders (1 L of volume) made of a plastic net with holes of 1 cm. Each experiment was composed of four treatments: control (environment irradiance and oligotrophic conditions), low irradiance and oligotrophic conditions, environment irradiance and eutrophic conditions, and low irradiance and eutrophic conditions. To simulate light reduction, a shading screen was placed on top of the cylinders, and the eutrophic condition was replicated by adding NPK nutrients of low and fast dilution. Each treatment had three replicas, and cylinders were randomly placed 1 m apart in a longline at open sea. To evaluate the physiological conditions of each species, measurements of effective quantum yield (Fv/Fm′) were taken with a pulse amplitude-modulated chlorophyll *a* fluorometer (Diving PAM—Walz Germany). Data was tested for normality with the Shapiro-Wilk test ($p < 0.05$), and homogeneity of variances was tested with Levene's test of the median ($p < 0.05$). An analysis of variance (ANOVA) of two factors was then made for each experiment. Post hoc test of Tukey separated statistically different groups ($p < 0.05$). All data analyses were made in R (version 3.6.0).

The ANOVA analysis showed a significant interaction ($p < 0.05$) between irradiance and nutrient concentrations in different species in the Trindade experiment (Fig. 8.2), but for the Fernando de Noronha experiment, only different concentrations of nutrients and species presented a significant interaction (Fig. 8.3). Although the rhodolith-forming species were the group with lower Fv/Fm′ in both experiments, when compared to other species, no significant difference was observed

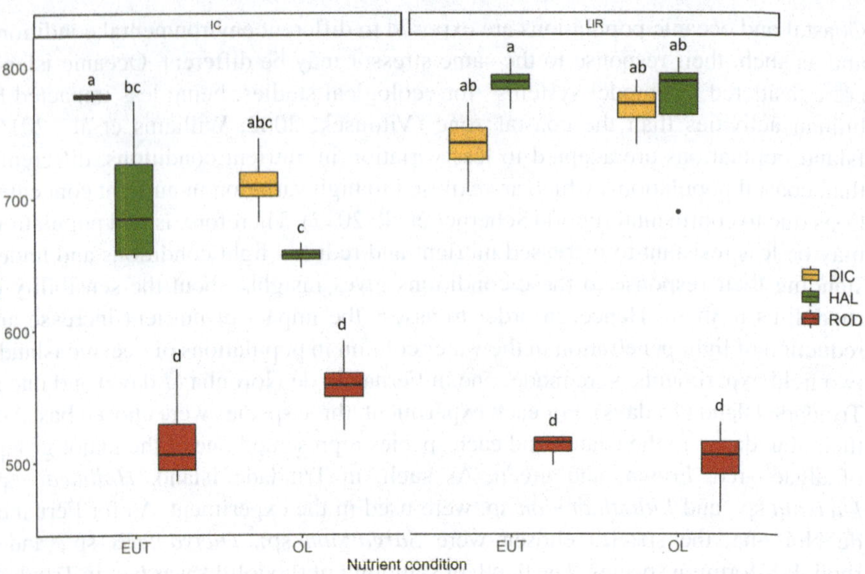

Fig. 8.2 Results from Trindade experiment. Effective quantum yield (Fv/Fm′) of *Dictyota* sp. (DIC), *Halimeda* sp. (HAL), and *Lithothamnion* sp. (ROD) in different light conditions, control (IC) and low light (LIR), and in different nutrient conditions, low nutrient concentration (OL) and high nutrient concentration (EUT). Letters represent different groups as identified by the Tukey test ($p < 0.05$), and straight lines in box plots represent the average

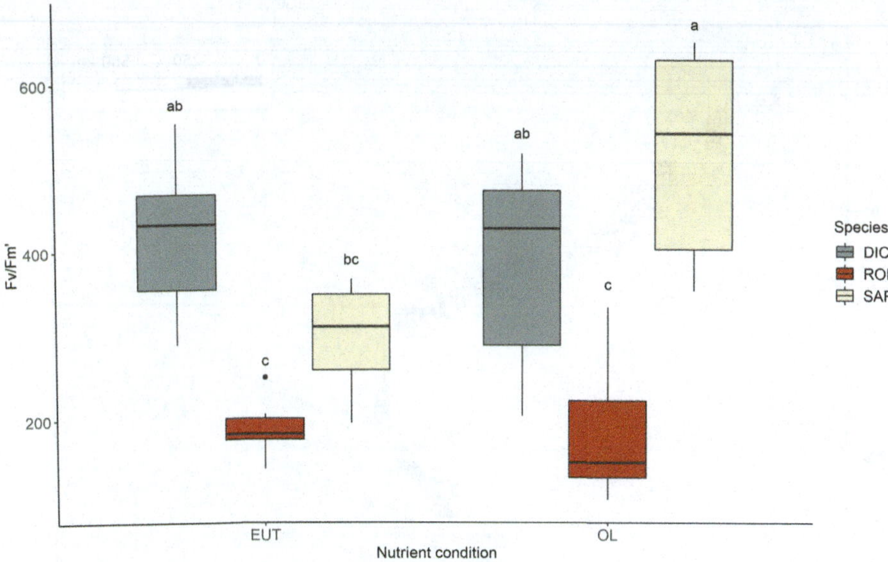

Fig. 8.3 Results from Fernando de Noronha experiment. Effective quantum yield (Fv/Fm′) of *Dictyopteris* sp. (DIC), a rhodolith-forming species. ROD and *Sargassum* sp. (SAR) in different nutrient conditions, low nutrient concentration (OL), and high nutrient concentration (EUT). Letters represent different groups as identified by the Tukey test ($p < 0.05$), and straight lines in box plots represent the average

between either different light or nutrient conditions in either experiment (Figs. 8.2 and 8.3). The concentration of nutrients used in the experiments was not detrimental to the photosynthetic efficiency of these rhodoliths, similar to what Qui-Minet et al. (2019) observed in a long-term experiment with three *mäerl* species. Moreover, neither condition of light (shaded or unshaded treatments) caused changes in the photosynthetic efficiency of rhodoliths from Trindade. This could be explained by the adaptation to the canopy column cover and depth in which the rhodoliths used in the experiments were collected (6 m), meaning that they were already adapted to lower light conditions.

8.2.3 *Economic Activities*

Rhodolith beds are spread throughout the Brazilian coast, and as such, they can be found very close to regions with economic interest (Fig. 8.4). That is the case of the deepwater rhodolith bed in the Peregrino oil field, one of the ten biggest oil fields in Brazil, according to the National Agency of Petroleum, Natural Gas and Biofuels (ANP). The Peregrino oil field is part of the Campos Basin, a region that has been intensively explored not only by activities of oil exploration but also by calcareous

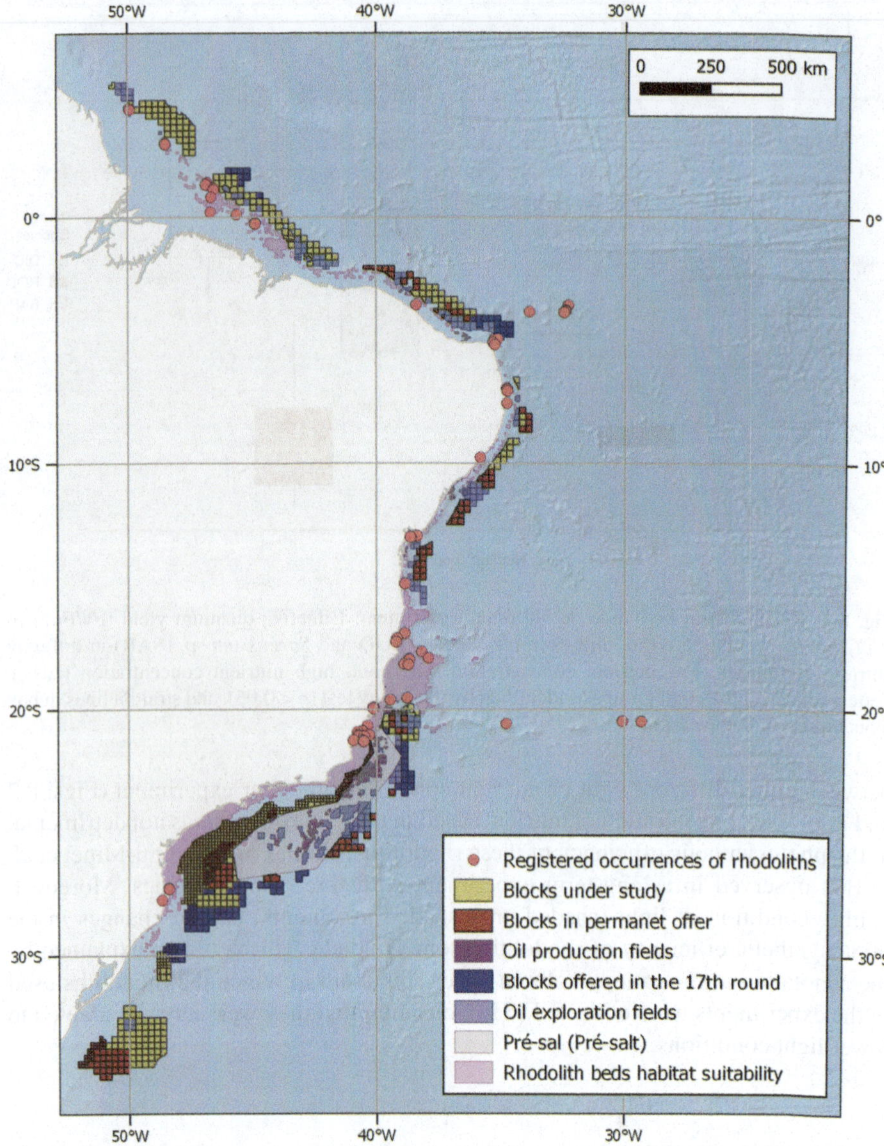

Fig. 8.4 Map showing the overlap between regions with habitat suitability and registered occurrences of rhodoliths in Brazil and fields of petroleum exploration. Oil production fields are fields with active petroleum extraction, while oil exploration fields are fields being explored in search of petroleum. Blocks are blocks for concession, which are available for auctioning in November of 2020. In the map are shown the blocks under study for auctioning, blocks in permanent offer for concessions, and blocks available in the 17th round of auctioning. Data of petroleum exploration/production was downloaded from the National Agency of Petroleum, Natural Gas and Biofuels (ANP—anp.gov.br, November 2020). Data of rhodolith bed habitat suitability and occurrences were extracted from Carvalho et al. (2019)

mining and fish dredging, which promote substantial sediment dislodgment (Villas-Bôas et al., 2014). The Peregrino oil field is one example of a conflict area, and another one is the most extensive rhodolith bed in the world, the Abrolhos bank (Amado-Filho et al., 2012). The auctioning of blocks close to the bank (as depicted in Fig. 8.4) and the oil production fields about 50 km from the reserve are of great concern for the health of the rhodolith beds (Moura et al., 2013). And more concern about petroleum activities has been raised recently due to the oil spill that affected almost the entire Brazilian coast (de Oliveira Soares et al., 2020). As brought to attention by Sissini et al. (2020), the impacts of this environmental disaster over rhodolith beds are still unknown. The Deepwater Horizon oil spill, which happened in 2010 in the Gulf of Mexico, caused a decrease of the biodiversity associated with rhodolith beds in this region (Felder et al., 2014) and also induced bleaching of rhodoliths (Fredericq et al., 2014). Therefore, it is reasonable to expect that rhodolith beds and associated fauna in regions affected by the oil spill in Brazil will endure impacts.

Rhodoliths are of economic interest as they can be used as a source of calcium carbonate and micronutrients, thus being used as fertilizers and correction of organic soils, and also in the construction sector, as raw material for cement production (Dias, 2000). Their extraction is done by dredging, and this is a practice that has been done in Europe since the 1970s (Grall & Hall-Spencer, 2003). Although rhodoliths are considered as a nonrenewable resource, as their growth rate is in average 1 mm year^{-1}, their extraction in many countries, such as Ireland (5000 t year $^{-1}$) and France (500,000 t year $^{-1}$), is a common practice (Coletti et al. 2017). In Brazil, extraction can be up to 18 t year^{-1} per company, and Brazilian law refers to rhodoliths as a nonliving resource (Santos et al., 2016). The reserves of rhodolith beds indicated for exploitation by the National Mining Agency are mostly distributed from southwest to north coast, with a total of 1,355,157,240 tons of $CaCO_3$ (Paiva et al., 2023). While Great Britain has ceased the licensing of rhodolith extraction and in Spain and the Mediterranean environmental laws regulate economic activities in some rhodolith beds (Salomidi et al., 2012), Brazil still does not have clear regulations regarding these ecosystems.

One of the many byproducts shared by activities of petroleum extraction, mining, and rhodolith extraction is the plume of sediments created, which can settle upon benthic assemblages. This load of sediments has been demonstrated to impair the photosynthetic activity of coralline algae (Wilson et al., 2004; Harrington et al., 2005), both shallow (Riul et al., 2008; Reynier et al., 2015) and deepwater species (Villas-Bôas et al., 2014; Reynier et al., 2015). The coverage of specimens *Lithothamnion* Heydrich sp. from a shallow-water coastal reef from Paraíba, Brazil, by a thin sediment layer reduced net production by 70% (Riul et al., 2008). Photosynthetic efficiency is also affected by sediment coverage, as demonstrated by Figueiredo et al. (2015). In this study, the two most dominant species of the deepwater (100 meters depth) rhodolith bed in the Peregrino oil field, *Mesophyllum engelhartii* (Foslie) W.H.Adey and *Lithothamnion* sp., presented a reduction in photosynthetic efficiency of 50% at 70% of sediment coverage, establishing a negative correlation between the added amount of sediment and photosynthetic rates. The reduction in photosynthesis efficiency is mostly due to the sediment coverage of the organisms, which affects not only light availability but also gas exchanges, which are important in the primary metabolism of these algae (Riul et al., 2008; Figueiredo et al., 2015). As a matter of fact, it is not only thin sediments that have been proven to affect rhodoliths. Natural coarse sediments from the Peregrino oil field and the complete burial of rhodoliths also caused reductions in photosynthetic efficiency, changes in coloration, and, in the case of burial, death (Villas-Bôas et al., 2014; Reynier et al., 2015).

Consequently, dredging and extraction activities close to or in rhodolith beds must be correctly regulated by environmental agencies. As highlighted by Paiva et al. (2023), the exploitation of rhodolith beds compromise the achievement of sustainable development goals, regarding the struggle against poverty, hunger, inequalities, climate change, and also for a decent work and healthy aquatic life (Paiva et al., 2023) besides other functions or contributions to people as described in Chap. 8. The Environmental Agency of the Brazilian Government (IBAMA) has, in the past, adopted the Precautionary Principle (Cooney, 2004), denying requests of oil or gas drilling near rhodolith beds (Santos et al., 2016). However, Brazil has historically focused on short-term targets to increase economic growth that often collides with sustainability needs and goals (Pereira, 2019). New practices, such as the unrestrained release of new pesticides [474 only in 2019—Oliveira and Tooge (2019)], or new legislations, like the proposition allowing mining in indigenous protected areas (Souza, 2020), are worsening this tradition. River runoff delivers these contaminants in the coastal zone, and the current mining and agricultural use of pesticides already implicate in polluted coastal zones in Brazil (Liebezeit et al., 2011; de Mahiques et al., 2013). Hydrocarbons from oil and gas exploration can also be found in coastal waters and species of economic interest (Santiago et al., 2016). To encourage the uncontrolled expansion of these activities, especially without a strong environmental agency monitoring and licensing them, IBAMA, the administrative arm of Brazil's Environmental Ministry, has been dismantled on all fronts (Tufanni, 2019) and is an allowance for the further devastation of marine ecosystems. The impacts of these hydrocarbons, metals from mining, and pesticides in rhodoliths are mostly still unassessed but have been demonstrated to impact other macroalgae (see Petsas & Vagi, 2017 for a review). For these reasons, there is an urgent need for clear and coherent regulations of these activities, both in land and ocean, to prevent the further degradation of rhodolith beds.

8.3 Combination of Global and Local Stressors

While investigating the effects of each of these stressors separately is important to understand individual mechanisms and is a cornerstone in studies of organismal physiology, assessing the impacts of multiple stressors is also important, considering that in the future they will happen simultaneously (Boyd et al., 2018). The interaction between stressors (or drivers) can have either an additive (the sum of their individual effects is either double-positive, double-negative, or opposing), antagonistic (less positive or negative than predicted additively), or synergistic effect (when the result is greater, positively or negatively, than the additive effect of individual drivers) (for further explanation of terms, see Piggott et al., 2015). As a result, assessing the combined effects of global and local stressors in rhodoliths is necessary to play out the possible scenarios of species and communities' responses to future environmental conditions.

Many studies have investigated the combined effects of two or three stressors in the physiology of coralline algae, some testing the hypothesis that one stressor may help them cope with another stressor (Hofmann et al., 2014; Qui-Minet et al., 2019). One of the most investigated interactions of stressors under this perspective is ocean acidification and warming. A few studies have demonstrated that ocean acidification exacerbates the negative effects of ocean warming in rhodoliths (Kroeker et al., 2013; Vásquez-Elizondo & Enríquez, 2016), increasing mortality, bleaching, and decreasing calcification and photosynthesis (Anthony et al., 2008; Diaz-Pulido et al., 2012). Conversely, the interaction of increased pCO_2 and temperatures in the physiology of *Phymatolithon lusitanicum* led to higher rates of photosynthesis, respiration, and calcification than when only temperatures were increased, indicating an intensification of the effects of temperatures under higher CO_2 concentrations (Sordo et al., 2019). The combination of global and local stressors, specifically nutrient enrichment, has also been investigated under this angle. Higher nitrogen concentration partially ameliorated the effects of higher pCO_2 by increasing calcification rates and photophysiology of *Porolithon onkodes* (Johnson & Carpenter, 2018). Qui-Minet et al. (2019) observed an antagonistic interaction between ocean warming, acidification, and nutrient enrichment in the physiology of three mäerl species, with nutrient enrichment partially alleviating the effects of climate change. In contrast, Schubert et al. (2019), when testing increasement of nutrients and temperature, found no effect (negative or positive) of nutrients in calcification in treatments with elevated temperature. Other aspects of coralline algae's life histories will also be affected, and their competition ability may also be impaired. For example, recruitment of CCA under higher nutrient conditions and CO_2 is highly disrupted, while turf increases their recruitment and coverage in the same conditions (Russell et al., 2009). As discussed previously, the individual effects of these stressors are species-specific, and their interaction appears to follow the same trend.

However, other factors also play an important role in mediating the response of coralline algae to changing environmental conditions. Seasonal fluctuations in temperature and irradiance (length and intensity) mediate and partially explain the ameliorated effects of ocean warming in rhodoliths, and the effects of warming will also be different between seasons (Qui-Minet et al., 2019). Moreover, coralline algae are exposed to substantial daily changes in pH and temperature, and thus, they may adapt to near-future conditions of ocean acidification (Qui-Minet et al., 2018). However, their response to long-term (> 12 months) increasement in temperature and CO_2 concentrations is likely different than their response to short-term exposure, as a result of these daily and seasonal fluctuations (Sordo et al., 2018), which implies that the mitigation of effects of one stressor by the other may also be a short-lived response. For this reason, longer studies exploring the effects of multiple stressors on rhodoliths are necessary to understand this complex dynamic.

Another important aspect that needs to be investigated regarding the impacts of multiple stressors is the effects of these on communities. Changes in community composition, which we can expect based on numerous studies demonstrating lower resistance of some species to altered environmental conditions, while others seem to cope with it, will also reflect in changed interactions. Other taxa, from other algae to fish, for example, will also be under the influence of these conditions, and experience shifts in their assemblages (Vergés et al., 2016). The expression of climate change in the interactions within the community may have a greater impact on populations than the effects of abiotic conditions and, thus, have larger effects on the resilience of ecosystems (Ockendon et al., 2014; Strona & Bradshaw, 2018). Therefore, future studies of rhodoliths will need to address the effects of changed community composition and interactions in order to more accurately predict the future of these ecosystems.

8.4 Final Remarks

In this "perfect storm" of ecological chaos, we cannot forget the underlying driver of change: human society. And human society is going to be heavily impacted by the collapse of ecosystems, as we depend greatly on services and functions provided by them (Worm et al., 2006). Rhodolith beds provide services of production through the provision of important fish habitats, of regulation by their importance in climate regulation, and of support by promoting a high biodiversity and complex ecological interactions, among others (Basso et al., 2016). The loss and/or diminishment of Brazilian rhodolith beds will result in loss of these ecosystem services, and the impacts on Brazil's economy and society are still to be assessed.

To avoid—or to at least attenuate—the loss of these important ecosystems, Brazil and other governments need to make a transformative change in their practices and processes across diverse sectors, especially the economic, business, and political ones (IPCC, 2019). Under the future climatic and environmental projections, the underlying socioeconomic and political scenario of Brazil appears to make the achievement of these transformative changes difficult. But still, initiatives around the world can give hope that these changes are achievable. Many countries are implementing plans to completely decarbonize their economies by 2030 or 2050 (Victor, 2020); 1840 local authorities in 32 countries have declared and acknowledged the climate emergency that we are in (CED, 2020); and investors initiatives, such as Climate Action 100+, are looking to ensure that the largest corporate greenhouse gas emitters take action in face of climate change (Climate Action 100+, 2020). We have the technology and the means to change the path we are in, as many other countries have realized. Brazil just needs to pick up the pace.

References

Amado-Filho, G. M., Moura, R. L., Bastos, A. C., Salgado, L. T., Sumida, P. Y., Guth, A. Z., et al. (2012). Rhodolith beds are major CaCO3 bio-factories in the tropical South West Atlantic. *PLoS One, 7*, e35171. https://doi.org/10.1371/journal.pone.0035171

Anthony, K. R. N., Kline, D. I., Diaz-Pulido, G., Dove, S., & Hoegh-Guldberg, O. (2008). Ocean acidification causes bleaching and productivity loss in coral reef builders. *Proceedings of the National Academy of Sciences of the United States of America, 105*, 17442–17446. https://doi.org/10.1073/pnas.0804478105

Basso, D., Babbini, L., Kaleb, S., Bracchi, V. A., & Falace, A. (2016). Monitoring deep Mediterranean rhodolith beds. *Aquatic Conservation: Marine and Freshwater Ecosystems, 26*, 549–561. https://doi.org/10.1002/aqc.2586

Berg, M. P., Toby Kiers, E., Driessen, G., van der Heijden, M., Kooi, B. W., Kuenen, F., et al. (2010). Adapt or disperse: Understanding species persistence in a changing world. *Global Change Biology, 16*, 587–598. https://doi.org/10.1111/j.1365-2486.2009.02014.x

Bjork, M., Mohammed, S. M., Bjorklund, M., & Semesi, A. (1995). Coralline algae, important coral-reef builders threatened by pollution. *Ambio, 24*, 502–505.

Boyd, P. W., Collins, S., Dupont, S., Fabricius, K., Gattuso, J. P., Havenhand, J., et al. (2018). Experimental strategies to assess the biological ramifications of multiple drivers of global ocean change—A review. *Global Change Biology, 24*, 2239–2261. https://doi.org/10.1111/gcb.14102

Büdenbender, J., Riebesell, U., & Form, A. (2011). Calcification of the Arctic coralline red algae Lithothamnion glaciale in response to elevated CO_2. *Marine Ecology Progress Series, 441*, 79–87. https://doi.org/10.3354/meps09405

Burdett, H., Perna, G., McKay, L., Broomhead, G., & Kamenos, N. (2018). Community-level sensitivity of a calcifying ecosystem to acute in situ CO_2 enrichment. *Marine Ecology Progress Series, 587*, 73–80. https://doi.org/10.3354/meps12421

Campestrini, I., & Jardim, W. F. (2017). Occurrence of cocaine and benzoylecgonine in drinking and source water in the São Paulo state region, Brazil. *Science of the Total Environmen, 576*, 374–380. https://doi.org/10.1016/j.scitotenv.2016.10.089

Carvalho, V. F., Assis, J., Serrão, E. A., Nunes, J. M., Batista, A. A., Batista, M. B., et al. (2019). Environmental drivers of rhodolith beds and epiphytes community along the South Western Atlantic coast. *Marine Environmental Research, 104827*, 104827. https://doi.org/10.1016/j.marenvres.2019.104827

CED, Climate Emergency Declaration. (2020). *Climate emergency declarations in 1,840 jurisdictions and local governments cover 820 million citizens*. Available at: https://climateemergencydeclaration.org/climate-emergency-declarations-cover-15-million-citizens/. Acessed 23 Nov 2020.

Celis-Plá, P. S. M., Hall-Spencer, J. M., Horta, P. A., Milazzo, M., Korbee, N., Cornwall, C. E., et al. (2015). Macroalgal responses to ocean acidification depend on nutrient and light levels. *Frontiers in Marine Science, 2*. https://doi.org/10.3389/fmars.2015.00026

Climate Action 100+ (2020). *Global investors driving business transition*. Available in: http://www.climateaction100.org/. Acessed 23 Nov 2020.

Coletti, G., Basso, D., & Frixa, A. (2017). Economic importance of coralline carbonates. In R. Riosmena-Rodríguez, W. Nelson, & J. Aguirre (Eds.), *Rhodolith/Maërl beds: A global perspective* (pp. 87–101). Springer International Publishing. https://doi.org/10.1007/978-3-319-29315-8_4

Cooney, R. (2004). *The precautionary principle in biodiversity conservation and natural resource management: An issues paper for policy-makers, researchers and practitioners*. Available at: http://data.iucn.org/dbtw-wpd/edocs/pgc-002.pdf.

Cornwall, C. E., Diaz-Pulido, G., & Comeau, S. (2019). Impacts of ocean warming on coralline algal calcification: Meta-analysis, knowledge gaps, and key recommendations for future research. *Frontiers in Marine Science, 6*, 1–10. https://doi.org/10.3389/fmars.2019.00186

de Mahiques, M. M., Figueira, R. C. L., Salaroli, A. B., Alves, D. P. V., & Gonçalves, C. (2013). 150 years of anthropogenic metal input in a biosphere reserve: The case study of the Cananéia-Iguape coastal system, southeastern Brazil. *Environment and Earth Science, 68*, 1073–1087. https://doi.org/10.1007/s12665-012-1809-6

De Sousa, A. C. A., & Barrocas, P. R. G. (2017). Privatizar ou não privatizar: Eis a questão. A única questão? A reedição da agenda liberal para o saneamento básico no Brasil. *Cadernos de Saúde Pública, 33*, e00048917. https://doi.org/10.1590/0102-311X00048917

Dias, G. T. M. (2000). Granulados bioclásticos-algas calcárias. *Revista Brasileira de Geofisica, 18*, 307–318. https://doi.org/10.1590/s0102-261x2000000300008

Diaz-Pulido, G., Anthony, K. R. N., Kline, D. I., Dove, S., & Hoegh-Guldberg, O. (2012). Interactions between ocean acidification and warming on the mortality and dissolution of coralline algae. *Journal of Phycology, 48*, 32–39. https://doi.org/10.1111/j.1529-8817.2011.01084.x

Doney, S. C., Fabry, V. J., Feely, R. A., & Kleypas, J. A. (2009). Ocean acidification: The other CO_2 problem. *Annual Review of Marine Science, 1*, 169–192. https://doi.org/10.1146/annurev.marine.010908.163834

FAO. (2017). *Water pollution from agriculture: A global review.* Available at: http://www.fao.org/3/a-i7754e.pdf.

Felder, D. L., Thoma, B. P., Schmidt, W. E., Sauvage, T., Self-Krayesky, S. L., Chistoserdov, A., et al. (2014). Seaweeds and decapod crustaceans on gulf deep banks after the macondo oil spill. *Bioscience, 64*, 808–819. https://doi.org/10.1093/biosci/biu119

Figueiredo, M. A. O., Eide, I., Reynier, M., Villas-Bôas, A. B., Tâmega, F. T. S., Ferreira, C. G., et al. (2015). The effect of sediment mimicking drill cuttings on deep water rhodoliths in a flow-through system: Experimental work and modeling. *Marine Pollution Bulletin, 95*, 81–88. https://doi.org/10.1016/j.marpolbul.2015.04.040

Fortes, M. D., & Lüning, K. (1980). Growth rates of North Sea macroalgae in relation to temperature, irradiance and photoperiod. *Helgoländer Meeresuntersuchungen, 34*, 15–29. https://doi.org/10.1007/BF01983538

Foster, M. S. (2001). Rhodoliths: Between rocks and soft places. *Journal of Phycology, 37*, 659–667. https://doi.org/10.1046/j.1529-8817.2001.00195.x

Fragkopoulou, E., Serrão, E. A., Horta, P. A., Koerich, G., & Assis, J. (2021). Bottom trawling threatens future climate refugia of rhodoliths globally. *Frontiers in Marine Science, 7*, 1246.

Fredericq, S., Arakaki, N., Camacho, O., Gabriel, D., Krayesky, D., Self-Krayesky, S., et al. (2014). A dynamic approach to the study of Rhodoliths: A case study for the northwestern Gulf of Mexico. *Cryptogamie Algologie, 35*, 77–98. https://doi.org/10.7872/crya.v35.iss1.2014.77

Frölicher, T. L., Fischer, E. M., & Gruber, N. (2018). Marine heatwaves under global warming. *Nature, 560*, 360–364. https://doi.org/10.1038/s41586-018-0383-9

Grall, J., & Hall-Spencer, J. M. (2003). Problems facing maerl conservation in Brittany. *Aquatic Conservation: Marine and Freshwater Ecosystems, 13*, 55–64. https://doi.org/10.1002/aqc.568

Harley, C. D. G., Anderson, K. M., Demes, K. W., Jorve, J. P., Kordas, R. L., Coyle, T. A., et al. (2012). Effects of climate change on global seaweed communities. *Journal of Phycology, 48*, 1064–1078. https://doi.org/10.1111/j.1529-8817.2012.01224.x

Harrington, L., Fabricius, K., Eaglesham, G., & Negri, A. (2005). Synergistic effects of diuron and sedimentation on photosynthesis and survival of crustose coralline algae. *Marine Pollution Bulletin, 51*, 415–427. https://doi.org/10.1016/j.marpolbul.2004.10.042

Hartmann, D. L. (2016). *Global physical climatology* (2nd ed.). Elsevier.

Harvey, B. P., Gwynn-Jones, D., & Moore, P. J. (2013). Meta-analysis reveals complex marine biological responses to the interactive effects of ocean acidification and warming. *Ecology and Evolution, 3*(4), 1016–1030.

Hoek, C. V. D. (1982). The distribution of benthic marine algae in relation to the temperature regulation of their life histories. *Biological Journal of the Linnean Society, 18*, 81–144. https://doi.org/10.1111/j.1095-8312.1982.tb02035.x

Hofmann, L. C., Heiden, J., Bischof, K., & Teichberg, M. (2014). Nutrient availability affects the response of the calcifying chlorophyte Halimeda opuntia (L.) J.V. Lamouroux to low pH. *Planta, 239*, 231–242. https://doi.org/10.1007/s00425-013-1982-1

Horta, P. A., Riul, P., Amado-Filho, G. M., Gurgel, C. F. D., Berchez, F., de Nunes, J. M., et al. (2016). Rhodoliths in Brazil: Current knowledge and potential impacts of climate change. *Brazilian Journal of Oceanography, 64*, 117–136. https://doi.org/10.1590/S1679-875920160870064sp2

Howarth, R. W. (2008). Coastal nitrogen pollution: A review of sources and trends globally and regionally. *Harmful Algae, 8*, 14–20. https://doi.org/10.1016/j.hal.2008.08.015

IPCC. (2014). Climate change 2014: Synthesis report. In *Contribution of working groups I, II and II to the fifht assessment report of the intergovernmental panel on climate change* (p. 151).

IPCC. (2019). *Special report: The ocean and cryosphere in a changing climate.*

Johnson, M. D., & Carpenter, R. C. (2018). Nitrogen enrichment offsets direct negative effects of ocean acidification on a reef-building crustose coralline alga. *Biology Letters, 14*, 20180371. https://doi.org/10.1098/rsbl.2018.0371

Kamenos, N. A., & Law, A. (2010). Temperature controls on coralline algal skeletal growth. *Journal of Phycology, 46*, 331–335. https://doi.org/10.1111/j.1529-8817.2009.00780.x

Kato, A., Hikami, M., Kumagai, N. H., Suzuki, A., Nojiri, Y., & Sakai, K. (2014). Negative effects of ocean acidification on two crustose coralline species using genetically homogeneous samples. *Marine Environmental Research, 94*, 1–6. https://doi.org/10.1016/j.marenvres.2013.10.010

Kroeker, K. J., Kordas, R. L., Crim, R., Hendriks, I. E., Ramajo, L., Singh, G. S., et al. (2013). Impacts of ocean acidification on marine organisms: Quantifying sensitivities and interaction with warming. *Global Change Biology, 19*, 1884–1896. https://doi.org/10.1111/gcb.12179

Liebezeit, G., Brepohl, D., Rizzi, J., Guebert, F., Krome, M., MacHado, E., et al. (2011). DDT in biota of Paranaguá Bay, southern Brazil: Recent input and rapid degradation. *Water, Air, and Soil Pollution, 220*, 181–188. https://doi.org/10.1007/s11270-011-0745-5

Lüning, K. (1990). *Seaweeds. Their environment, biogeography, and ecophysiology*. Wiley Interscience.

Lüthi, D., Le Floch, M., Bereiter, B., Blunier, T., Barnola, J. M., Siegenthaler, U., et al. (2008). High-resolution carbon dioxide concentration record 650,000-800,000 years before present. *Nature, 453*, 379–382. https://doi.org/10.1038/nature06949

Martin, S., Cohu, S., Vignot, C., Zimmerman, G., & Gattuso, J. P. (2013). One-year experiment on the physiological response of the Mediterranean crustose coralline alga, Lithophyllum cabiochae, to elevated pCO2 and temperature. *Ecology and Evolution, 3*, 676–693. https://doi.org/10.1002/ece3.475

Martin, S., & Hall-Spencer, J. M. (2017). Effects of ocean warming and acidification on Rhodolith/Maerl beds. In R. Riosmena-Rodríguez, W. Nelson, & J. Aguirre (Eds.), *Rhodolith/mäerl beds: A global perspective*. Springer International Publishing. https://doi.org/10.1007/978-3-319-29315-8

Martins, C. D. L., Arantes, N., Faveri, C., Batista, M. B., Oliveira, E. C., Pagliosa, P. R., et al. (2012). The impact of coastal urbanization on the structure of phytobenthic communities in southern Brazil. *Marine Pollution Bulletin, 64*, 772–778. https://doi.org/10.1016/j.marpolbul.2012.01.031

Mccoy, S. J., & Kamenos, N. A. (2015). Coralline algae (Rhodophyta) in a changing world: Integrating ecological, physiological, and geochemical responses to global change. *Journal of Phycology, 51*, 6–24. https://doi.org/10.1111/jpy.12262

Moura, R. L., Secchin, N. A., Amado-Filho, G. M., Francini-Filho, R. B., Freitas, M. O., Minte-Vera, C. V., et al. (2013). Spatial patterns of benthic megahabitats and conservation planning in the Abrolhos Bank. *Continental Shelf Research, 70*, 109–117. https://doi.org/10.1016/j.csr.2013.04.036

Muñoz, P. T., Sáez, C. A., Martínez-Callejas, M. B., Flores-Molina, M. R., Bastos, E., Fonseca, A., et al. (2018). Short-term interactive effects of increased temperatures and acidification on the calcifying macroalgae *Lithothamnion crispatum* and Sonderophycus capensis. *Aquatic Botany, 148*, 46–52. https://doi.org/10.1016/j.aquabot.2018.04.008

Nash, M. C., Cornwall, C. E., Diaz-pulido, G., & Adela, S. (2019). *Coralline algal calcification: A morphological and process-based understanding.*

Ngatia, L., Grace, J. M., III, Moriasi, D., & Taylor, R. (2019). Nitrogen and phosphorus eutrophication in marine ecosystems. *Monitoring of Marine Pollution, 1*, 1–17. https://doi.org/10.5772/intechopen.81869

Noisette, F., Duong, G., Six, C., Davoult, D., & Martin, S. (2013a). Effects of elevated pCO_2 on the metabolism of a temperate rhodolith Lithothamnion corallioides grown under different temperatures. *Journal of Phycology, 49*, 746–757. https://doi.org/10.1111/jpy.12085

Noisette, F., Egilsdottir, H., Davoult, D., & Martin, S. (2013b). Physiological responses of three temperate coralline algae from contrasting habitats to near-future ocean acidification. *Journal of Experimental Marine Biology and Ecology, 448*, 179–187. https://doi.org/10.1016/j.jembe.2013.07.006

Ockendon, N., Baker, D. J., Carr, J. A., White, E. C., Almond, R. E. A., Amano, T., et al. (2014). Mechanisms underpinning climatic impacts on natural populations: Altered species interactions are more important than direct effects. *Global Change Biology, 20*, 2221–2229. https://doi.org/10.1111/gcb.12559

Paiva, S. V., Carneiro, P. B. M., Garcia, T. M., Tavares, T. C. L., de Souza Pinheiro, L., Neto, A. R. X., et al. (2023). Marine carbonate mining in the southwestern Atlantic: Current status, potential impacts, and conservation actions. *Marine Policy, 148*, 105435. https://doi.org/10.1016/j.marpol.2022.105435

Pereira, J. C. (2019). Climate change governance in the Atlantic Basin: The cases of the United States, the European Union, and Brazil. In N. S. Teixeira & D. Marcos (Eds.), *Evolving human security challenges in the Atlantic space* (pp. 207–236). Foreign Policy Institute. https://doi.org/10.5305/amerjin-telaw.110.2.0288.3

Petsas, A. S., & Vagi, M. C. (2017). Effects on the photosynthetic activity of algae after exposure to various organic and inorganic pollutants. *Chlorophyll, 37*. https://doi.org/10.5772/67991

Piggott, J. J., Townsend, C. R., & Matthaei, C. D. (2015). Reconceptualizing synergism and antagonism among multiple stressors. *Ecology and Evolution, 5*, 1538–1547. https://doi.org/10.1002/ece3.1465

Poloczanska, E. S., Brown, C. J., Sydeman, W. J., Kiessling, W., Schoeman, D. S., Moore, P. J., et al. (2013). Global imprint of climate change on marine life. *Nature Climate Change, 3*, 919–925. https://doi.org/10.1038/nclimate1958

Qui-Minet, Z. N., Coudret, J., Davoult, D., Grall, J., Mendez-Sandin, M., Cariou, T., et al. (2019). Combined effects of global climate change and nutrient enrichment on the physiology of three temperate maerl species. *Ecology and Evolution, 1–21*, 13787. https://doi.org/10.1002/ece3.5802

Qui-Minet, Z. N., Delaunay, C., Grall, J., Six, C., Cariou, T., Bohner, O., et al. (2018). The role of local environmental changes on maerl and its associated non-calcareous epiphytic flora in the Bay of Brest. *Estuarine, Coastal and Shelf Science, 208*, 140–152. https://doi.org/10.1016/j.ecss.2018.04.032

Ragazzola, F., Foster, L. C., Form, A., Anderson, P. S. L., Hansteen, T. H., & Fietzke, J. (2012). Ocean acidification weakens the structural integrity of coralline algae. *Global Change Biology, 18*, 2804–2812. https://doi.org/10.1111/j.1365-2486.2012.02756.x

Ragazzola, F., Foster, L. C., Form, A. U., Büscher, J., Hansteen, T. H., & Fietzke, J. (2013). Phenotypic plasticity of coralline algae in a high CO_2 world. *Ecology and Evolution, 3*, 3436–3446. https://doi.org/10.1002/ece3.723

Reynier, M. V., Tâmega, F. T. S., Daflon, S. D. A., Santos, M. A. B., Coutinho, R., & Figueiredo, M. A. O. (2015). Long- and short-term effects of smothering and burial by drill cuttings on calcareous algae in a static-renewal test. *Environmental Toxicology and Chemistry, 34*, 1572–1577. https://doi.org/10.1002/etc.2938

Rich, W. A., Schubert, N., Schläpfer, N., Carvalho, V. F., Horta, A. C., & Horta, P. A. (2018). Physiological and biochemical responses of a coralline alga and a sea urchin to climate change: Implications for herbivory. *Marine Environmental Research, 142*, 100–107.

Rilov, G., Mazaris, A. D., Stelzenmüller, V., Helmuth, B., Wahl, M., Guy-Haim, T., et al. (2019). Adaptive marine conservation planning in the face of climate change: What can we learn from physiological, ecological and genetic studies? *Global Ecology and Conservation, 17*, e00566. https://doi.org/10.1016/j.gecco.2019.e00566

Riul, P., Targino, C. H., Farias, J. D. N., Visscher, P. T., & Horta, P. A. (2008). Decrease in Lithothamnion sp. (Rhodophyta) primary production due to the deposition of a thin sediment layer. *Journal of the Marine Biological Association of the United Kingdom, 88*, 17–19. https://doi.org/10.1017/S0025315408000258

Ruddiman, W. F. (2008). *Earth's climate: Past and future* (2nd ed.). Freeman, W. H. & Company.

Russell, B. D., Thompson, J. A. I., Falkenberg, L. J., & Connell, S. D. (2009). Synergistic effects of climate change and local stressors: CO_2 and nutrient-driven change in subtidal rocky habitats. *Global Change Biology, 15*, 2153–2162. https://doi.org/10.1111/j.1365-2486.2009.01886.x

Salomidi, M., Katsanevakis, S., Borja, Á., Braeckman, U., Damalas, D., Galparsoro, I., et al. (2012). Assessment of goods and services, vulnerability, and conservation status of European seabed biotopes: A stepping stone towards ecosystem-based marine spatial management. *Mediterranean Marine Science, 13*, 49–88. https://doi.org/10.12681/mms.23

Santiago, I. U., Molisani, M. M., Nudi, A. H., Scofield, A. L., Wagener, A. D. L. R., & Limaverde Filho, A. M. (2016). Hydrocarbons and trace metals in mussels in the Macaé coast: Preliminary assessment for a coastal zone under influence of offshore oil field exploration in southeastern Brazil. *Marine Pollution Bulletin, 103*, 349–353. https://doi.org/10.1016/j.marpolbul.2015.12.034

Santos, C. S. G., Lino, J. B., de Cerqueira Veras, P., Amado-Filho, G. M., Francini-Filho, R. B., Motta, F. S., et al. (2016). Environmental licensing on rhodolith beds: Insights from a worm. *Natureza & Conservação, 14*, 137–141. https://doi.org/10.1016/j.ncon.2016.06.002

Scherner, F., Bonomi Barufi, J., & Horta, P. A. (2012). Photosynthetic response of two seaweed species along an urban pollution gradient: Evidence of selection of pollution-tolerant species. *Marine Pollution Bulletin, 64*, 2380–2390. https://doi.org/10.1016/j.marpolbul.2012.08.012

Scherner, F., Horta, P. A., de Oliveira, E. C., Simonassi, J. C., Hall-Spencer, J. M., Chow, F., et al. (2013). Coastal urbanization leads to remarkable seaweed species loss and community shifts along the SW Atlantic. *Marine Pollution Bulletin, 76*, 106–115. https://doi.org/10.1016/j.marpolbul.2013.09.019

Scherner, F., Pereira, C. M., Duarte, G., Horta, P. A., e Castro, C. B., Barufi, J. B., et al. (2016). Effects of ocean acidification and temperature increases on the photosynthesis of tropical reef calcified macroalgae. *PLoS One, 11*, e0154844. https://doi.org/10.1371/journal.pone.0154844

Schubert, N., Salazar, V. W., Rich, W. A., Vivanco Bercovich, M., Almeida Saá, A. C., Fadigas, S. D., et al. (2019). Rhodolith primary and carbonate production in a changing ocean: The interplay of warming and nutrients. *Science of the Total Environment, 676*, 455–468. https://doi.org/10.1016/j.scitotenv.2019.04.280

Sgrò, C. M., Lowe, A. J., & Hoffmann, A. A. (2011). Building evolutionary resilience for conserving biodiversity under climate change. *Evolutionary Applications, 4*, 326–337. https://doi.org/10.1111/j.1752-4571.2010.00157.x

Short, J., Foster, T., Falter, J., Kendrick, G. A., & McCulloch, M. T. (2015). Crustose coralline algal growth, calcification and mortality following a marine heatwave in Western Australia. *Continental Shelf Research, 106*, 38–44. https://doi.org/10.1016/j.csr.2015.07.003

Short, J., Kendrick, G. A., Falter, J., & McCulloch, M. T. (2014). Interactions between filamentous turf algae and coralline algae are modified under ocean acidification. *Journal of Experimental Marine Biology and Ecology, 456*, 70–77. https://doi.org/10.1016/j.jembe.2014.03.014

Simkiss, K. (1964). Phosphates as crystal poisons of calcification. *Biological Reviews, 39*, 487–504. https://doi.org/10.1111/j.1469-185x.1964.tb01166.x

Sissini, M. N., Berchez, F., Hall-Spencer, J., Ghilardi-Lopes, N., Carvalho, V. F., Schubert, N., et al. (2020). Brazil oil spill response: Protect rhodolith beds. *Science, 367*, 156. https://doi.org/10.1126/science.aba2582

de Oliveira Soares, M., Teixeira, C. E. P., Bezerra, L. E. A., Paiva, S. V., Tavares, T. C. L., Garcia, T. M., et al. (2020). Oil spill in South Atlantic (Brazil): Environmental and governmental disaster. *Marine Policy, 115*, 103879. https://doi.org/10.1016/j.marpol.2020.103879

Sordo, L., Santos, R., Barrote, I., & Silva, J. (2018). High CO_2 decreases the long-term resilience of the free-living coralline algae Phymatolithon lusitanicum. *Ecology and Evolution, 8*, 4781–4792. https://doi.org/10.1002/ece3.4020

Sordo, L., Santos, R., Barrote, I., & Silva, J. (2019). Temperature amplifies the effect of high CO_2 on the photosynthesis, respiration, and calcification of the coralline algae Phymatolithon lusitanicum. *Ecology and Evolution, 9*(19), 11000–11009. https://doi.org/10.1002/ece3.5560
Sordo, L., Santos, R., Reis, J., Shulika, A., & Silva, J. (2016). A direct CO2 control system for ocean acidification experiments: Testing effects on the coralline red algae Phymatolithon lusitanicum. *PeerJ, 2016*, e2503. https://doi.org/10.7717/peerj.2503
Steller, D. L., Hernández-Ayón, J. M., Riosmena-Rodríguez, R., & Cabello-Pasini, A. (2007). Efecto de la temperature sobre las tasas de fotosíntesis, crecimiento y calcificación del alga coralina de vida libre Lithophyllum margaritae. *Ciencias Marinas, 33*, 441–456. https://doi.org/10.7773/cm.v33i4.1255
Strona, G., & Bradshaw, C. J. A. (2018). Co-extinctions annihilate planetary life during extreme environmental change. *Scientific Reports, 8*, 1–12. https://doi.org/10.1038/s41598-018-35068-1
Vásquez-Elizondo, R. M., & Enríquez, S. (2016). Coralline algal physiology is more adversely affected by elevated temperature than reduced pH. *Scientific Reports, 6*, 1–14. https://doi.org/10.1038/srep19030
Vergés, A., Doropoulos, C., Malcolm, H. A., Skye, M., Garcia-Pizá, M., Marzinelli, E. M., et al. (2016). Long-term empirical evidence of ocean warming leading to tropicalization of fish communities, increased herbivory, and loss of kelp. *Proceedings of the National Academy of Sciences of the United States of America, 113*, 13791–13796. https://doi.org/10.1073/pnas.1610725113
Vergés, A., Steinberg, P. D., Hay, M. E., Poore, A. G. B., Campbell, A. H., Ballesteros, E., et al. (2014). The tropicalization of temperate marine ecosystems: Climate-mediated changes in herbivory and community phase shifts. *Proceedings of the Royal Society B: Biological Sciences, 281*, 20140846. https://doi.org/10.1098/rspb.2014.0846
Victor, D. G. (2020). *Deep decarbonization: A realistic way forward on climate change*. Available in: https://e360.yale.edu/features/deep-decarbonization-a-realistic-way-forward-on-climate-change. Acessed 23 Nov 2020.
Villas-Bôas, A. B., Tâmega, F. T. D. S., Andrade, M., Coutinho, R., & Figueiredo, M. A. D. O. (2014). Experimental effects of sediment burial and light attenuation on two coralline algae of a deep water Rhodolith bed in Rio de Janeiro, Brazil. *Cryptogamie, Algologie, 35*, 67–76. https://doi.org/10.7872/crya.v35.iss1.2014.67
Vitousek, P. M. (2002). Oceanic islands as model systems for ecological studies. *Journal of Biogeography, 29*, 573–582. https://doi.org/10.1046/j.1365-2699.2002.00707.x
Wernberg, T., Bennett, S., Babcock, R. C., De Bettignies, T., Cure, K., Depczynski, M., et al. (2016). Climate-driven regime shift of a temperate marine ecosystem. *Science, 353*, 169–172. https://doi.org/10.1126/science.aad8745
Wernberg, T., Smale, D. A., Tuya, F., Thomsen, M. S., Langlois, T. J., De Bettignies, T., et al. (2013). An extreme climatic event alters marine ecosystem structure in a global biodiversity hotspot. *Nature Climate Change, 3*, 78–82. https://doi.org/10.1038/nclimate1627
Wethey, D. S., Woodin, S. A., Hilbish, T. J., Jones, S. J., Lima, F. P., & Brannock, P. M. (2011). Response of intertidal populations to climate: Effects of extreme events versus long term change. *Journal of Experimental Marine Biology and Ecology, 400*, 132–144. https://doi.org/10.1016/j.jembe.2011.02.008
Williams, I. D., Baum, J. K., Heenan, A., Hanson, K. M., Nadon, M. O., & Brainard, R. E. (2015). Human, oceanographic and habitat drivers of central and western pacific coral reef fish assemblages. *PLoS One, 10*, 1–19. https://doi.org/10.1371/journal.pone.0120516
Wilson, S., Blake, C., Berges, J. A., & Maggs, C. A. (2004). Environmental tolerances of free-living coralline algae (maerl): Implications for European marine conservation. *Biological Conservation, 120*, 279–289. https://doi.org/10.1016/j.biocon.2004.03.001
Worm, B., Barbier, E. B., Beaumont, N., Duffy, J. E., Folke, C., Halpern, B. S., et al. (2006). Impacts of biodiversity loss on ocean ecosystem services. *Science, 80*(314), 787–790. https://doi.org/10.1126/science.1132294

Chapter 9
Contributions of Rhodolith Beds to People

Tainá Luchese Gaspar, Marina Nasri Sissini, Jorge Luiz Rodrigues Filho, and Paulo A. Horta

Abstract As presented in the other chapters of the book, calcareous algae are of paramount importance in several ecosystemic aspects. In this chapter, we address the importance of assessing the value of this ecosystem in an integrated way that goes according to different perspectives and contexts around the world. The pluralist understanding and evaluation of the value of these ecosystems should lead policymakers to maintain their sustainability. Understanding, therefore, aspects of various rhodolith beds should help in understanding the importance of these ecosystems in providing contributions to people. Here, we briefly present the processes that occur within rhodolith beds that lead to the contributions of this ecosystem to people. These contributions are known within regulatory categories, such as oxygen production and carbon storage; materials, such as nutrient cycling and the food chain; in addition to nonmaterial contributions, such as scenic beauty landscapes that attract tourists; or even as inspiration for art and science.

Keywords Coralline red algae · Benefits to society · Ecosystem contributions

T. L. Gaspar (✉)
Programa de Pós-Graduação em Ecologia, Universidade Federal de Santa Catarina, Florianópolis, SC, Brazil

Laboratório de Ecologia de Ambientes Recifais, Departamento de Ecologia e Zoologia, Universidade Federal de Santa Catarina, Florianópolis, SC, Brazil

M. N. Sissini
Programa de Pós-Graduação em Ecologia, Universidade Federal de Santa Catarina, Florianópolis, SC, Brazil

J. L. R. Filho
Laboratório de Ecologia Aplicada e Conservação, Universidade do Estado de Santa Catarina, Laguna, SC, Brazil
e-mail: jorge.rodrigues@udesc.br

P. A. Horta
Laboratório de Ficologia, Departamento de Botânica, Universidade Federal de Santa Catarina, Florianópolis, SC, Brasil

P. A. Horta, M. N. Sissini (eds.), *Brazilian Rhodolith Beds*, Brazilian Marine Biodiversity, https://doi.org/10.1007/978-3-031-61449-1_9

9.1 Introduction

The high diversity and the ecological importance of coralline red algae (CRA) grounds and its free-living form, the rhodoliths, were highlighted in previous chapters (see Chap. 1). Beyond the importance of these ecosystems for marine wildlife, their influence on humankind's well-being must be taken into account in decision-making about the future of these environments and all related coastal marine ecosystems. A better understanding of the ecological functions of this structuring organisms and the consequent ecosystem processes that culminate in benefits for society is essential to more fairly assess the importances of these environments, enabling society, stakeholders, and policymakers awareness, with the necessary and expected maintenance of their sustainability (Bennett et al., 2015). In this way, improving the knowledge about the composition of rhodolith beds, the richness, and abundance of species that compose them and their functions in the ecosystem, as well as the connections and interdependencies among them and other ecosystems or environments, allows us to better understand the benefits generated for society. In addition, this multidisciplinary approach enables their valuation and fair decision-making about these ecosystems regarding human needs and nature as a subject of rights. In this perspective, each species plays a role in the environment in which it is found, which is defined as a function of the species (Cadotte et al., 2011), with consequences that extrapolate their territories and lifetime. The functions of different species occurring interactively and concurrently generate processes within and beyond the ecosystem, which will culminate in contributions from nature to people.

Taking into account that nature is perceived in completely different ways depending on the observer (Daily et al., 2000; Martinez-Alier, 2003), regarding different cultures and perspectives of life, it is essential to substantiate and highlight the importance of these contributions to broad categories of the stakeholders, enriching arguments and dialectic. Acknowledging the socioecological significance of ecosystems can lead to developing effective strategies for inclusive management and conservation of these environments, ultimately achieving sustainable development goals for all related socio-ecosystems (Bennett et al., 2015). Here, the benefits to society arising from ecosystem functioning will be treated as "contributions of nature to people," as recommended by Pascual et al. (2017), rather than "ecosystem services or benefits." The selected approach seeks to encompass concepts associated with other worldviews on the relationship between man-nature and knowledge systems (Díaz et al., 2015), and it is utilized by Intergovernmental Platform on Biodiversity and Ecosystem Services (IPBES).

Nature's contributions to people (NCP) are categorized into regulatory, material, and nonmaterial (Hill et al., 2021). Regulatory contributions are related to the functional aspects of the ecosystem that alter environmental conditions and regulate the generation of material and immaterial benefits (e.g., climate regulation). Material contributions refer to substances and elements that support the physical existence of society (e.g., provision of food). Finally, the nonmaterial category is associated with the contributions of nature under a more subjective aspect to the quality of human life (e.g., landscapes as artistic inspiration). In an alternative classification system, the Millennium Ecosystem Assessment (MEA), the NCP can be divided into four major clusters: supporting, provisioning, regulating, and cultural (Carpenter et al., 2006). The integrated utilization of these complementary perspectives (Fig. 9.1) reinforces the interaction and interdependencies of these concepts, as well as biodiversity components and functioning, in an eco-centric or at least a less anthropocentric perspective. Regarding empirical evidence and theoretical analyses, non-anthropocentric evaluations are essential for nature conservation and eventual selection of management strategies, during the development of policies and actions (Himes & Muraca, 2018).

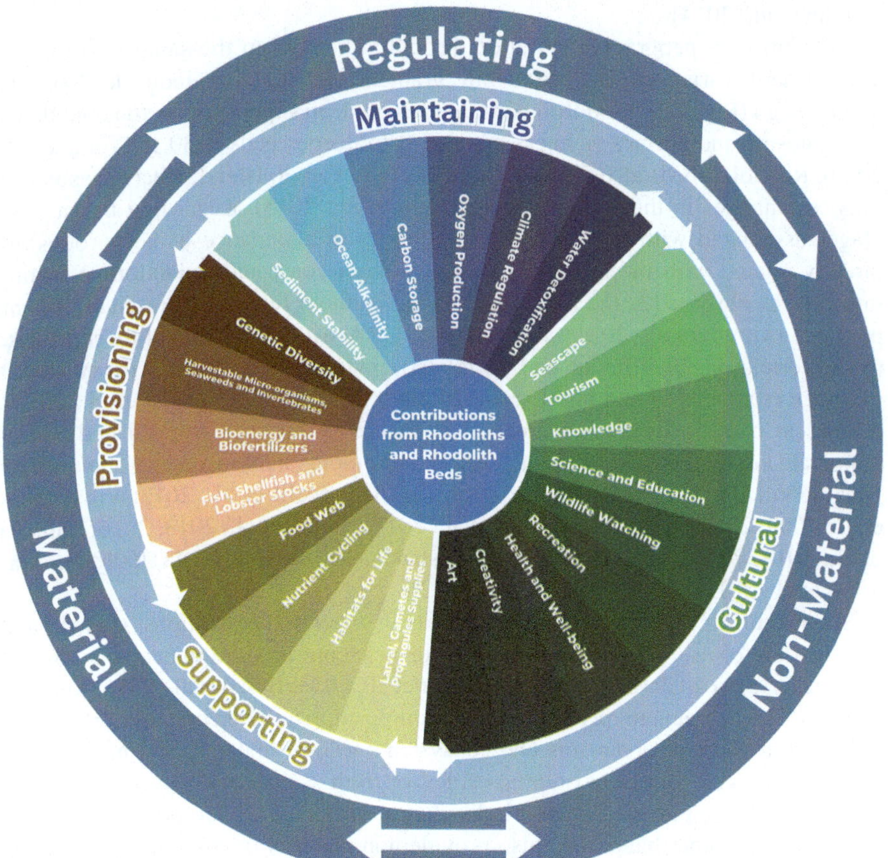

Fig. 9.1 Rhodolith and rhodolith bed contributions to people are categorized into different clusters regarding the perspective of the Intergovernmental Platform on Biodiversity and Ecosystem Services (IPBES), in the external ring and, Millennium Ecosystem Assessment (MEA) in the internal ring with the interrelated contributions. (Adapted from Carpenter et al., 2006; Hill et al., 2021)

Rhodolith beds are a widely distributed and diverse ecosystem in the marine realm (see Chap. 3). There are a high complexity in the configuration of rhodolith beds, since they are influenced by macroscale factors, as latitudinal gradients, and also by microscale factors, such as distance from the coast or depths (Amado-Filho et al., 2017; Veras et al., 2020). Taking into account the latitudinal gradient, for example, Sissini et al. (2022) found a variation of 5–50 phylogenetic species of CRA distributed in different ecoregions (Spalding et al., 2007) of the Brazilian coast. Eastern Brazil (between Bahia and Rio de Janeiro) is the ecoregion with the highest number of species (50), while the Amazon ecoregion had five species. The associated fauna and flora of these beds also vary significantly according to their location; in Arvoredo Island, Santa Catarina, 238 taxa were found (Bouzon & Freire, 2007; Metri & Rocha, 2008; Scherner et al., 2010; Pascelli et al., 2013), and on the coast of Paraíba, 102 taxa were found (Riul et al., 2009; Santos et al., 2011; Gondin et al., 2014).

In addition to geographic variation, rhodolith beds from the same location can also present variations due to depth gradient. In the Fernando de Noronha Archipelago (Fig. 9.2), rhodolith beds up to 30 m deep support twice the abundance of organisms and species richness compared to deeper beds (>30 m, Veras et al., 2020). Rhodolith beds also vary over time, as in Arvoredo Island, where seasonality imposes changes in the assemblage (Pascelli et al., 2013). Regarding fish assemblages associated with rhodolith beds, increases in abundance of key groups are directly related with the increased distances from coast and reduction of fishing intensity, as observed at the Abrolhos region, southern region of Brazilian tropical province (Moura et al., 2021). Furthermore, CRA with other life habits contribute to the formation of different and correlated environments, such as rocky shores (Tâmega et al., 2015) and coral reefs (Mariath et al., 2012; Tâmega et al., 2014), besides to present different degrees of connectivity between tropical and warm-temperate ecoregions (Sissini et al., 2022) and among shallow (Bahia et al., 2010; Dias & Villaça, 2012; Dominguez et al., 2013; Pascelli et al., 2013; Villas-Boas et al., 2009, 2014), deep (Figueiredo et al., 2012; Moura et al., 2013, 2016; Tâmega et al., 2013, 2014), coastal and oceanic islands (Villas-Boas et al., 2005; Pereira-Filho et al., 2012; Amado-Filho et al., 2012).

As a result of this ecological distinctness in species compositions, growth, and aggregation forms, rhodolith beds provide diverse contributions to people across all the categories of NCP (Nelson, 2009; Foster et al., 2013; Basso et al., 2016; Rendina et al., 2022). In a general perspective, the composition and functioning of rhodolith beds assign to these organisms and ecosystems values, material, and nonmaterial values. From symbolic to existential values, or from use or emotive values (Gunton et al., 2017), their related socio-biodiversity contributes to key processes on the planet. On the other hand, they also provide non material or cultural contributions building seascapes, enriching the underwater tourism; inspiring scientific questions, education, and knowledge; and learning opportunities, as well as health and well-being, or creativity and art (Fig. 9.1). In the following sections, we will look in detail at the main potential contributions.

Fig. 9.2 Rhodolith beds in the Fernando de Noronha Archipelago, Brazil (**a**–**e**), are different in species composition and associated fauna and flora, resulting in particular ecosystem contributions. These contributions are known within regulatory categories, such as oxygen production and carbon storage, materials, such as nutrient cycling and the food chain, in addition to nonmaterial contributions, such as scenic beauty landscapes that attract tourists, or even as inspiration for art and science

9.2 Contributions of Rhodolith Beds to People

9.2.1 *Regulatory Contributions*

Rhodolith beds might provide regulatory contributions in marine environments, by changing the physical and chemical environments above and below ground. Photosynthetic organisms, such as calcareous algae, produce oxygen and absorb/fix carbon dioxide through photosynthesis, in addition to incorporating it into their cells during growth (Gattuso et al., 1998; Kamenos et al., 2013), mechanism also responsible for the water detoxification. However, much of this carbon is released back into the atmosphere through respiration or decomposition (Farrelly et al., 2013). Even so, even if carbon dioxide is released, the rate between respiration and photosynthesis will determine whether environments built by calcareous algae will serve as a CO_2 sink or not (Grace, 2004). The direction and magnitude of carbon fluxes from rhodolith beds are associated with community productivity and balance between the abundance of heterotrophic and autotrophic organisms (Martin et al., 2007; Qui-Minet et al., 2022).

Added to the photosynthetic capacity, the calcification of algae in limestone highlights the importance of environments such as rhodolith beds in the carbon and calcium carbonate cycle and storage (Martin et al., 2005, 2007; Schwarz et al., 2005; Nelson, 2009; Amado-Filho et al., 2012; Basso & Granier, 2012; James et al., 2022). Because they are very abundant and spatially distributed, calcareous algae, even with slow growth, have high carbon and calcium carbonate production rates (Foster, 2001). Calcite production by calcareous algae depends not only on the species but also on their growth and photosynthesis rates (Schubert et al., 2019) and environmental conditions such as irradiance, temperature, and $CaCO_3$ saturation (van der Heijden & Kamenos, 2015; Qui-Minet et al., 2021). Understanding, therefore, the specificity of the calcareous algae that form the rhodolith beds and their relationship with environmental conditions is essential to bring about their regulatory contribution to the carbon balance, as well as to maintain this provision over time.

In the Capitalocene, when anthropogenic activities such as burning fossil fuels and deforestation significantly increased the concentration of greenhouse gasses in the atmosphere (Bala, 2013; IPCC, 2021), the contribution of macroalgae to climate adaptation (Horta et al., 2016) and mitigation objectives has been highlighted (Krause-Jensen & Duarte, 2016; Queirós et al., 2019). Rhodolith beds act in carbon storage and in climate regulation, mainly by the formation of carbon stocks when the rate of photosynthetic CO_2 gain exceeds the offset of respiration losses (Grace, 2004). Such environments represent one of the largest carbonate deposits and potential sinks in the world (Testa & Boscence, 1999), being able to deposit up to 4×10^8 t C $year^{-1}$ (van der Heijden & Kamenos, 2015). Since rhodolith beds and calcareous algae have a wide distribution, occurring from the tropics to the poles and can be found at depths greater than 200 meters (Foster, 2001), their importance in the global carbon cycle and their viability in the mitigation of CO_2 emissions are a fact

(Duarte et al., 2013; Singh & Ahluwalia, 2013). Other key characteristics of rhodolith beds, in addition to their wide distribution, such as their representative primary and secondary productivity, further reinforce their importance in the biogeochemical balance of carbon, whose representativeness has so far not been well evaluated (Laffoley et al., 2020).

Calcareous algae ecosystems, in addition to presenting high rates of carbonate production and sequestration (Amado-Filho et al., 2012; van der Heijden & Kamenos, 2015; Teed et al., 2020), represent blocked biogenic carbonate deposits, through high individual longevity-community of these environments. Nonliving carbonates are slowly dissolved through microbial respiration, increasing alkalinity and capturing CO_2, a process accelerated by ocean acidification (<10×; Kamenos et al., 2013; Burdett et al., 2018). In addition, the three-dimensionalization provided by rhodoliths considerably facilitates the amount of particulate organic carbon accumulated and settled (Stelzer et al., 2021) between and below the rhodoliths, which is stored for millennia (Mao et al., 2020). This latter ecological function is associated with sediment stability in marine grounds.

9.2.2 *Material Contributions*

Like corals in reefs, rhodoliths are recognized ecosystem engineers (Foster et al., 2007) and might provide habitats and influence the niche conditions for different species. In unconsolidated substrate sites, rhodolith beds develop under limiting conditions and significantly increase habitat structural complexity and primary production. The increase in the three-dimensionality of habitats, provided by the interstitial space within and between rhodoliths, provides different substrates and niches for an extensive diversity of organisms, such as macroalgae (see Chap. 4) and invertebrates (e.g., Otero-Ferrer et al., 2019; Veras et al., 2020). This supporting contribution of habitat provision is associated with significant changes in the richness and abundance of benthic species and end up harboring several species of commercial interest, either in their adult stage or still serving as a nursery or refuge (Steller et al., 2003; Kamenos et al., 2004; Costa et al., 2020; Moura et al., 2021). The aforementioned property rights, and the extension of the community of justice to nonhuman entities diversity and increase in the stocks of harvestable invertebrates and fishes (Fig. 9.1).

The rhodolith beds might enhance food web complexity, since the rhodolith-forming algae, in addition to macroalgae and microphytobenthos, are consumed by herbivores (Scherner et al., 2010; Carvalho et al., 2020a, b). The predatory species feed on invertebrates present in the three-dimensionalization provided by the rhodoliths and by the marine forest to them associated (Carvalho et al., 2020a, b), and the detritivores and decomposers recycle the dead matter present in the beds (Costa et al., 2020). In this way, rhodoliths are fundamental for the maintenance of marine biodiversity, being often considered as hotspots (Scherner et al., 2010; Peña et al., 2014; Horta et al., 2016; Riosmena-Rodriguez, 2017, de Araújo Costa et al., 2021).

Rhodolith beds therefore harbor a biodiversity of invertebrates capable of sustaining an ichthyofauna equivalent or as abundant as those found on coral reefs (Moura et al., 2021). In addition, rhodolith beds are important regions of genetic and ecological connectivity with other habitats, providing migration corridors from spawning areas to other life cycle habitats. This connectivity allows the dispersion of gametes, larvals, and other reproductive forms in seascapes, resulting in an increase of the reef species abundances (Costa et al., 2020; Moura et al., 2021; Carneiro et al., 2022).

In recent decades, rhodolith beds have also been the target of bioprospecting studies, given the presence of several compounds of pharmacological interest, such as hypolipidemic (Ara et al., 2002), antioxidant (Yuan & Walsh, 2006; Matloub et al., 2015), immunological (Saker et al., 2004), antitumoral (Harada et al., 1997; Lee & Sung, 2003; Aslam et al., 2009, 2010; Gheda et al., 2018), antiviral (Rinehart et al., 1983; Matloub et al., 2015), and antibacterial activities (Lima-Filho et al., 2002). In addition to material contributions through their biological resources and bioprospecting, rhodolith beds have also been the target of mining to obtain mineral resources in Brazil (Riul et al., 2008). However, calcareous algae are nonrenewable resources due to their slow growth, around 1–1.5 mm/year (Amado-Filho et al., 2012), which is not consistent with the pace of economic exploration activities, and may jeopardize the contributions of these environments to the whole socio-ecosystem.

In Brazil, it is estimated that 96,000–120,000 tons of calcium carbonate are extracted per year (Riul et al., 2008). The dredging process, the result of exploration, directly impacts these ecosystems through the removal of rhodoliths and indirectly with the suspension of sediments and the availability of stored carbon for remineralization (Hill et al., 2015). In France, for example, the unsustainable exploitation of a rhodolith bed promoted severe changes in the biodiversity of the benthic community (Cabioch, 1969; Biomaerl Team, 1999; Hall-Spencer & Moore, 2000). Despite the advances in potential applications for rhodolith extraction, there are still gaps in knowledge about these environments, especially regarding their ability to recover from extractive activities and regarding the scenario of growing impacts of multiple stressors (Horta et al., 2016), from global to local perspective, that already threaten these formations. In this context, the Decade of the Oceans offers the scientific community the chance to stimulate ocean science and the valuation of their environments based on sustainable development (Ryabinin et al., 2019). Therefore, it is first necessary to build a global information system about restoring and maintaining the health of ocean ecosystems (Duarte et al., 2018) in order to achieve global sustainable development (Heymans et al., 2020).

9.2.3 *Nonmaterial Contributions*

In addition to the material and regulatory contributions explained above, rhodolith beds and calcareous algae also contribute to social well-being through nonmaterial contributions. Such nonmaterial contributions are related to a good quality of life through the construction of relationships received from people with nature, which integrate the sense of identity and spirituality of communities (Pascual et al., 2017). In this way, nonmaterial contributions reflect elements of cultural identity, social cohesion, and social and moral responsibilities toward nature (UNEP, 2014), which are related to quality of life, learning and artistic inspiration, and the cultural identity of the associated communities. The contributions of rhodolith beds to people reflect the fact that these ecosystems are unique environments in the seascape that generate and sustain human experiences and end up supporting diverse livelihoods and identity, as well as providing opportunities for research, tourism, and education (Motuhi et al., 2016; Cinner, 2014).

Taking into account that rhodolith beds are configured, in terms of biodiversity, as reef environments (Riding, 2002), the nonmaterial benefits arising from these ecosystems are also very similar. Rhodolith beds have inspiring scenic beauty, which attracts visitors through recreational and wildlife watching activities such as snorkeling and scuba diving (Fig. 9.2c), in addition to recreational fishing (Brander et al., 2007). The value of tourism in coral reefs, for example, is estimated at US$ 35.8 billion dollars globally per year, through the direct contribution of tourism (diving, snorkeling and boat trips) and indirect (calm and clear waters, beaches, views with scenic beauty, seafood, and advertising) (Spalding et al., 2017). It is worth mentioning that monetization is just one of the ways to value these ecosystems and their contributions, and that given the complexity of contributions to people, it becomes necessary to expand the way society recognizes the values of ecosystems and adopt approaches to pluralist appreciation (Pascual et al., 2017). Another example of the nonmaterial contribution of reef environments to people is in Hawaii, where fisheries encompass a range of cultural values, including livelihoods (physical and cultural), knowledge, and social cohesion (Grafeld et al., 2017). Considering that traditional fisheries activities are associated with several cultural ecosystem services (Pellowe & Leslie, 2021), it is possible to infer that the contributions of rhodolith beds generate a well-being to fisheries communities worldwide.

One way to enhance people's understanding of the value of nonmaterial contributions from ecosystems such as rhodolith beds to people is through integrated approaches to transdisciplinary issues such as environmental education (Godemann, 2008; Reyers et al., 2013). Environmental education consists of forming citizens who learn about the environment, its importance, and its associated problems (Stapp, 1969). Such citizens become aware of the value of ecosystems and how to help solve their problems. In addition, they are motivated to solve them by the symbolic relationship between citizens and the environment created by environmental education. Linked to educational actions, management, and sustainability actions regarding social responsibility, they are also effective in raising the community's environmental awareness (Powell & Ham, 2008; Lucrezi et al., 2017; Mutanga et al., 2017; Smit et al., 2017; Sutcliffe & Barnes, 2018).

The tool of environmental education is one of the pillars for the preservation of the oceans (SDG, 2020; Ojala, 2013), with the aim of stimulating social well-being in parallel with the viability of the natural environment and economic activities (Ciegis et al., 2009; Griggs et al., 2013). However, in addition to traditional teaching, nonformal environmental education experiences, which include recreational activities and classes outside the school environment, can facilitate learning and direct behavioral changes toward ecologically conscious biases (Rios & Brewer, 2014). The development of ecological awareness, finally, results in communities that, in addition to understanding environmental issues, act in a non-predatory manner and create sustainable practices for the environment and life (Dickmann & Carneiro, 2012; Freire, 2018).

In this context, Brazil, with 8500 km of coastal territory (MMA, 2020), where CRA are present in practically all environments (Amado-Filho et al., 2012) and rhodolith beds in almost every coastline, presents an immense potential for nontraditional environmental education. On the northeastern coast of Brazil, raising awareness for the protection of marine environments through environmental education can be applied to rhodolith beds accessible from the coast (Costa et al., 2020). In addition, it is worth mentioning several other places in Brazil, whose rhodolith beds also present enormous nonmaterial contributions and potential for environmental education, such as Fernando de Noronha Archipelago (Amado-Filho et al., 2012), Arvoredo Island (Horta et al., 2016), Abrolhos (Moura et al., 2013), and coastal rhodolith beds accessible, as in Rio Grande do Norte (Testa & Bosence, 1999), Bahia (Bahia et al., 2010; Dominguez et al., 2013), and Espírito Santo (in which there is a Conservation Unit called "Costa das Algas" referring to the abundance of these organisms, Amado-Filho et al., 2007; Dias & Villaça, 2012; Villas-Boas et al., 2014).

On Arvoredo Island, whose rhodolith bed is located within the Arvoredo Marine Biological Reserve, the MAArE Project was carried out, a condition indicated by the Chico Mendes Institute for Biodiversity Conservation (ICMBio) within the scope of the environmental licensing process for the Brazilian Institute for the Environment and Renewable Natural Resources (IBAMA). The information collected by the project contributed to the expansion of knowledge about the local marine community. Such results were palpable, not only in the scientific community, but extending it to the community, which can download the book for free in a language accessible to the population (link for download: https://www.maare.ufsc.br/realizacoes/#o-livro). Using this book as an environmental education tool, the community was able to become aware of the ecosystems of Arvoredo Island, including its rhodolith bed. The result of this is the artistic inspiration promoted by the scenic beauty of the rhodolith beds of Arvoredo Island (Fig. 9.3a), which was portrayed in a watercolor painting made by the mother of one of the collaborators of the book (Fig. 9.3b). This scene represents an opportunity to reinforce the interrelation or interdependencies of contributions. At the same moment and place besides the nonmaterial perspective, we can highlight the role of rhodolith beds as a provisioning contribution, feeding a fish population, and also the supporting contribution of cycling nutrients.

Fig. 9.3 (**a**) Shoal of Garapoás (*Pseudocaranx dentex*) feeding on a rhodolith bed at Arvoredo Island. (Photo: João Paulo Krajewski—MAArE). (**b**) Painting made in watercolor inspired by the photo of the rhodolith bed on Arvoredo Island. (Art by Bernardete Wrublevski Aued and photo by Marcelo Crivellaro)

9.3 Final Remarks

Given the wide range of contributions of rhodolith beds discussed above, it is coherent that these habitats are considered one of the most valuable to preserve, integrating different categories of Marine Protected Areas around the world, and especially on Brazilian coast where such environments are found in practically all marine territory (Barbera et al., 2003; OSPAR Commission, 2010; Turra & Denadai, 2015; Riosmena-Rodríguez, 2017; de Araújo Costa et al., 2021). Valuing these environments, on the other hand, must be carried out in a pluralistic way, in order to minimize conflicts over interpretations of their values, which affect decision-making about them (Benett et al., 2015). Identifying the diversity of values of individuals and social groups toward the ecosystem is often a challenge, but not doing so can produce unsustainable results (Martin et al., 2016; Pascual et al., 2017). We must therefore make use of available tools that enable the recognition of the values of contributions and their diversity, such as guides developed by IPBES (UNEP, 2014), in addition to methodologies for practical approaches to valuing and evaluating nature's contributions (Pascual et al., 2017). We claim that all these efforts, as fostered by Muradian and Gómez-Baggethun (2021), should be conducted by a paradigm shift, from a morality and ethic of utility to a morality and ethic of care (Curtin, 1991), a reallocation of property rights, and the extension of the community of justice to nonhuman entities.

References

Amado-Filho, G. M., Maneveldt, G., Manso, R. C. C., Marins-Rosa, B. V., Pacheco, M. R., & Guimarães, S. M. P. B. (2007). Structure of rhodolith beds from 4 to 55 meters deep along the southern coast of Espírito Santo State, Brazil. *Ciencias Marinas, 33*(4), 399–410.

Amado-Filho, G. M., Pereira-Filho, G. H., Bahia, R. G., Abrantes, D. P., Veras, P. C., & Matheus, Z. (2012). Occurrence and distribution of rhodolith beds on the Fernando de Noronha Archipelago of Brazil. *Aquatic Botany, 101*, 41–45.

Amado-Filho, G. M., Bahia, R. G., Pereira-Filho, G. H., & Longo, L. L. (2017). South Atlantic rhodolith beds: Latitudinal distribution, species composition, structure and ecosystem functions, threats and conservation status. In *Rhodolith/Maërl Beds: A Global Perspective* (pp. 299–317).

Ara, J., Sultana, V., Qasim, R., & Ahmad, V. U. (2002). Hypolipidaemic activity of seaweed from Karachi Coast. *Phytotherapy Research, 16*, 479–483.

Aslam, M. N., Bhagavathula, N., Paruchuri, T., Hu, X., Chakrabarty, S., & Varani, J. (2009). Growth-inhibitory effects of a mineralized extract from the red marine algae, Lithothamnion calcareum, on Ca2+-sensitive and Ca2+-resistant human colon carcinoma cells. *Cancer Letters, 283*(2), 186–192.

Aslam, M. N., Kreider, J. M., Paruchuri, T., Bhagavathula, N., DaSilva, M., Zernicke, R. F., et al. (2010). A mineral-rich extract from the red marine algae Lithothamnion calcareum preserves bone structure and function in female mice on a Western-style diet. *Calcified Tissue International, 86*, 313–324.

Bahia, R. G., Abrantes, D. P., Brasileiro, P. S., Pereira-Filho, G. H., & Amado-Filho, G. M. (2010). Rhodolith bed structure along a depth gradient on the northern coast of Bahia State, Brazil. *Brazilian Journal of Oceanography, 58*, 323–337.

Barbera, C., Bordehore, C., Borg, J. A., Glémarec, M., Grall, J., Hall-Spencer, J. M., et al. (2003). Conservation and management of northeast Atlantic and Mediterranean maerl beds. *Aquatic Conservation: Marine and Freshwater Ecosystems, 13*(S1), S65–S76.

Basso, D., & Granier, B. (2012). Calcareous algae in changing environments. *Geodiversitas, 34*(1), 5–11.

Basso, D., Babbini, L., Kaleb, S., Bracchi, V., & Falace, A. (2016) *Monitoring deep Mediterranean rhodolith beds*.

Bennett, E. M., Cramer, W., Begossi, A., Cundill, G., Díaz, S., Egoh, B. N., et al. (2015). Linking biodiversity, ecosystem services, and human well-being: Three challenges for designing research for sustainability. *Current Opinion in Environmental Sustainability, 14*, 76–85.

BIOMAERL Team. (1999). BIOMAERL: Maerl biodiversity; functional structure and anthropogenic impacts, EC Con- tract No. MAS3-CT95-0020, p. 973.

Bouzon, J. L., & Freire, A. S. (2007). The Brachyura and Anomura fauna (Decapoda; Crustacea) in the Arvoredo Marine Biological Reserve on the southern brazilian coast. *Brazilian Journal of Biology, 67*, 321–325.

Brander, L. M., Van Beukering, P., & Cesar, H. S. (2007). The recreational value of coral reefs: A meta-analysis. *Ecological Economics, 63*(1), 209–218.

Burdett, H. L., Perna, G., McKay, L., Broomhead, G., & Kamenos, N. A. (2018). Community-level sensitivity of a calcifying ecosystem to acute in situ CO_2 enrichment. *Marine Ecology Progress Series, 587*, 73–80.

Cabioch, J. (1969). Les fonds de maërl de la baie de Morlaix et leur peuplement végétal. *Cahiers de Biologie Marine, 10*, 139–161.

Cadotte, M. W., Carscadden, K., & Mirotchnick, N. (2011). Beyond species: Functional diversity and the maintenance of ecological processes and services. *Journal of Applied Ecology, 48*, 1079–1087.

Carneiro, P. B., Ximenes Neto, A. R., Jucá-Queiroz, B., Teixeira, C. E., Feitosa, C. V., Barroso, C. X., et al. (2022). Interconnected marine habitats form a single continental-scale reef system in South America. *Scientific Reports, 12*(1), 17359.

Carpenter, S. R., DeFries, R., Dietz, T., Mooney, H. A., Polasky, S., Reid, W. V., & Scholes, R. J. (2006). Millennium ecosystem assessment: research needs. *Science, 314*(5797), 257–258.

Carvalho, V. F., Assis, J., Serrao, E. A., Nunes, J. M., Anderson, A. B., Batista, M. B., et al. (2020a). Environmental drivers of rhodolith beds and epiphytes community along the South Western Atlantic coast. *Marine Environmental Research, 154*, 104827.

Carvalho, V. F., Silva, J., Kerr, R., Anderson, A. B., Bastos, E. O., Cabral, D., et al. (2020b). When descriptive ecology meets physiology: a study in a South Atlantic rhodolith bed. *Journal of the Marine Biological Association of the United Kingdom, 100*(3), 347–360.

Ciegis, R., Ramanauskiene, J., & Martinkus, B. (2009). The concept of sustainable development and its use for sustainability scenarios. *Engineering Economics, 62*(2).

Cinner, J. (2014). Coral reef livelihoods. *Current Opinion in Environmental Sustainability, 7*, 65–71.

Costa, A. C. P., Garcia, T. M., Paiva, B. P., Neto, A. R. X., & de Oliveira Soares, M. (2020). Seagrass and rhodolith beds are important seascapes for the development of fish eggs and larvae in tropical coastal areas. *Marine Environmental Research, 161*, 105064.

Curtin, D. (1991). Toward an ecological ethic of care. *Hypatia, 6*(1), 60–74.

Daily, G. C., Söderqvist, T., Aniyar, S., Arrow, K., Dasgupta, P., Ehrlich, P. R., et al. (2000). The value of nature and the nature of value. *Science, 289*(5478), 395–396.

de Araújo Costa, D., de Lucena, R. F. P., da Silva, F. D. A., da Silva, G. M. B., Massei, K., Christoffersen, M. L., & Dolbeth, M. (2021). Importance of rhodoliths as habitats for benthic communities in impacted environments. *Regional Studies in Marine Science, 48*, 102055.

Dias, G. T., & Villaça, R. C. (2012). Coralline algae depositional environments on the Brazilian central–south-eastern shelf. *Journal of Coastal Research, 28*(1), 270–279.

Díaz, S., Demissew, S., Carabias, J., Joly, C., Lonsdale, M., Ash, N., et al. (2015). The IPBES Conceptual Framework—Connecting nature and people. *Current Opinion in Environmental Sustainability, 14*, 1–16.

Dickmann, I., & Carneiro, S. M. M. (2012). Paulo Freire e Educação ambiental: Contribuições a partir da obra Pedagogia da Autonomia. *Revista de Educação Pública, 21*(45), 87–102.
Dominguez, J. M. L., da Silva, R. P., Nunes, A. S., & Freire, A. F. M. (2013). The narrow, shallow, low-accommodation shelf of central Brazil: Sedimentology, evolution, and human uses. *Geomorphology, 203*, 46–59.
Duarte, C. M., Hendriks, I. E., Moore, T. S., Olsen, Y. S., Steckbauer, A., Ramajo, L., et al. (2013). Is ocean acidification an open-ocean syndrome? Understanding anthropogenic impacts on seawater pH. *Estuaries and Coasts, 36*, 221–236.
Duarte, C. M., Poiner, I., & Gunn, J. (2018). Perspectives on a global observing system to assess ocean health. *Frontiers in Marine Science, 5*, 265. https://doi.org/10.3389/fmars.2018.00265
Farrelly, D., Everard, C. D., Fagan, C. C., & McDonnell, K. P. (2013). Carbon sequestration and the role of biological carbon mitigation: A review. *Renewable and Sustainable Energy Reviews, 21*, 712–727. https://doi.org/10.1016/j.rser.2012.12.038
Figueiredo, M. A. O., Coutinho, R., Villas-Boas, A. B., Tâmega, F. T. S., & Mariath, R. (2012). Deep-water rhodolith productivity and growth in the southwestern Atlantic. *Journal of Applied Phycology, 24*, 487–493.
Foster, M. S. (2001). Rhodoliths: between rocks and soft places. *Journal of Phycology, 37*(5), 659–667.
Foster, M. S., McConnico, L. M., Lundsten, L., Wadsworth, T., Kimball, T., Brooks, L. B., et al. (2007). Diversity and natural history of a Lithothamnion muelleri-Sargassum horridum community in the Gulf of California. *Ciencias Marinas, 33*(4), 367–384.
Foster, M. S., Amado Filho, G. M., Kamenos, N. A., Riosmena-Rodríguez, R., & Steller, D. L. (2013). Rhodoliths and rhodolith beds. *Research and discoveries: The revolution of science through SCUBA*.
Freire, P. (2018). *Pedagogy of the oppressed*. Bloomsbury Publishing.
Gattuso, J. P., Frankignoulle, M., & Wollast, R. (1998). Carbon and carbonate metabolism in coastal aquatic ecosystems. *Annual Review of Ecology and Systematics, 29*(1), 405–434. https://doi.org/10.1146/annurev.ecolsys.29.1.405
Gheda, S., El-Sheekh, M., & Abou-Zeid, A. (2018). In vitro anticancer activity of polysaccharide extracted from red alga Jania rubens against breast and colon cancer cell lines. *Asian Pacific Journal of Tropical Medicine, 11*(10), 583.
Godemann, J. (2008). Knowledge integration: A key challenge for transdisciplinary cooperation. *Environmental Education Research, 14*(6), 625–641.
Gondim, A. I., Dias, T. P., de Souza, C., Duarte, R., Riul, P., Lacouth, P., & Christoffersen, M. L. (2014). Filling a knowledge gap on the biodiversity of rhodolith-associated Echinodermata from northeastern Brazil. *Tropical Conservation Science, 7*(1), 87–99.
Grace, J. (2004). Understanding and managing the global carbon cycle. *Journal of Ecology, 92*, 189–202. https://doi.org/10.1111/j.0022-0477.2004.00874.x
Grafeld, S., Oleson, K. L., Teneva, L., & Kittinger, J. N. (2017). Follow that fish: Uncovering the hidden blue economy in coral reef fisheries. *PLoS One, 12*(8), e0182104.
Griggs, D., Stafford-Smith, M., Gaffney, O., Rockström, J., Öhman, M. C., Shyamsundar, P., et al. (2013). Sustainable development goals for people and planet. *Nature, 495*(7441), 305–307.
Gunton, R. M., van Asperen, E. N., Basden, A., Bookless, D., Araya, Y., Hanson, D. R., et al. (2017). Beyond ecosystem services: Valuing the invaluable. *Trends in Ecology & Evolution, 32*(4), 249–257.
Hall-Spencer, J. M., & Moore, P. G. (2000). Scallop dredging has profound, long-term impacts on maerl habitats. *ICES Journal of Marine Science, 57*(5), 1407–1415.
Harada, H., NoRo, T., & Kamei, Y. (1997). Selective antitumor activity in vitro from marine algae from Japan coasts. *Biological and Pharmaceutical Bulletin, 20*(5), 541–546.
Heymans, J. J., Bundy, A., Christensen, V., Coll, M., De Mutsert, K., Fulton, E. A., et al. (2020). The ocean decade: a true ecosystem modeling challenge. *Frontiers in Marine Science, 7*, 554573.

Hill, R., Bellgrove, A., Macreadie, P. I., Petrou, K., Beardall, J., Steven, A., & Ralph, P. J. (2015). Can macroalgae contribute to blue carbon? An A ustralian perspective. *Limnology and Oceanography, 60*(5), 1689–1706.

Hill, R., Díaz, S., Pascual, U., Stenseke, M., Molnár, Z., & Van Velden, J. (2021). Nature's contributions to people: Weaving plural perspectives. *One Earth, 4*(7), 910–915.

Himes, A., & Muraca, B. (2018). Relational values: the key to pluralistic valuation of ecosystem services. *Current Opinion in Environmental Sustainability, 35*, 1–7.

Horta, P. A., Riul, P., Amado Filho, G. M., Gurgel, C. F. D., Berchez, F., Nunes, J. M. D. C., et al. (2016). Rhodoliths in Brazil: Current knowledge and potential impacts of climate change. *Brazilian Journal of Oceanography, 64*, 117–136.

James, K., Macreadie, P. I., Burdett, H., Davies, I., Baxter, J., & Kamenos, N. (2022). Non-classical blue carbon ecosystems are key stores of marine organic carbon. *Available at SSRN 4301771.*

Kamenos, N. A., Moore, P. G., & Hall-Spencer, J. M. (2004). Maerl grounds provide both refuge and high growth potential for juvenile queen scallops (Aequipecten opercularis L.). *Journal of Experimental Marine Biology and Ecology, 313*(2), 241–254.

Kamenos, N. A., Burdett, H. L., Aloisio, E., Findlay, H. S., Martin, S., Longbone, C., Dunn, J., Widdicombe, S., & Calosi, P. (2013). Coralline algal structure is more sensitive to rate, rather than the magnitude, of ocean acidification. *Global Change Biology, 19*, 3621–3628. https://doi.org/10.1111/gcb.12351

Krause-Jensen, D., & Duarte, C. M. (2016). Substantial role of macroalgae in marine carbon sequestration. *Nature Geoscience, 9*, 737–742.

Laffoley, D., Baxter, J. M., Amon, D. J., Currie, D. E., Downs, C. A., Hall-Spencer, J. M., et al. (2020). Eight urgent, fundamental and simultaneous steps needed to restore ocean health, and the consequences for humanity and the planet of inaction or delay. *Aquatic Conservation: Marine and Freshwater Ecosystems, 30*(1), 194–208.

Lee, E. J., & Sung, M. K. (2003). Chemoprevention of azoxymethaneinduced rat colon carcinogenesis by seatangle, a fiber-rich seaweed. *Plant Food for Human Nutrition, 58*, 1–8.

Lima-Filho, J. V. M., Carvalho, A. F., Freitas, S. M., & Melo, V. M. (2002). Antibacterial activity of extracts of six macroalgae from the northeastern brazilian coast. *Brazilian Journal of Microbiology, 33*, 311–314.

Lucrezi, S., Milanese, M., Markantonatou, V., Cerrano, C., Sarà, A., Palma, M., & Saayman, M. (2017). Scuba diving tourism systems and sustainability: Perceptions by the scuba diving industry in two marine protected areas. *Tourism Management, 59*, 385–403.

Mao, J., Burdett, H. L., McGill, R. A., Newton, J., Gulliver, P., & Kamenos, N. A. (2020). Carbon burial over the last four millennia is regulated by both climatic and land use change. *Global Change Biology, 26*(4), 2496–2504.

Mariath, R., Riosmena-Rodriguez, R., & Figueiredo, M. (2012). Lithothamnion steneckii sp. nov. and Pneophyllum conicum: New coralline red algae (Corallinales, Rhodophyta) for coral reefs of Brazil. *Algae, 27*(4), 249–258.

Martin, S., Clavier, J., Guarini, J. M., Chauvaud, L., Hily, C., Grall, J., et al. (2005). Comparison of Zostera marina and maerl community metabolism. *Aquatic Botany, 83*(3), 161–174.

Martin, S., Clavier, J., Chauvaud, L., & Thouzeau, G. (2007). Community metabolism in temperate maerl beds. I. Carbon and carbonate fluxes. *Marine Ecology Progress Series, 335*, 19–29.

Martin, A., Coolsaet, B., Corbera, E., Dawson, N. M., Fraser, J. A., Lehmann, I., & Rodriguez, I. (2016). Justice and conservation: The need to incorporate recognition. *Biological Conservation, 197*, 254–261.

Martinez-Alier, J. (2003). *The Environmentalism of the poor: A study of ecological conflicts and valuation*. Edward Elgar Publishing.

Matloub, A. A., Elsouda, S. S. M., El-Senousy, W. M., Hamed, M., Aly, H., Ali, S. A., et al. (2015). In vitro antiviral, cytotoxic, antioxidant and hypolipidemic activities of polysaccharide isolated from marine algae. *Journal of Pharmacognosy and Phytochemistry Research, 7*, 1099–1111.

Metri, R., & Rocha, R. D. (2008). Bancos de algas calcárias, um ecossistema rico a ser preservado. [Calcareous algae banks, a rich ecosystem to be preserved]. *Natureza & Conservaçao (Brasil), 6*, 8–17.

Motuhi, S. E., Mehiri, M., Payri, C. E., La Barre, S., & Bach, S. (2016). Marine natural products from New Caledonia—A review. *Marine Drugs, 14*(3), 58.

Moura, R. L., Secchin, N. A., Amado-Filho, G. M., Francini-Filho, R. B., Freitas, M. O., Minte-Vera, C. V., & Bastos, A. C. (2013). Spatial patterns of benthic megahabitats and conservation planning in the Abrolhos Bank. *Continental Shelf Research, 70*, 109–117.

Moura, R. L., Amado-Filho, G. M., Moraes, F. C., et al. (2016). An extensive reef system at the Amazon River mouth. *Science Advances, 2*, 1–12.

Moura, R. L., Abieri, M. L., Castro, G. M., et al. (2021). Tropical rhodolith beds are a major and belittled reef fish habitat. *Scientific Reports, 11*, 794.

Muradian, R., & Gómez-Baggethun, E. (2021). Beyond ecosystem services and nature's contributions: Is it time to leave utilitarian environmentalism behind? *Ecological Economics, 185*, 107038.

Mutanga, C. N., Vengesayi, S., Chikuta, O., Muboko, N., & Gandiwa, E. (2017). Travel motivation and tourist satisfaction with wildlife tourism experiences in Gonarezhou and Matusadona National Parks, Zimbabwe. *Journal of Outdoor Recreation and Tourism, 20*, 1–18.

Nelson, W. A. (2009). Calcified macroalgae—Critical to coastal ecosystems and vulnerable to change: A review. *Marine and Freshwater Research, 60*, 787–801. https://doi.org/10.1071/MF08335

Ojala, M. (2013). Coping with climate change among adolescents: Implications for subjective well-being and environmental engagement. *Sustainability, 5*(5), 2191–2209.

OSPAR Commission. (2010).

Otero-Ferrer, F., Mannarà, E., Cosme, M., Falace, A., Montiel-Nelson, J. A., Espino, F., Haroun, R., & Tuya, F. (2019). Early-faunal colonization patterns of discrete habitat units: A case study with rhodolith-associated vagile macrofauna. *Estuarine, Coastal and Shelf Science, 218*, 9–22.

Pascelli, C., Riul, P., Riosmena-Rodríguez, R., Scherner, F., Nunes, M., Hall-Spencer, J. M., et al. (2013). Seasonal and depth-driven changes in rhodolith bed structure and associated macroalgae off Arvoredo Island (southeastern Brazil). *Aquatic Botany, 111*, 62–65.

Pascual, U., Balvanera, P., Díaz, S., Pataki, G., Roth, E., Stenseke, M., et al. (2017). Valuing nature's contributions to people: the IPBES approach. *Current Opinion in Environmental Sustainability, 26*, 7–16.

Pellowe, K. E., & Leslie, H. M. (2021). Ecosystem service lens reveals diverse community values of small-scale fisheries. *Ambio, 50*(3), 586–600.

Peña, V., Bárbara, I., Grall, J., Maggs, C. A., & Hall-Spencer, J. M. (2014). The diversity of seaweeds on maerl in the NE Atlantic. *Marine Biodiversity, 44*, 533–551.

Pereira-Filho, G. H., Amado-Filho, G., de Moura, R. L., Bastos, A. C., Guimaraes, S. M., Salgado, L. T., et al. (2012). Extensive rhodolith beds cover the summits of southwestern Atlantic Ocean seamounts. *Journal of Coastal Research, 28*(1), 261–269.

Powell, R. B., & Ham, S. H. (2008). Can ecotourism interpretation really lead to pro-conservation knowledge, attitudes and behaviour? Evidence from the Galapagos Islands. *Journal of Sustainable Tourism, 16*(4), 467–489.

Queirós, A. M., Stephens, N., Widdicombe, S., Tait, K., McCoy, S. J., Ingels, J., et al. (2019). Connected macroalgal-sediment systems: Blue carbon and food webs in the deep coastal ocean. *Ecological Monographs, 89*(3), e01366.

Qui-Minet, Z. N., Davoult, D., Grall, J., Delaunay, C., Six, C., Cariou, T., & Martin, S. (2021). Physiology of maerl algae: Comparison of inter-and intraspecies variations. *Journal of Phycology, 57*(3), 831–848.

Rendina, F., Buonocore, E., di Montanara, A. C., & Russo, G. F. (2022). The scientific research on rhodolith beds: A review through bibliometric network analysis. *Ecological Informatics*, 101738.

Reyers, B., Biggs, R., Cumming, G. S., Elmqvist, T., Hejnowicz, A. P., & Polasky, S. (2013). Getting the measure of ecosystem services: A social–ecological approach. *Frontiers in Ecology and the Environment, 11*(5), 268–273.

Riding, R. (2002). Structure and composition of organic reefs and carbonate mud mounds: Concepts and categories. *Earth-Science Reviews, 58*(1–2), 163–231.

Rinehart, K. L., Gloer, J. B., Wilson, G. R., Hughes, R. G., Li, L. H., Renis, H. E., & McGovren, J. P. (1983). Antiviral and antitumor compounds from tunicates. *Federation Proceedings, 42*, 87–90.

Rios, J. M., & Brewer, J. (2014). Outdoor education and science achievement. *Applied Environmental Education & Communication, 13*(4), 234–240.

Riosmena-Rodríguez, R. (2017). Natural history of rhodolith/maërl beds: their role in near-shore biodiversity and management. In *Rhodolith/maërl beds: A global perspective* (pp. 3–26). Springer.

Riul, P., Lacouth, P., Pagliosa, P. R., Christoffersen, M. L., & Horta, P. A. (2009). Rhodolith beds at the easternmost extreme of South America: Community structure of an endangered environment. *Aquatic Botany, 90*(4), 315–320.

Riul, P., Targino, C. H., Farias, J. D. N., Visscher, P. T., & Horta, P. A. (2008). Decrease in Lithothamnion sp. (Rhodophyta) primary production due to the deposition of a thin sediment layer. *Journal of the Marine Biological Association of the United Kingdom, 88*(1), 17–19.

Ryabinin, V., Barbière, J., Haugan, P., Kullenberg, G., Smith, N., McLean, C., Troisi, A., Fischer, A., Aricò, S., Aarup, T., Pissierssens, P., Visbeck, M., Enevoldsen, H. O., & Rigaud, J. (2019). The UN decade of ocean science for sustainable development. *Frontiers in Marine Science, 6*, 470. https://doi.org/10.3389/fmars.2019.00470

Saker, K. E., Fike, J. H., Veit, H., & Ward, D. L. (2004). Brown seaweed-(Tasco) treated conserved forage enhances antioxidant status and immune function in heat-stressed weather lambs. *Journal of Animal Physiology and Animal Nutrition, 88*, 122–130.

Santos, A. S., Riul, P., Brasil, A. C. D. S., & Christoffersen, M. L. (2011). Encrusting Sabellariidae (Annelida: Polychaeta) in rhodolith beds, with description of a new species of Sabellaria from the Brazilian coast. *Journal of the Marine Biological Association of the United Kingdom, 91*(2), 425–438.

Scherner, F., Riul, P., Bastos, E., Bouzon, Z. L., Pagliosa, P. R., Blankensteyn, A., et al. (2010). Herbivory in a rhodolith bed: a structuring factor. *Pan-American Journal of Aquatic Sciences, 5*(3), 358–366.

Schubert, N., Salazar, V. W., Rich, W. A., Bercovich, M. V., Saá, A. A., Fadigas, S. D., et al. (2019). Rhodolith primary and carbonate production in a changing ocean: The interplay of warming and nutrients. *Science of the Total Environment, 676*, 455–468.

Schwarz, A. M., Hawes, I., Andrew, N., Mercer, S., Cummings, V., & Thrush, S. (2005). Primary production potential of non-geniculate coralline algae at Cape Evans, Ross Sea, Antarctica. *Marine Ecology Progress Series, 294*, 131–140.

SDGs. (2020). Sustainable development goals.

Singh, U. B., & Ahluwalia, A. S. (2013). Microalgae: A promising tool for carbon sequestration. *Mitigation and Adaptation Strategies for Global Change, 18*(1), 73–95.

Sissini, M. N., Koerich, G., Barros-Barreto, M. B., Coutinho, L. M., Gomes, F. P., Oliveira, W., et al. (2022). Diversity, distribution, and environmental drivers of coralline red algae: the major reef builders in the Southwestern Atlantic. *Coral Reefs, 41*(3), 711–725.

Smit, I. P., Roux, D. J., Swemmer, L. K., Boshoff, N., & Novellie, P. (2017). Protected areas as outdoor classrooms and global laboratories: Intellectual ecosystem services flowing to-and-from a National Park. *Ecosystem Services, 28*, 238–250.

Spalding, M. D., Fox, H. E., Allen, G. R., Davidson, N., Ferdaña, Z. A., Finlayson, M., et al. (2007). Marine ecoregions of the world: A bioregionalization of coastal and shelf areas. *Bioscience, 57*(7), 573–583.

Spalding, M., Burke, L., Wood, S. A., Ashpole, J., Hutchison, J., & Zu Ermgassen, P. (2017). Mapping the global value and distribution of coral reef tourism. *Marine Policy, 82*, 104–113.

Stapp, W. B. (1969). The concept of environmental education. *Environmental Education, 1*, 30–31.
Steller, D. L., Riosmena-Rodríguez, R., Foster, M. S., & Roberts, C. (2003). Rhodolith bed diversity in the Gulf of California: The importance of rhodolith structure and consequences of anthropogenic disturbances. *Aquatic Conservation, 13*, S5–S20.
Stelzer, P. S., Mazzuco, A. C. A., Gomes, L. E., Martins, J., Netto, S., & Bernardino, A. F. (2021). Taxonomic and functional diversity of benthic macrofauna associated with rhodolith beds in SE Brazil. *PeerJ, 9*, e11903.
Sutcliffe, S. R., & Barnes, M. L. (2018). The role of shark ecotourism in conservation behaviour: Evidence from Hawaii. *Marine Policy, 97*, 27–33.
Teed, L., Bélanger, D., Gagnon, P., & Edinger, E. (2020). Calcium carbonate (CaCO3) production of a subpolar rhodolith bed: Methods of estimation, effect of bioturbators, and global comparisons. *Estuarine, Coastal and Shelf Science, 242*, 106822.
Tâmega, F.T.S., Spotorno-Oliveira, P., & Figueiredo, M. A. O. (2013). *Catalogue of the benthic marine life from Peregrino oil field* (p. 140). Campos Basin, Brazil. Instituto Biodiversidade Marinha, Rio de Janeiro.
Tâmega, F. T., Bassi, D., Figueiredo, M. A., & Cherkinsky, A. (2014). Deep-water rhodolith bed from central Brazilian continental shelf, Campos Basin: Coralline algal and faunal taxonomic composition. *Galaxea, Journal of Coral Reef Studies, 16*(1), 21–31.
Tâmega, F. T. S., Riosmena-Rodríguez, R., Sportono-Oliveira, P., et al. (2015). Taxonomy and distribution of non-geniculate coralline red algae (Corallinales, Rhodophyta) on rocky reefs from Ilha Grande Bay, Brazil. *Phytotaxa, 192*, 267–278.
Testa, V., & Bosence, D. W. (1999). Physical and biological controls on the formation of carbonate and siliciclastic bedforms on the northeast Brazilian shelf. *Sedimentology, 46*, 279–301.
Turra, A., & Denadai, M. R. (2015). Protocolos para o monitoramento de habitats bentônicos costeiros.
UNEP: IPBES-2/4. (2014). Conceptual framework for the intergovernmental science-policy platform on biodiversity and ecosystem services. In *Report of the second session of the plenary of the intergovernmental science-policy platform on biodiversity and ecosystem services.*
Van Der Heijden, L. H., & Kamenos, N. (2015). Reviews and syntheses: Calculating the global contribution of coralline algae to total carbon burial. *Biogeosciences, 12*(21), 6429–6441.
Veras, P., Pierozzi-Jr, I., Lino, J. B., Amado-Filho, G. M., de Senna, A. R., Santos, C. S. G., et al. (2020). Drivers of biodiversity associated with rhodolith beds from euphotic and mesophotic zones: Insights for management and conservation. *Perspectives in Ecology and Conservation, 18*(1), 37–43.
Villas Bôas, A. B., Figueiredo, M. A. D. O., & Villaça, R. C. (2005). Colonization and growth of crustose coralline algae (Corallinales, Rhodophyta) on the Rocas Atoll. *Brazilian Journal of Oceanography, 53*, 147–156.
Villas-Bôas, A. B., Riosmena-Rodriguez, R., Amado-Filho, G. M., et al. (2009). Rhodolith-forming species of *Lithophyllum* (Corallinales; Rhodophyta) from Espírito Santo State, Brazil, including the description of *L. depressum* sp. nov. *Phycologia, 48*, 237–248.
Villas-Bôas, A. B., Riosmena-Rodriguez, R., & Figueiredo, M. A. O. (2014). Community structure of rhodolith-forming beds on the central Brazilian continental shelf. *Helgoland Marine Research, 68*, 27–35.
Yuan, Y. V., & Walsh, N. A. (2006). Antioxidant and antiproliferative activities of extracts from a variety of edible seaweeds. *Food and Chemical Toxicology, 44*, 1144–1150.

Chapter 10
The Legacy of Gilberto M. Amado-Filho to South Atlantic Rhodolith Beds Knowledge: Memories of His Students

Guilherme H. Pereira-Filho, Ricardo da Gama Bahia, Fernando C. Moraes, and Leonardo T. Salgado

Abstract *"I seem to have been only like a boy playing on the seashore, and diverting myself in now and then finding a smoother pebble or a prettier shell than ordinary, whilst the great ocean of truth lay all undiscovered before me."* This quote is attributed to one of the most important scientists of our society, Sir Isaac Newton. However, it could easily be attributed to Gilberto M. Amado-Filho by those who have had the opportunity to know him. Pebbles (or gravels) were how extensive areas of the Brazilian Exclusive Economic Zone (EEZ) had been classified before Dr. Amado-Filho took his diver gears down into several turbid waters of this portion of the Western Atlantic.

Keywords Brazilian Exclusive Economic Zone (EEZ) · Rhodolith Beds · Southwestern Atlantic · Marine biodiversity · Coraline algae · Marine conservation

"I seem to have been only like a boy playing on the seashore, and diverting myself in now and then finding a smoother pebble or a prettier shell than ordinary, whilst the great ocean of truth lay all undiscovered before me." This quote is attributed to one of the most important scientists of our society, Sir Isaac Newton. However, it could easily be attributed to Gilberto M. Amado-Filho by those who have had the opportunity to know him. Pebbles (or gravels) were how extensive areas of the Brazilian Exclusive Economic Zone (EEZ) had been classified before Dr.

G. H. Pereira-Filho (✉)
Instituto de Ciências do Mar, Universidade Federal de São Paulo, Santos, São Paulo, Brazil
e-mail: pereira.filho@unifesp.br

R. da Gama Bahia · L. T. Salgado
Instituto de Pesquisas Jardim Botânico do Rio de Janeiro, Rio de Janeiro, Brazil

F. C. Moraes
Instituto de Pesquisas Jardim Botânico do Rio de Janeiro, Rio de Janeiro, Brazil

Museu Nacional, Universidade Federal do Rio de Janeiro, Rio de Janeiro, Brazil
e-mail: fmoraes@mn.ufrj.br

P. A. Horta, M. N. Sissini (eds.), *Brazilian Rhodolith Beds*, Brazilian Marine Biodiversity, https://doi.org/10.1007/978-3-031-61449-1_10

Amado-Filho took his diver gears down into several turbid waters of this portion of the Western Atlantic.

Interestingly, his journey revealing the magnitude of Southwestern Atlantic Rhodolith Beds began accidentally around 2004, when he was studying Brazilian unpretending kelp beds ~21°S (Fig. 10.1). During those dives, he realized that the kelp *Laminaria abyssalis* was always attached on carbonate nodules built by coralline algae, that is, rhodoliths (e.g., Marins et al., 2012; Marins et al., 2013). He absorbed himself into the extensive areas covered by rhodolith beds, a completely new habitat to him, with its amazing associated marine biodiversity. It did not take too long to new species and new occurrences of algae begin to be described associated with rhodolith beds (e.g., Guimaraes & Amado-Filho, 2008, 2009; Villas-Boas et al., 2009). The seasonal pattern observed to *L. abyssalis* (i.e., alternation of small filamentous phase with large foliose one, according local/regional oceanographic features) also instigated him to undercover the dynamic of the benthic community associated with the hard, yet unattached, bottom provided by rhodoliths (Amado-Filho et al., 2007). Experimental designs soon revealed a pattern of seaweeds detachment during winter storms when rhodoliths roll (Amado-Filho et al., 2010). Thus, these habitats were also providing resources for other adjacent ones (e.g.,

Fig. 10.1 *Laminaria abyssalis* associated with rhodolith beds ~21°S. (**a**) "Facilities" used by Amado-Filho at the beginning of his rhodolith beds explorations. This image was taken, in 2004, by Gilberto Amado-Filho before diving down to 53 m depth. (**b**) *Laminaria abyssalis* on this same rhodolith bed. Image taken by Amado-Filho in 2004. (**c**) (Detail of a recovered sample of *L. abyssalis* attached to a rhodolith. (**d**) Gilberto Amado-Filho after a diving day during an expedition to sample *L. abyssalis* associated with rhodolith beds in 2010 (Photos from Gilberto's Laboratory Archive)

Andrades et al., 2014). If too many questions have emerged from a punctual rhodolith bed, what could come if others Southwestern Atlantic "pebbles" bottoms began to be explored, especially by a marine biologist willing to challenge the belittled impression that most researchers had about scientific diving (see Foster et al. 2013)?

Coincidence or not, dives carried by Amado-Filho and his colleagues in 2009 on the seamounts of the Vitoria-Trindade Chain (Fig. 10.2) revealed that rhodolith beds were not only covering large areas of the Brazilian Continental Shelf but were also the major bottom feature on oceanic shallow and mesophotic reefs (Pereira-Filho et al., 2011; Pereira-Filho et al., 2012; Meirelles et al., 2015). Amado-Filho had accessed an early vast range of literature that had suggested the occurrence of rhodolith beds along the Brazilian EEZ, but most of them called this habitat as "pebbles" or "gravels." He challenged himself to dive wherever he could to prove that those bottoms were alive, harboring high diversity of marine life, and thus were highly productive sites rather than just "pebbles/gravels" bottoms, as described in most environmental licensing and/or Brazilian geological department reports (see Amado-Filho & Pereira-Filho, 2012; Santos et al., 2016). His efforts had no limit! Amado-Filho was 51 years old when, in 2011, he became one of the few people to have ever dived in all Brazilian oceanic islands and atoll during the same year. Besides the oceanic islands, he also sampled many other sites in that year, totalizing more than 160 days diving and over 1300 miles navigated. As would be expected from him, rhodoliths were mapped surrounding most of Brazilian oceanic islands and atoll (Amado-Filho et al. 2012a, 2016; Pereira-Filho et al., 2015,) as well as many other areas of the Brazilian continental shelf (see Amado-Filho et al., 2017), including the largest rhodolith bed in the world (Amado-Filho et al., 2012b) and an extensive (~9500 km^2), carbonate system off the Amazon River mouth, including rhodolith beds beneath the plume of the world's largest river (Fig. 10.3) (Moura et al., 2016; Vale et al., 2018). Gilberto stood out for his ability to lead multidisciplinary research networks, permeating various topics, from cell biology and calcareous algae ecology and conservation of marine environments. In this journey, he established national and international collaborations to advance in the hitherto incipient knowledge about rhodoliths and other marine habitats from the continental

Fig. 10.2 Field surveys on the Vitoria-Trindade Chain. (**a**) Gilberto Amado-Filho starting a technical dive at Trindade Island shelf, 2009. (**b**) Amado-Filho sampling rhodoliths on one of the Vitória-Trindade Seamounts, 2011. (Photos: Ronaldo Francini-Filho)

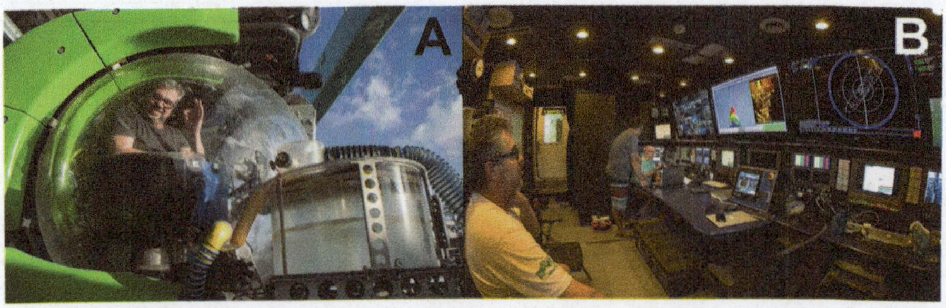

Fig. 10.3 Gilberto Amado-Filho during an emblematic survey of the Amazon carbonate reefs. (**a**) Amado-Filho before dive down to Amazon reefs in the submersible. (**b**) Amado-Filho watching the bottom survey from the @@Alluccia bridge command. (Photos: Fernando Moraes/Rede Abrolhos)

Fig. 10.4 Gilberto Amado-Filho (06/out/1959–5/mar/2019)

shelf and oceanic islands. This was done through intensive collaborations with proficient marine scientists in Brazil and from the Americas, Europe, Africa, and Asia. He contributed to the development of a network of scientists including dozens of experts from different fields of knowledge. Gilberto, along with the research group that he led, has unraveled much of the coralline algae diversity and taxonomy in Brazil, including the description of several new rhodolith-forming species (e.g., Villas-Boas et al., 2009; Bahia et al. 2011, 2014a, b, 2015a, b; Jesionek et al., 2016, 2020).

It is no surprise that, having witnessed the magnitude of many Brazilian rhodolith beds, Amado-Filho did not hesitate to claim that the ecological importance of those habitats would be comparable with the Australian Great Barrier Reef (GBR). Indeed, he showed that the $CaCO_3$ production by rhodoliths from Abrolhos Bank rivals that of the GBR (Amado-Filho et al., 2012b) and changed the way in which the Brazilian rhodolith beds were internationally known (Fig. 10.4) This was Amado-Filho! He was intense in advocating his ideas and results, becoming a researcher that would not save efforts to make science, to argue in favor of marine conservation and to proudly represent his institution, the Instituto de Pesquisas Jardim Botânico do Rio de Janeiro.

Amado-Filho left us in March 2019, leaving his immeasurable legacy to knowledge development, formation of young marine scientists, and science diffusion about the Brazilian rhodolith beds. Besides pioneering the mapping and sampling of these habitats in situ, he influenced a generation of marine biologists with his amazing explorations. Amado-Filho supervised more than 30 researchers that are now employed in different Brazilian institutions, all of them pursuing the mission to protect and disseminate the ecological importance of rhodolith beds. Well done Gilberto! Thank you for sharing your intensity with us! Rest In Peace!

References

Amado Filho, G. M., Maneveldt, G., Marins, B. V., & Guimarães, S. M. P. (2007). Structure of rhodolith beds from a depth gradient of 4 to 55 meters at the south of Espírito Santo state coast, Brazil. *Ciencias Marinas, 33*, 399–410.

Amado Filho, G. M., Maneveldt, G., Pereira-Filho, G., Manso, R. C., Bahia, R., Barreto, M. B. B., & Guimarães, S. M. P. B. (2010). Seaweed diversity associated with a Brazilian tropical rhodolith bed. *Ciencias Marinas, 36*, 371–391.

Amado-Filho, G. M., & Pereira-Filho, G. H. (2012). Rhodolith beds in Brazil: A new potential habitat for marine bioprospection. *Revista Brasileira de Farmacognosia, 22*, 782–7882.

Amado-Filho, G. M., Pereira-Filho, G. H., Bahia, R. G., Abrantes, D. P., Veras, P. C., & Matheus, Z. (2012a). Occurrence and distribution of rhodolith beds on the Fernando de Noronha archipelago of Brazil. *Aquatic Botany, 101*, 41–45.

Amado-Filho, G. M., Moura, R. L., Bastos, A. C., Salgado, L. T., Sumida, P. Y., Guth, A. Z., Francini-Filho, R. B., Pereira-Filho, G. H., Abrantes, D. P., Brasileiro, P. S., Bahia, R. G., Leal, R. N., Kaufman, L., Kleypas, J. A., Farina, M., & Thompson, F. L. (2012b). Rhodolith beds are major CaCO3 bio-factories in the tropical South West Atlantic. *PLoS One, 7*, e35171.

Amado-Filho, G. M., Moura, R. L., Bastos, A. C., Francini-Filho, R. B., Pereira-Filho, G. H., Moraes, F. C., & Motta, F. S. (2016). Mesophotic ecosystems of the unique South Atlantic atoll are composed by rhodolith beds and scattered consolidated reefs. *Marine Biodiversity, 46*, 933–936.

Amado-Filho, G. M., Bahia, R., Pereira-Filho, G. H., & Longo, L. L. (2017). South Atlantic Rhodolith beds: Latitudinal distribution, species composition, structure and ecosystem functions, threats and conservation status. In R. Riosmena-Rodríguez, W. Nelson, & J. Aguirre (Eds.), *Rhodolith/Maërl beds: A global perspective* (Vol. 15, pp. 299–317). Springer.

Andrades, R., Gomes, M. P., Pereira-Filho, G. H., Souza-Filho, J., Alburquerque, C. Q., & Martins, A. S. (2014). *The influence of allochthonous macroalgae on the fish communities of tropical sandy beaches*. Estuarine, Coastal and Shelf Science.

Bahia, R. G., Riosmena-Rodriguez, R., Maneveldt, G. W., & Amado-Filho, G. M. (2011). Research note: First report of *Sporolithon ptychoides* (Sporolithales, Corallinophycidae,) Rhodophyta for the Atlantic Ocean. *Phycological Research, 59*, 64–69.

Bahia, R. G., Maneveldt, G. M., & Amado-Filho, G. M. (2014a). *Porolithon improcerum* (Porolithoideae, Corallinaceae) and *Mesophyllum macroblastum* (Melobesioideae, Hapalidiaceae): New records of crustose coralline red algae for the Southwest Atlantic Ocean. *Phytotaxa, 190*, 38.

Bahia, R. G., Amado-Filho, G. M., & Maneveldt, G. M. (2014b). *Sporolithon molle* (Heydrich) Heydrich (Sporolithales, Corallinophycidae, Rhodophyta): An addition to the Atlantic Flora found on a remote Oceanic Island. *Cryptogamie, Algologie, 35*, 7–14.

Bahia, R. G., Maneveldt, G. M., Amado-Filho, G. M., & Yoneshigue-Valentin, Y. (2015a). New diagnostic characters for the order Sporolithales (Corallinophycidae, Rhodophyta). *Journal of Phycology*.

Bahia, R. G., Amado-Filho, G. M., Maneveldt, G. M., Adey, W. H., Johnson, G., Jesionek, M. B., & Longo, L. L. (2015b). Sporolithon yoneshigueae sp. nov. (Sporolithales, Corallinophycidae, Rhodophyta), a new rhodolith-forming coralline alga from the Southwest Atlantic. *Phytotaxa, 224*, 140–158.

Foster, M., Amado-Filho, G. M., Steller, D., Riosmena-Rodriguez, R., & Kamenos, N. (2013). *Rhodoliths and rhodoliths beds. Contribution of SCUBA diving to research and discovery in marine environments* (Vol. 39, 39th ed., pp. 143–156). Smithsonian Institution Scholarly Press.

Guimarães, S. M. P. B., & Amado-Filho, G. M. (2008). Deep-water gelatinous rhodophytes from southern Espírito Santo state, Brazil. *Botanica Marina, 51*, 378–387.

Guimarães, S. M. P. B., & Amado-Filho, G. M. (2009). First record of Reticulocaulis mucosissimus I. A. Abbott (Naccariaceae, Rhodophyta) for the western Atlantic Ocean. *Revista Brasileira de Botânica, 32*, 671–675.

Jesionek, M. B., Bahia, R. G., Hernández-Kantún, J. J., Adey, W. H., Yoneshigue-Valentin, Y., Longo, L. L., & Amado-Filho, G. M. (2016). A taxonomic account of non-geniculate coralline algae (Corallinophycidae, Rhodophyta) from shallow reefs of the Abrolhos Bank, Brazil. *Algae, 31*, 317–340.

Jesionek, M.B., Bahia, R.G., Lyra, M.B., Leão, L.A.B., Oliveira, M.C., Amado-Filho, G.M. (2020) Newly discovered coralline algae in Southeast Brazil: Tectolithon fluminense gen. et sp. nov. and Crustaphytum atlanticum sp. nov. (Hapalidiales, Rhodophyta).. Phycologia.

Marins, B. V., Amado-Filho, G. M., Barreto, M. B. B., & Longo, L. L. (2012). Taxonomy of the southwestern Atlantic endemic kelp: *Laminaria abyssalis* and *Laminaria brasiliensis* (Phaeophyceae, Laminariales) are not different species. *Phycological Research, 60*, 51–60.

Marins, B. V., Amado-Filho, G. M., Barbarino, E., Pereira-Filho, G. H., & Longo, L. L. (2013). Seasonal changes in population structure of the tropical deep-water kelp *L aminaria abyssalis*. *Phycological Research, 62*.

Meirelles, P. M., Amado-Filho, G. M., Pereira-Filho, G. H., Pinheiro, H. T., Moura, R. L., Joyeux, J. C., Mazzei, E. F., Bastos, A. C., Edwards, R. A., Dinsdale, E., Paranhos, R., Santos, E. O., Iida, T., Gotoh, K., Nakamura, S., Sawabe, T., Rezende, C. E., Gadelha, L. M. R., Francini-Filho, R. B., Thompson, C., & Thompson, F. L. (2015). Baseline assessment of Mesophotic reefs of the Vitória-Trindade seamount chain based on water quality, microbial diversity, benthic cover and fish biomass data. *PLoS One, 10*, e0130084.

Moura, R. L., Amado-Filho, G. M., Moraes, F. C., Brasileiro, P. S., Salomon, P. S., Mahiques, M. M., Bastos, A. C., Almeida, M. G., Silva, J. M., Araujo, B. F., Brito, F. P., Rangel, T. P., Oliveira, B. C. V., Bahia, R. G., Paranhos, R. P., Dias, R. J. S., Siegle, E., Figueiredo, A. G., Pereira, R. C., Leal, C. V., Hajdu, E., Asp, N. E., Gregoracci, G. B., Neumann-Leitao, S., Yager, P. L., et al. (2016). An extensive reef system at the Amazon River mouth. *Science Advances, 2*, e1501252–e1501252.

Pereira-Filho, G., Amado-Filho, G. M., Guimarães, S. M. P. B., Moura, R. L., Sumida, P. Y. G., Abrantes, D. P., Bahia, R., Guth, A. Z., Jorge, R., & Francini-Filho, R. (2011). Reef fish and benthic assemblages of the Trindade and Martin Vaz Island group, southwestern Atlantic. *Brazilian Journal of Oceanography, 59*, 201–212.

Pereira-Filho, G. H., Amado-Filho, G. M., Moura, R. L., Bastos, A. C., Guimarães, S. M. P. B., Salgado, L. T., Francini-Filho, R. B., Bahia, R. G., Abrantes, D. P., Guth, A. Z., & Brasileiro, P. S. (2012). Extensive Rhodolith beds cover the summits of southwestern Atlantic Ocean seamounts. *Journal of Coastal Research, 279*, 261–269.

Pereira-Filho, G. H., Veras, P. C., Francini-Filho, R. B., Moura, R. L., Pinheiro, H. T., Gibran, F. Z., Matheus, Z., Neves, L. M., & Amado-Filho, G. M. (2015). Effects of the sand tilefish

Malacanthus plumieri on the structure and dynamics of a rhodolith bed in the Fernando de Noronha Archipelago, tropical West Atlantic. *Marine Ecology Progress Series, 541*, 65–73.
Santos, C. S. G., Lino, J. B., Veras, P. C., Amado-Filho, G. M., Francini-Filho, R. B., Motta, F. S., Moura, R. L., & Pereira-Filho, G. H. (2016). Environmental licensing on rhodolith beds: Insights from a worm. *Natureza & Conservação, 14*, 137–141.
Vale, N. F., Amado-Filho, G. M., Braga, J. C., Brasileiro, P. S., Karez, C. S., Moraes, F. C., Bahia, R. G., Bastos, A. C., & Moura, R. L. (2018). Structure and composition of rhodoliths from the Amazon River mouth, Brazil. *Journal of South American Earth Sciences, 84*, 149–159.
Villas-Boas, A. B., Riosmena-Rodriguez, R., Amado Filho, G. M., Maneveldt, G., & Figueiredo, M. A. (2009). Rhodolith-forming species of Lithophyllum (Corallinales; Rhodophyta) from Espírito Santo state, Brazil, including the description of L. depressum sp. nov. *Phycologia, 48*, 237–248.

Index

A
Anthropogenic impacts, 156, 164, 167

B
Benefits to society, 186
Benthic communities, 22, 68, 132, 149, 170, 192, 204
Biodiversity, 3, 7–9, 24, 68–96, 106–108, 115, 121, 122, 166, 175, 178, 187, 188, 192, 193
Biogenic reefs, 2, 146
Brazil, 16, 17, 24, 30, 36–40, 68–96, 106, 115, 121, 128, 130–134, 137, 147, 149, 151, 153, 154, 156, 165, 167–171, 173–176, 178, 188, 189, 192, 194, 206
Brazilian rhodolith beds, 16–24, 72, 73, 86, 156, 169, 178, 206, 207

C
Contributions, 2, 6, 7, 9, 168, 169, 176, 186–196
Corallinales, 2, 30, 33, 35–37, 39, 44, 46, 73
Coralline red algae (CRA), 2–9, 16, 34, 38, 68, 73, 165, 168, 186, 188, 194
Corallinophycidae, 2, 30, 33, 35–36, 40, 42, 46
Cretaceous, 4–6, 128–131, 137

E
Ecological cascade, 69
Ecosystem contributions, 189
Ecosystem services and functions, 2
Eutrophication, 170

F
Fauna, 5, 84–86, 146, 148, 175, 188, 189
Flora, 5, 7, 17, 37, 39, 45, 71–83, 86, 146, 188, 189
Fossil coralline algae, 128

H
Hapalidiales, 2, 30, 33, 35, 36, 39, 42–45, 50, 52, 73
Holocene, 128, 131–133, 135, 138

M
Macrofauna, 86
Marine biodiversity, 5, 147, 156, 191, 204
Marine ecology, 3, 17, 23, 24, 67–96, 106, 121, 145–156, 164, 166, 171, 178, 186, 188, 191, 192, 194, 205–207

O
Ocean Literacy, 24

P
Physiological responses, 164

R
Reef builders, 5, 6
Reef evolution, 2

P. A. Horta, M. N. Sissini (eds.), *Brazilian Rhodolith Beds*, Brazilian Marine Biodiversity, https://doi.org/10.1007/978-3-031-61449-1

Resilience, 96, 170, 178
Rhodolith beds, 3, 16, 30, 68, 106, 128, 146, 164, 186, 204
Rhodoliths, 3, 16, 30, 68, 107, 131, 146, 164, 186, 204

S

Sporolithales, 2, 30, 33, 35, 36, 42, 44, 49, 73
Systematics, 2, 16, 35–36, 73

SPRINGER NATURE

GPSR Compliance

The European Union's (EU) General Product Safety Regulation (GPSR) is a set of rules that requires consumer products to be safe and our obligations to ensure this.

If you have any concerns about our products, you can contact us on ProductSafety@springernature.com

In case Publisher is established outside the EU, the EU authorized representative is:

Springer Nature Customer Service Center GmbH
Europaplatz 3
69115 Heidelberg, Germany

The manufacturer's authorised representative in the EU is Springer Nature Customer Service Centre GmbH, Europaplatz 3, 69115 Heidelberg, Germany. If you have any concerns regarding our products, please contact ProductSafety@springernature.com

Printed and bound by CPI Group (UK) Ltd, Croydon, CR0 4YY
28/09/2026
02224081-0008